New Frontiers in Veterinary Virology

New Frontiers in Veterinary Virology

Editor: Travis Schroeder

AMERICAN
MEDICAL PUBLISHERS
www.americanmedicalpublishers.com

AMERICAN
MEDICAL PUBLISHERS
www.americanmedicalpublishers.com

Cataloging-in-Publication Data

New frontiers in veterinary virology / edited by Travis Schroeder.
 p. cm.
Includes bibliographical references and index.
ISBN 978-1-63927-520-5
 1. Veterinary virology. 2. Veterinary medicine. 3. Animals--Diseases. 4. Virus diseases. I. Schroeder, Travis.
SF780.4 .N49 2022
636.089 601 94--dc23

American Medical Publishers,
41 Flatbush Avenue,
1st Floor, New York,
NY 11217, USA

ISBN 978-1-63927-520-5 (Hardback)

Contents

Preface

The world is advancing at a fast pace like never before. Therefore, the need is to keep up with the latest developments. This book was an idea that came to fruition when the specialists in the area realized the need to coordinate together and document essential themes in the subject. That's when I was requested to be the editor. Editing this book has been an honour as it brings together diverse authors researching on different streams of the field. The book collates essential materials contributed by veterans in the area which can be utilized by students and researchers alike.

There are many viruses that affect the health of animals, and the field of study related to viruses in animals is known as veterinary virology. Some of these viruses have their own range and infect particular species. There are also some viruses which can affect different species. Veterinary virology focuses on the study of viruses like rhabdoviruses, foot and mouth disease viruses, circoviruses, herpes viruses and retroviruses. Rhabdoviruses are single stranded, negative sense RNA viruses inheriting six genera which infect a wide variety of plants and animals. Foot and mouth disease viruses are positive strand, non-enveloped, RNA viruses that can cause foot and mouth diseases in animals. Herpes viruses consist of ubiquitous pathogens which can infect both animals and humans. Retroviruses are a type of viruses that can cause cancer or immune deficiency. This book provides comprehensive insights into the field of veterinary virology. Its aim is to present researches that have transformed this discipline and aided its advancement. This book aims to serve as a resource guide for students and experts alike and contribute to the growth of the discipline.

Each chapter is a sole-standing publication that reflects each author's interpretation. Thus, the book displays a multi-facetted picture of our current understanding of application, resources and aspects of the field. I would like to thank the contributors of this book and my family for their endless support.

Editor

Feline panleukopenia virus in cerebral neurons of young and adult cats

Mutien Garigliany[1][*][†], Gautier Gilliaux[1][†], Sandra Jolly[1], Tomas Casanova[1], Calixte Bayrou[1], Kris Gommeren[2], Thomas Fett[3], Axel Mauroy[3], Etienne Lévy[1], Dominique Cassart[1], Dominique Peeters[2], Luc Poncelet[4] and Daniel Desmecht[1]

Abstract

Background: Perinatal infections with feline panleukopenia virus (FPV) have long been known to be associated with cerebellar hypoplasia in kittens due to productive infection of dividing neuroblasts. FPV, like other parvoviruses, requires dividing cells to replicate which explains the usual tropism of the virus for the digestive tract, lymphoid tissues and bone marrow in older animals.

Results: In this study, the necropsy and histopathological analyses of a series of 28 cats which died from parvovirus infection in 2013 were performed. Infections were confirmed by real time PCR and immunohistochemistry in several organs. Strikingly, while none of these cats showed cerebellar atrophy or cerebellar positive immunostaining, some of them, including one adult, showed a bright positive immunostaining for viral antigens in cerebral neurons (diencephalon). Furthermore, infected neurons were negative by immunostaining for p27^{Kip1}, a cell cycle regulatory protein, while neighboring, uninfected, neurons were positive, suggesting a possible re-entry of infected neurons into the mitotic cycle. Next-Generation Sequencing and PCR analyses showed that the virus infecting cat brains was FPV and presented a unique substitution in NS1 protein sequence. Given the role played by this protein in the control of cell cycle and apoptosis in other parvoviral species, it is tempting to hypothesize that a cause-to-effect between this NS1 mutation and the capacity of this FPV strain to infect neurons in adult cats might exist.

Conclusions: This study provides the first evidence of infection of cerebral neurons by feline panleukopenia virus in cats, including an adult. A possible re-entry into the cell cycle by infected neurons has been observed. A mutation in the NS1 protein sequence of the FPV strain involved could be related to its unusual cellular tropism. Further research is needed to clarify this point.

Keywords: Feline panleukopenia virus, Cat, Neurons, Neurological disorders

Background

Feline panleukopenia virus (FPV) and canine parvovirus (CPV) both belong to the *Protoparvovirus* genus within the *Parvovirinae* subfamily of the *Parvoviridae* family of single-stranded DNA viruses [1]. CPV-1 and 2 infect *Canidae* and CPV-2 emerged as a new host range variant in the mid-1970s (CPV-2) and spread worldwide in the canine population in 1978 [2]. Then, antigenic variants CPV-2a, b, c have gained infectivity for other species such as cats. FPV and FPV-like strains (such as mink enteritis

virus, MEV) are unable to infect *Canidae* [1, 3]. Although most FPV and CPV strains have been isolated from cats and dogs, a broad range of alternative hosts have been identified within the *Carnivora* order [1].

Parvovirus genome replication takes place in the nucleus and requires cells in S phase, since it relies on host cell machinery for the formation of double-stranded replication intermediates [4, 5]. This requirement limits the tropism of FPV and CPV to highly dividing cells such as those found in the intestine, bone marrow or lymphoid tissues. In kittens during the perinatal period, the infection of neuroblasts of the external granular layer is thought to be responsible for the cerebellar hypoplasia typically associated with such infections. However, viral

* Correspondence: mmgarigliany@ulg.ac.be
[†]Equal contributors
[1]Department of Morphology and Pathology, University of Liège, Liège, Belgium

proteins are expressed in some Purkinje cells despite the fact that these neurons are post-mitotic at this development stage [6, 7]. Nervous tissue infection by FPV has never been described in adult cats, although positive CPV immunostaining of feline cerebral neurons has been reported [8], which raises questions about the possible re-entry of some neurons into the S phase of the cell cycle, making them susceptible to infection.

In the present study, we show strong evidence of infection of cerebral neurons by FPV in young and adult cats, some of which with a history of neurological signs, associated with a unique mutation in the NS1 (nonstructural protein 1) amino acid sequence. Besides, one affected cat showed a co-infection by feline bocavirus type 1, which is the first evidence of nervous system infection by a bocavirus.

Results

Twenty eight parvovirus-positive cats, aged from 6 weeks to 5 years (mean: 12.5 months +/- 17.5 months; Table 1) were investigated in this study. Real time PCR revealed the presence of parvovirus DNA in most organs tested, with the highest concentrations in the spleen, small intestine and mesenteric lymph node (mean C_T (Cycle Threshold) for these three organs: 19,3 +/- 2,9). Interestingly, brain tissues were positive for most cats, with relatively low C_T values (20–25) in several of them (cats No 5, 9, 10, 11, 14, 15 and 16). Especially, the differences in C_T value between brain tissues and the ileum (small intestine) were highly variable, ranging from 2.2 to 14.9. This difference was the lowest (<6) in cats No 5, 10, 11, 14 and 15. Several of these cats were reported by referring veterinarians to have shown neurological disorders before death, mostly ataxia and/or dysphagia.

Affected cats were characterized on gross examination by mild to severe fibrinous to fibrinomucoid enteritis with thickened mucosa and highlighted Peyer's patches and mesenteric lymphadenomegaly.

The histopathological analyses showed classical lymphocyte necrosis and depletion in lymphoid organs (Peyer's patches and mesenteric lymph nodes) and severe intestinal villous blunting. Intestinal crypt cells were necrotic. Intestinal villi were depleted of their enterocytes and covered with a thick fibrinonecrotic exudate. Focal neuronal satellitosis and neuronophagic pictures (Fig. 1) were observed in the brain of cats No 5, 10, 11 and 15 (Table 1). None of the cats, even the youngest, showed any evidence of cerebellar atrophy.

Immunohistochemical (IHC) staining of the different sampled organs revealed the presence of parvoviral antigens in most locations. In particular, a bright staining was observed in cells from ileal crypts and follicular dendritic cells in the spleen and mesenteric lymph nodes. Staining was negative in the cerebellum but 10 cats (No

Table 1 Details of the 28 FPV-positive cats included in the study

Cat number	Sex	Breed	Age
1	F	Persian	22 weeks
2	M	European shorthair	16 weeks
3	M	European shorthair	5 years
4	F	Maine Coon	4 years
5	**M**	**Maine Coon**	**6 weeks**
6	F	European shorthair	22 weeks
7	F	European shorthair	2.5 years
8	M	European shorthair	14 weeks
9	M	European shorthair	10 weeks
10	**F**	**European shorthair**	**10 weeks**
11	**M**	**European shorthair**	**4.5 years**
12	F	European shorthair	24 weeks
13	F	Siamese cross	20 weeks
14	M	European shorthair	14 weeks
15	**F**	**European shorthair**	**12 weeks**
16	M	European shorthair	11 weeks
17	M	European shorthair	5 months
18	M	European shorthair	1.5 years
19	F	European shorthair	2 years
20	F	European shorthair	4 years
21	F	European shorthair	14 weeks
22	M	European shorthair	1.5 years
23	F	European shorthair	24 weeks
24	F	European shorthair	14 weeks
25	F	Siamese	10 weeks
26	F	European shorthair	8 weeks
27	F	European shorthair	8 weeks
28	M	European shorthair	1 year

The four cats with positive immunostaining for FPV antigens in cerebral neurons are bolded

5, 10, 11, 12, 14, 15, 16, 22, 23, 26; Table 1) showed strong positivity for parvoviral antigens in other brain regions, especially the interthalamic adhesion of the diencephalon (Fig. 2). Glial cells (mostly microglial cells) but also neurons (in four cats, No 5, 10, 11 and 15) showed bright cytoplasmic staining. Infection was associated in some but not all infected neurons by signs of neuronal degeneration and glial reaction (Fig. 1). The four cats bearing infected neurons were 6 weeks, 10 weeks, 12 weeks and 4.5 years-old (cats No 5, 10, 11 and 15, Table 1). Further immunostaining of the brain from cat number 15 for p27^{Kip1} antigen revealed an absence of nuclear p27^{Kip1} expression in FPV-infected neurons, while it was clearly expressed in the nucleus of uninfected neurons (Fig. 3). A relatively strong background reaction could not be avoided with the antibody

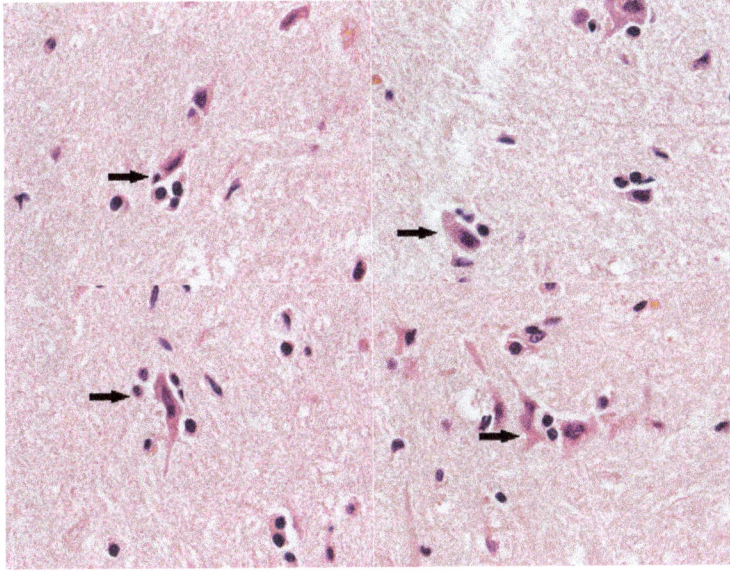

Fig. 1 Histopathological features of brain sections (diencephalic region) from a 12-week-old parvovirus-infected cat presented in this study (original magnification x400). Satellitosis around neurons with condensed chromatine (neuronophagia) is observed

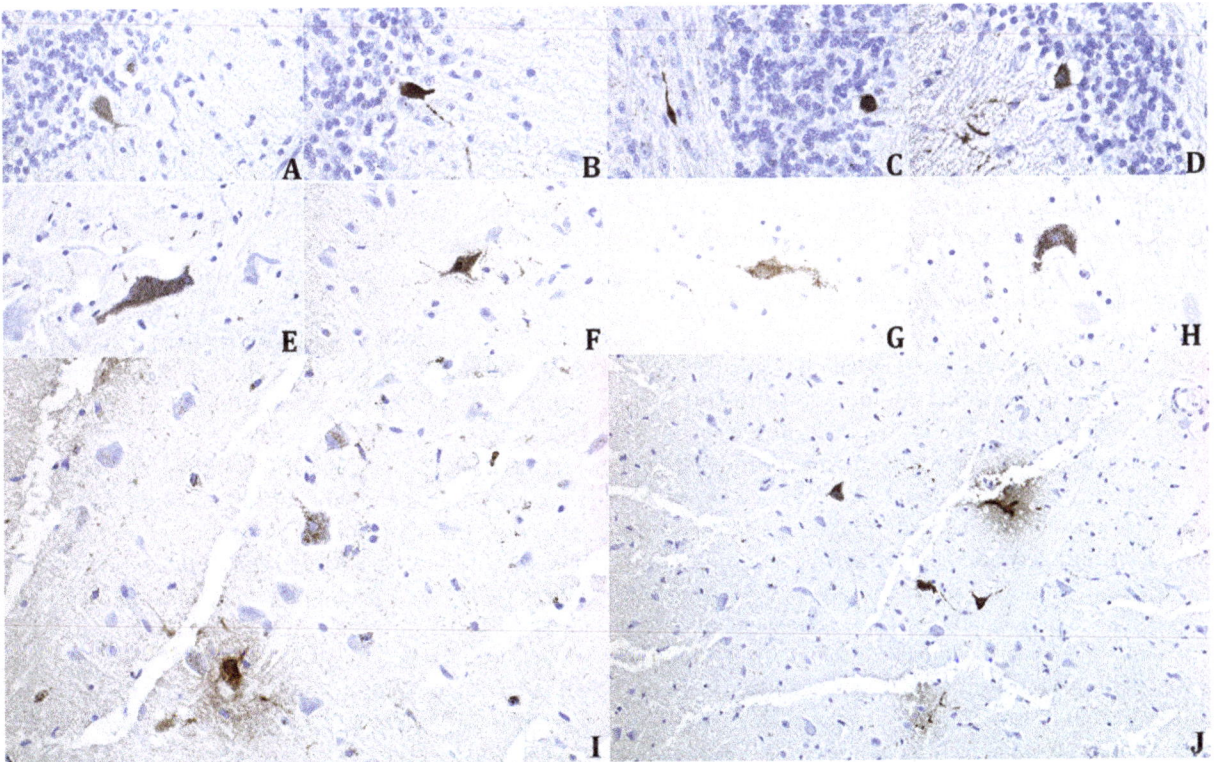

Fig. 2 Immunohistochemical staining for FPV antigens in parvovirus-positive cat brains. **a** to **d** Intense cytoplasmic or nuclear immunostaining of Purkinje cells from an infected kitten cerebellum used as a positive control (original magnification x400). **e** to **j** Bright staining of neuronal bodies and processes and (**i**) several microglial cells in the diencephalon (interthalamic adhesion) from a 12-week-old cat presented in this study (original magnification x400; (**j**) original magnification x200)

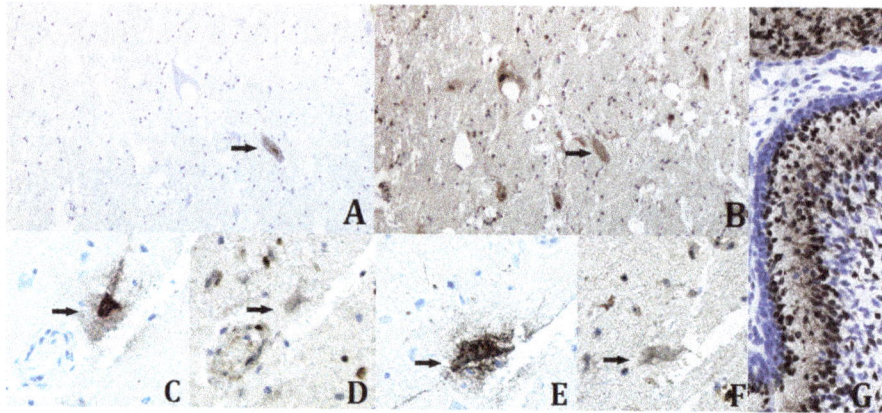

Fig. 3 Immunohistochemical staining for FPV and p27[Kip1] antigens. Immunostaining for FPV (**a, c, e**) and p27[Kip1] (**b, d, f**) antigens on adjacent sections showing the absence of nuclear staining for p27[Kip1] in FPV-infected neurons (**c-f**), while uninfected neurons still express nuclear p27 (**a, b**) (original magnification x100 (**a, b**) or x400 (**c** to **f**)). **g** Immunostaining for p27[Kip1] antigen of cerebellar cortex from a feline fetus (estimated gestational age: 54 days) used as a positive control (original magnification x400)

dilution (1/50) necessary to obtain a frank nuclear staining in p27[Kip1] positive cells (Fig. 3).

The full genome of the parvovirus present in the brain of the cat with the strongest IHC staining (cat No 5) was obtained using next-generation sequencing (GenBank accession number: KP769859). Sequence analysis allowed its classification as a feline panleukopenia virus, which was confirmed by a phylogenetic analysis (Fig. 4) and a sequence identity matrix (Fig. 5) based on the VP2 nucleotide coding sequence. A unique L → S substitution was observed at position 582 in the amino acid sequence of the NS1 protein (Table 2), resulting from a T → C nucleotide substitution at position 1745 of the NS1 coding sequence. Further sequence comparisons confirmed that this substitution had never been observed in FPV and CPV sequences available in GenBank to date. Sequences of VP1 and VP2 proteins were typical of FPV (Table 2).

A PCR targeting the region of interest of NS1 coding sequence confirmed the unique substitution in the KP769859 strain. The same PCR was applied to DNA extracted from the other three cats (No 10, 11, 15) with neuronal IHC staining and revealed the presence of the same substitution (Table 2). Subsequent next-generation sequencing of the full genome of the FPV genome from two of these cats (No 10 and 11) showed an identical amino acid sequence of NS1, VP1 and VP2 to that of KP769859 strain (data not shown). The NS1 substitution was not found in intestinal FPV strains infecting three cats with a negative cerebral immunostaining (cats number 14, 18 and 23; Table 2). It has to be noted here that these three cats had a positive FPV PCR in brain tissues, but with high C_T values, especially when compared to C_T values in the ileum.

Further analysis of the sequencing data from the cat brain infected by KP769859 FPV showed the presence of feline bocavirus. Subsequent PCR amplification with specific primers allowed the sequencing of around 80 % of the full genome (Genbank accession number: KP769860). Phylogenetic analysis based on the available partial genome allowed classification of this strain as a type 1 feline bocavirus (Fig. 6).

Discussion

Feline panleukopenia virus has long been known to cause cerebellar hypoplasia in neonatal kittens through in utero or perinatal infection of the external germinal epithelium of the cerebellum [6, 9, 10]. Parvoviruses typically target highly mitotic cells such as those from intestinal crypts, bone marrow and lymphoid tissues [10]. This requirement explains why cerebellar hypoplasia is only observed in kittens infected perinatally or in utero and not in older cats. Cell cycle re-entry might be involved in the productive infection of Purkinje cells in kittens [11].

Extra-cerebellar lesions of the central nervous system associated with parvoviral infections were described in cats with demyelination of the spinal cord [12] and in cats with canine parvovirus replication in cerebral neurons [8, 13]. Such productive infections of cat cerebral neurons were strikingly associated with old CPV-2 strains, which seem to infect cats while they do not circulate in dog populations anymore, and have never been identified with CPV-2a and -2b variants, nor with FPV strains [8].

In the present study, we showed the presence of FPV proteins, likely VP1/VP2, given the antigens targeted by the CPV1-2A1 antibody used [7], in glial cells and neurons from diencephalic region of four cats, mostly

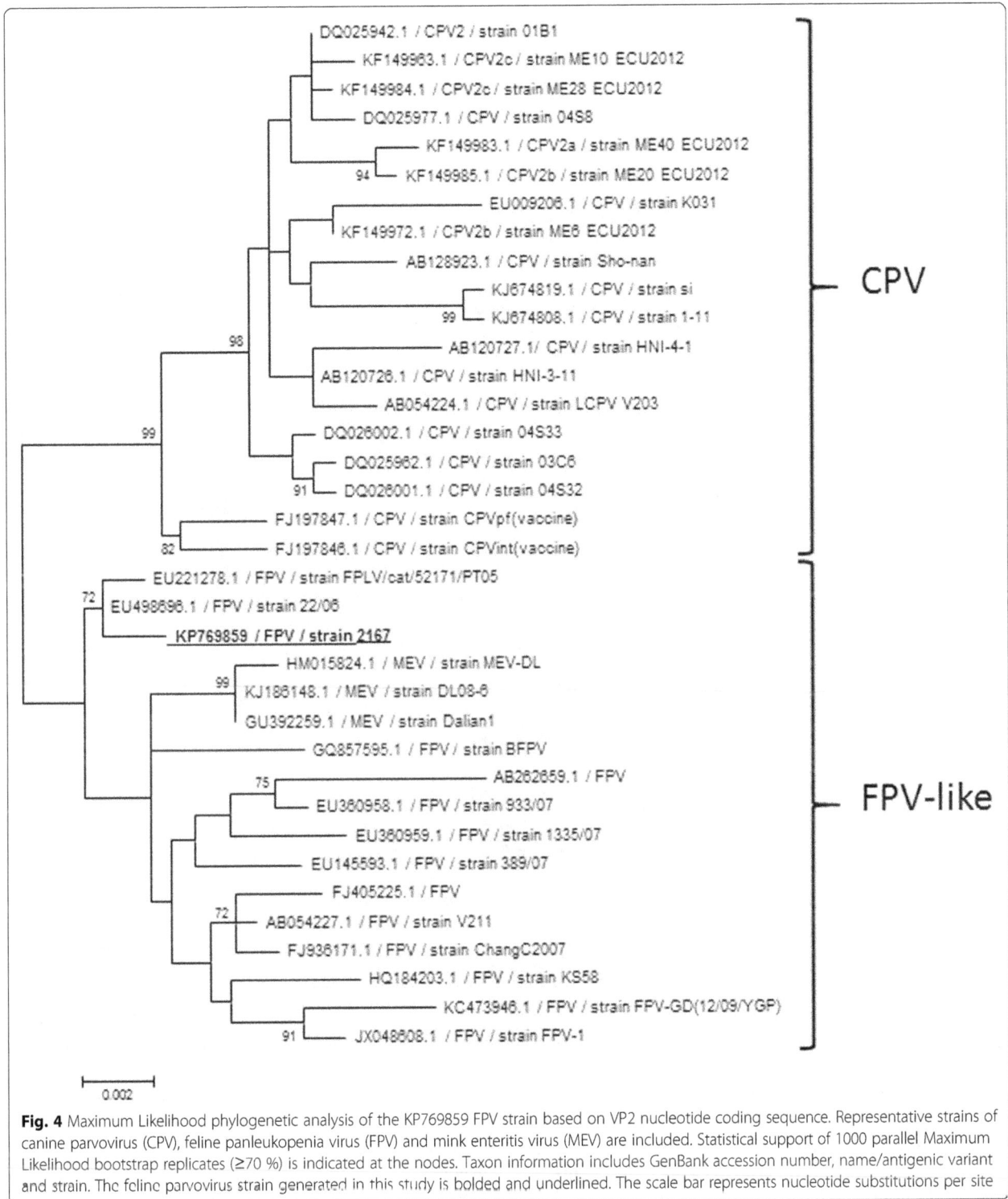

Fig. 4 Maximum Likelihood phylogenetic analysis of the KP769859 FPV strain based on VP2 nucleotide coding sequence. Representative strains of canine parvovirus (CPV), feline panleukopenia virus (FPV) and mink enteritis virus (MEV) are included. Statistical support of 1000 parallel Maximum Likelihood bootstrap replicates (≥70 %) is indicated at the nodes. Taxon information includes GenBank accession number, name/antigenic variant and strain. The feline parvovirus strain generated in this study is bolded and underlined. The scale bar represents nucleotide substitutions per site

the interthalamic adhesion. Similarly to observations reported with CPV-2 strains [8], the infection of cerebral neurons occurred in the absence of visible cerebellar infection or lesion. This is of utmost interest since some of the cats had a history of neurological signs before death. Moreover, one affected cat was adult, this last point raising questions about the possibility of a virus-induced re-entry of post-mitotic neurons into cell cycle [11]. To specifically address this point, we assessed the p27^{Kip1} expression of FPV-infected neurons by IHC. The cyclin-dependent kinase inhibitor p27^{Kip1} is expressed in cells that have exited the mitotic cycle [14]. Anti-p27^{Kip1}

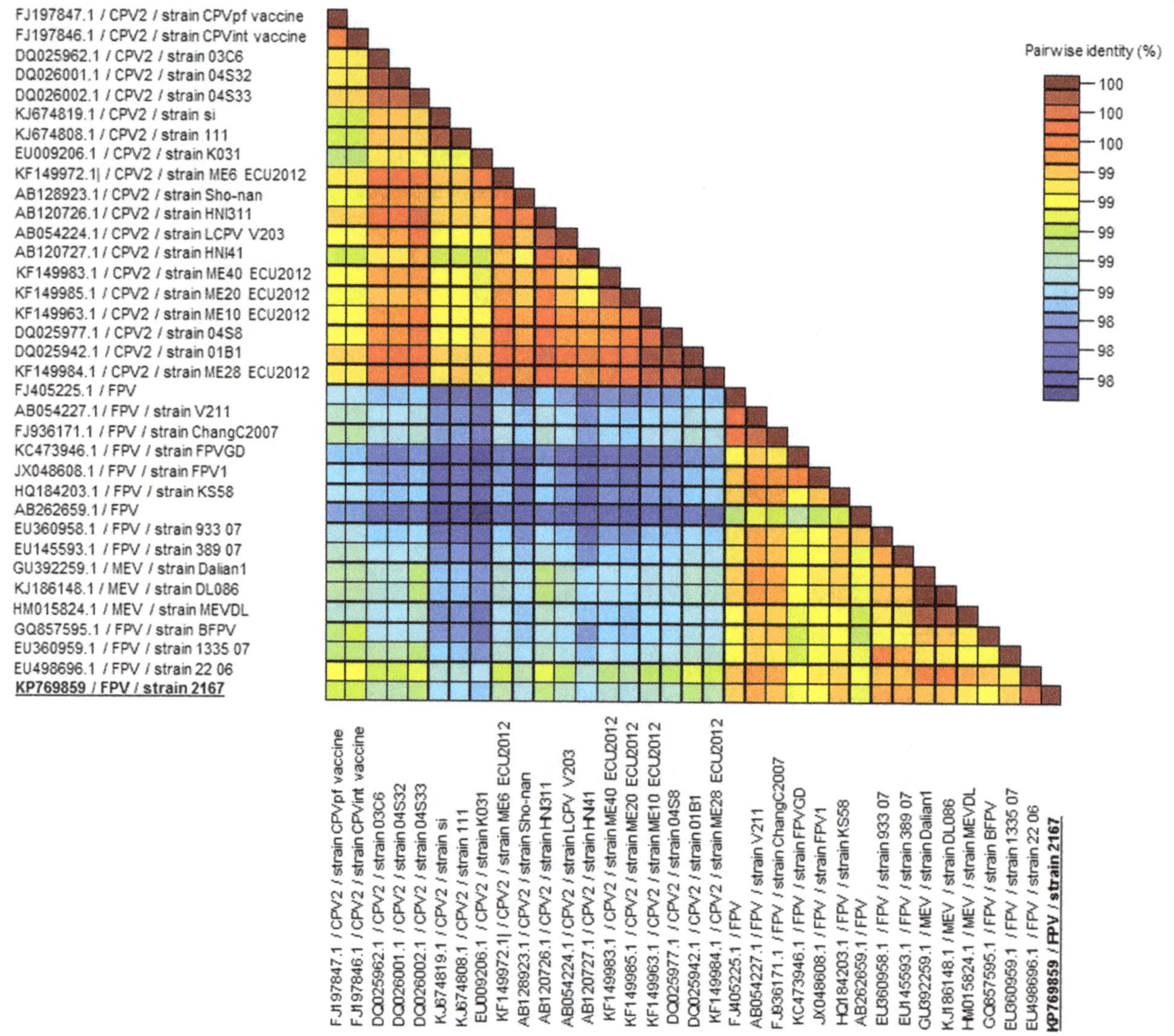

Fig. 5 Pairwise identity matrix based on the viral VP2 gene sequence. A color-coded pairwise identity matrix based on the VP2 nucleotide coding sequence from representative strains of canine parvovirus (CPV), feline panleukopenia virus (FPV) and mink enteritis virus (MEV) reveals the KP769859 strain generated in this study belongs to the FPV group. The KP769859 feline parvovirus strain is bolded and underlined. A color key indicates the correspondence between pairwise identities and the colors displayed in the matrix

and anti-FPV immunostainings were realized on serial sections from the same samples and revealed that while most neurons were p27-positive in the nucleus, those infected by FPV were p27-negative, meaning a possible re-entry into the mitotic cycle of some neurons including those expressing FPV proteins [14]. The ability of the anti-human p27^{Kip1} antibody used in this study to efficiently and specifically bind the feline p27^{Kip1} protein had been demonstrated previously [11]. Further, relatively high concentrations of the antibody with some background reaction were necessary to get a significant nuclear staining. Recent evidence suggest that cell cycle re-entry in post-mitotic neurons may occur under specific circumstances [15]. Although the number of cases

included in the present study is too low to be conclusive, this observation deserves to be further investigated. Besides, whether the lack of p27 expression is a cause or a consequence of the infection of these neurons by FPV remains to be determined.

Sequences analyses revealed that a unique L → S substitution was present in the NS1 protein from brain-infecting strains and not in other strains presented in this study. This substitution has thus far never been identified, neither in FPV nor in CPV strains. NS1 protein of human B19 parvovirus is known to cause cell cycle arrest at late S phase, which favors viral DNA replication [16] and is pro-apoptotic [17]. Similarly, Minute virus of Mice NS1 was associated with cell cycle arrest

Table 2 Parvovirus sequences from tissues of cats investigated (cats No 5, 10, 11, 15, 14, 18 and 23) in comparison with reference feline panleukopenia and canine parvovirus strains

Cat No or virus type/accession No	Origin	Amino acid position				
		VP2 aa 96	VP2 aa 106	VP2 aa 578	NS1 aa 248	NS1 aa 582
Cat No 5/KP769859	Cat - brain	K	V	A	T	S
Cat No 10	Cat - brain	K	V	A	T	S
Cat No 11	Cat - brain	K	V	A	T	S
Cat No 15	Cat - brain	K	V	A	ND	S
FPV/EU221279	Cat - fecal sample	K	V	A	ND	ND
FPV/BAA19024	Cat - ND	ND	ND	ND	T	L
CPV-2/FJ197847	Dog - fecal sample	N	A	G	ND	ND
CPV-2/NC_001539	Dog - cell culture	N	A	G	I	L
Cat No 14	Cat- ileum	ND	ND	ND	ND	L
Cat No 18	Cat- ileum	ND	ND	ND	ND	L
Cat No 23	Cat- ileum	ND	ND	ND	ND	L

ND: not determined

and p53 activation [18] and CPV-2 NS1 was shown to cause caspase-3 activation [19]. It is tempting to hypothesize that FPV NS1 could also be able to manipulate the cell cycle. The putative cause to effect of the NS1 mutation described for FPV strains in association with productive infection of cerebral neurons in this study deserves additional research. The potential link between this NS1 mutation and the lack of p27^{Kip1} expression in infected neurons should also be specifically addressed.

Lastly, the next-generation sequencing analysis of one parvovirus-infected cat's brain tissue showed the co-infection by type 1 feline bocavirus. Bocaviruses are enteric viruses of the *Parvoviridae* family which have

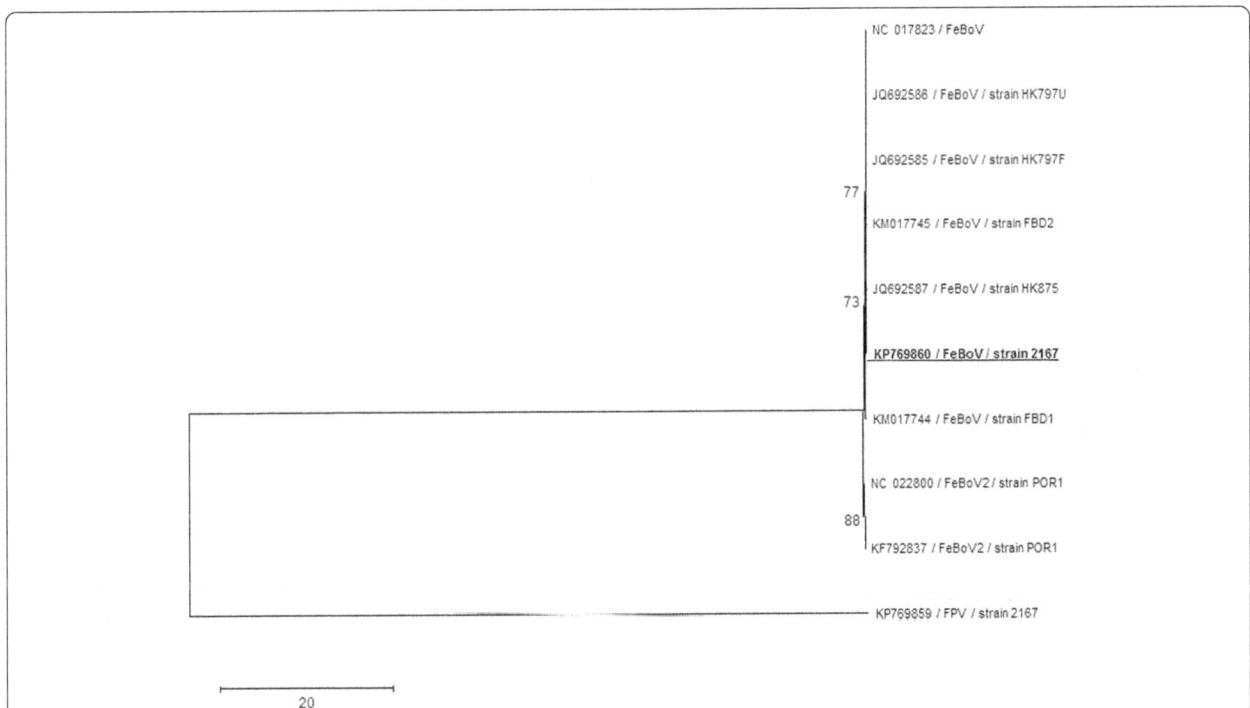

Fig. 6 Maximum Likelihood phylogenetic analysis of feline bocavirus strain KP769860 based on partial genomic sequence. Currently available type 1 and type 2 feline bocavirus strains are included. A partial genome (4,001 nucleotides) was used for analysis. Statistical support of 1000 parallel Maximum Likelihood bootstrap replicates (≥70 %) is indicated at the nodes. FPV strain 2167 (Genbank accession number KP769859, also described in this study) is used as outgroup. Taxon information includes GenBank accession number, name/antigenic variant and strain. The feline bocavirus strain generated in this study is bolded and underlined. The scale bar represents nucleotide substitutions per site

been described in several species and only recently in cats [20]. The pathogenic potential of these viruses remains to be determined [20]. Even if not confirmed by in situ histological techniques, the present study is the first report of nervous system infection by a bocavirus. The potential effects of the co-infection with FPV in the affected cat remains to be assessed.

Conclusions

Overall, the results presented in this study show that, like CPV-2, FPV is able to infect cerebral neurons of young or adult cats without involvement of the cerebellum. The identification of a unique substitution in NS1 protein might be related to this unusual tropism but this hypothesis would require further investigations. This study also provides the first evidence of nervous tissue infection by a bocavirus.

Methods

Sample collection and real time PCR

During the year 2013, a total of 28 cats referred for necropsy to the Veterinary Pathology Department of the University of Liège, Belgium, were confirmed as cases of feline panleukopenia. Samples of brain, cerebellum, spleen, small intestine, mesenteric lymph node, liver, kidney, lung and myocardium were collected and stored at –80 °C for subsequent PCR analysis or fixed in 10 % formalin, routinely processed and embedded in paraffin for histopathological evaluation.

The feline panleukopenia virus genome was detected using a Taqman real time PCR assay previously described [21]. In the Taqman probe, BHQ1 was used as a quencher.

All studies were in accordance with the guidelines of the Institutional Animal Care and Use Committees of the University of Liège.

Immunohistochemical staining

The organs examined for histopathology were subsequently submitted to immunohistochemical analysis. A commercial mouse monoclonal anti-parvovirus antibody (clone CPV1-2A1, sc57961, Santa Cruz, Dallas, Texas,

USA) [7] and a polyclonal rabbit anti-human p27[Kip1] antibody (ab7961, Abcam, Cambridge, United Kingdom) [11] were used as primary antibodies (dilution 1:50, for both). Goat anti-mouse or goat anti-rabbit serum (dilution 1/160, 323-005-024 and 323-005-024, respectively, Jackson ImmunoResearch, West Grove, USA) was used as secondary antibody. Binding was visualized using the peroxidase anti-peroxidase method with diaminobenzidine as the chromogen; sections were counterstained with Mayer hematoxylin. Cerebellar sections from an FPV-infected kitten with cerebellar hypoplasia were used as a positive control for the anti-parvovirus immunohistochemistry [7, 11], and sections from a fetal cerebellum (estimated gestational age, 54 days) were used as a positive control for the anti-p27[Kip1] immunohistochemistry [11].

Next-generation sequencing, confirmation PCRs and sequence analysis

Brain samples from three cats (5, 10, 11) for which sufficient amounts of tissue were available were submitted to next-generation sequencing. Briefly, 500 mg of tissue were homogenized in 1 ml of 1x DNase buffer (Life Technologies, Ghent, Belgium) for 5 min at 30Hz using a TissueLyser II device (Qiagen, Hilden, Germany). After centrifugation for 10 min at 11,000 g, supernatant was collected and filtered using a 0.2 µm filter (Pall Corporation, Newquay, United Kingdom). Viral particles were concentrated using a Microcon-100 kDa column (Merck Millipore, Billerica, USA) for 30 min at 500 g. Viral particles were washed once with 100 µl of 1x DNase buffer for 20 min at 500 g and eluted by inverting the column for 3 min at 1000 g. Turbo™Dnase (Life Technologies, Ghent, Belgium) and RNase A/T1 (Thermo Scientific, Waltham, USA) were added to the elution at a 1:50 dilution. The mixture was stored for 1 h at 37 °C. Viral particles were then digested using proteinase K and total DNA was extracted using NucleoSpin Tissue kit (Macherey-Nagel, Düren, Germany) according to manufacturer's instructions. Libraries were prepared using Ion Plus Fragment Library Kit and sequencing was performed with the Ion Torrent PGM technology (Life

Table 3 Primers used for amplification and sequencing of feline bocavirus (design/numbering based on GenBank accession number KM017745) and for confirmation of the substitution in NS1 coding sequence of FPV (design/numbering based on GenBank accession number KP769859)

Forward Primer	Sequence	Position	Reverse primer	Sequence	Position
BocaF1	ACAATGCCTGGACCTGAATC	481 → 500	BocaR1	TTTTACTCATTCTGGCATTCACA	1,276 → 1,254
BocaF2	CTAGTCAGGGACGAACCAA	1,141 → 1,159	BocaR2	CCATTGAGTATGAAAGCCAACT	2,103 → 2,082
BocaF3	CACGTCTAGAGAGCACCTTTGG	1,994 → 2,015	BocaR3	ATCAACCTCCATGGCAACC	3,118 → 3,100
BocaF4	DGGCATTTTAGCATGGGCGAA	2,966 → 2,986	BocaR4	GAAAATACCCGTATTGCGGAAGT	4,113 → 4,091
BocaF5	TACAGCGGGTGTACACATTT	3,9864,005	BocaR5	GTAGCAGTGTGGAGGGTGT	5,368 → 5,350
ParvF3	CAACCAATAAGAGACAGAATGTTGA	1,785 → 1,809	ParvR3	CCCCCACTTTACTAACACACC	2,342 → 2,322

Technologies, Ghent, Belgium). Full genome assembly and sequence analysis were performed with Geneious 8.0.5 (Biomatters, Auckland, New Zealand). Phylogenetic analyses were realized using MEGA 6.0 [22]. A nucleotide sequence identity matrix was produced with Sequence Demarcation Tool version 1.2 [23].

A confirmation PCR was used to verify a point mutation in NS1 coding sequence. Primers (Table 3) were designed using Primer3 [24]. PCRs were performed using HotStarTaq Plus Master Mix (Qiagen, Hilden, Germany) and conditions were as follows: a denaturation step at 95 °C for 5 min, followed by 45 cycles of 95 °C for 20 s, 55 °C for 45 s, 72 °C for 2 min, and final elongation 5 min at 72 °C. Sanger sequencing of the amplicons was performed by GATC Biotech (Konstanz, Germany).

Besides, primers were designed (Table 3) to amplify the genome of a feline bocavirus identified in the brain of one cat by next-generation sequencing. PCR/sequencing conditions were as described for panleukopenia virus.

Ethical consideration
According to the law of the Walloon region in Belgium [25], standard diagnostic procedures performed on dead animals or animal tissues do not require permission from the Ethical Board. Cats included in this study were specifically referred by field veterinarians to the Faculty, in agreement with the owners, for pathological evaluation (necropsy and histopathology) and diagnosis of feline panleukopenia.

Competing interests
The authors declare that they have no competing interest.

Authors' contributions
MG, GG, SJ, CB, TC, KG, TF, AM, EL, DC, DP and LP participated in the sample collection and experimental part. MG, LP and DD designed the study and drafted the manuscript. All authors read and approved the final manuscript.

Acknowledgements
We thank Jérôme Wayet and Michaël Sarlet for excellent technical support.

Author details
[1]Department of Morphology and Pathology, University of Liège, Liège, Belgium. [2]Department of Clinical Sciences, University of Liège, Liège, Belgium. [3]Department of Infectious and Parasitic Diseases, Centre for Fundamental and Applied Research for Animals & Health, Faculty of Veterinary Medicine, University of Liège, Liège, Belgium. [4]Laboratory of Anatomy, Biomechanics and Organogenesis, Faculty of Medicine, Free University of Brussels, Brussels, Belgium.

References
1. Allison AB, Kohler DJ, Ortega A, Hoover EA, Grove DM, Holmes EC, Parrish CR. Host-specific parvovirus evolution in nature is recapitulated by in vitro adaptation to different carnivore species. PLoS Pathog. 2014;10, e1004475.
2. Siegl G, Bates RC, Berns KI, Carter BJ, Kelly DC, Kurstak E, Tattersall P. Characteristics and taxonomy of Parvoviridae. Intervirology. 1985;23:61–73.
3. Parrish CR, O'Connell PH, Evermann JF, Carmichael LE. Natural variation of canine parvovirus. Science. 1985;230:1046–8.
4. Berns KI. Parvovirus replication. Microbiol Rev. 1990;54:316–29.
5. Deleu L, Pujol A, Faisst S, Rommelaere J. Activation of promoter P4 of the autonomous parvovirus minute virus of mice at early S phase is required for productive infection. J Virol. 1999;73:3877–85.
6. Resibois A, Coppens A, Poncelet L. Naturally occurring parvovirus-associated feline hypogranular cerebellar hypoplasia– A comparison to experimentally-induced lesions using immunohistology. Vet Pathol. 2007;44:831–41.
7. Poncelet L, Heraud C, Springinsfeld M, Ando K, Kabova A, Beineke A, Peeters D, Op De Beeck A, Brion J-P. Identification of feline panleukopenia virus proteins expressed in Purkinje cell nuclei of cats with cerebellar hypoplasia. Vet J. 2013; 196:381–7.
8. Url A, Truyen U, Rebel-Bauder B, Weissenbock H, Schmidt P. Evidence of parvovirus replication in cerebral neurons of cats. J Clin Microbiol. 2003;41: 3801–5.
9. Kilham L, Margolis G, Colby ED. Congenital infections of cats and ferrets by feline panleukopenia virus manifested by cerebellar hypoplasia. Lab Invest. 1967;17:465–80.
10. Pedersen NC. Feline panleukopenia virus. In: Appel MJG, editor. Virus infection of carnivores. New York: Elsevier Science Publishers; 1987. p. 247–54.
11. Poncelet L, Springinsfeld M, Ando K, Heraud C, Kabova A, Brion J-P. Expression of transferrin receptor 1, proliferating cell nuclear antigen, p27(Kip1) and calbindin in the fetal and neonatal feline cerebellar cortex. Vet J. 2013;196:388–93.
12. Csiza CK, Scott FW, De Lahunta A, Gillespie JH. Respiratory signs and central nervous system lesions in cats infected with panleukopenia virus. A case report. Cornell Vet. 1972;62:192–5.
13. Url A, Schmidt P. Do canine parvoviruses affect canine neurons? An immunohistochemical study. Res Vet Sci. 2005;79:57–9.
14. Miyazawa K, Himi T, Garcia V, Yamagishi H, Sato S, Ishizaki Y. A role for p27/Kip1 in the control of cerebellar granule cell precursor proliferation. J Neurosci. 2000;20:5756–63.
15. Frade JM, Ovejero-Benito MC. Neuronal cell cycle: the neuron itself and its circumstances. Cell Cycle. 2015;14:712–20.
16. Luo Y, Kleiboeker S, Deng X, Qiu J. Human parvovirus B19 infection causes cell cycle arrest of human erythroid progenitors at late S phase that favors viral DNA replication. J Virol. 2013;87:12766–75.
17. Poole BD, Zhou J, Grote A, Schiffenbauer A, Naides SJ. Apoptosis of liver-derived cells induced by parvovirus B19 nonstructural protein. J Virol. 2006; 80:4114–21.
18. Op De Beeck A, Sobczak-Thepot J, Sirma H, Bourgain F, Brechot C, Caillet-Fauquet P. NS1- and minute virus of mice-induced cell cycle arrest: involvement of p53 and p21(cip1). J Virol. 2001;75:11071–8.
19. Saxena L, Kumar GR, Saxena S, Chaturvedi U, Sahoo AP, Singh LV, Santra L, Palia SK, Desai GS, Tiwari AK. Apoptosis induced by NS1 gene of Canine Parvovirus-2 is caspase dependent and p53 independent. Virus Res. 2013; 173:426–30.
20. Lau SKP, Woo PCY, Yeung HC, Teng JLL, Wu Y, Bai R, Fan RYY, Chan K-H, Yuen K-Y. Identification and characterization of bocaviruses in cats and dogs reveals a novel feline bocavirus and a novel genetic group of canine bocavirus. J Gen Virol. 2012;93(Pt 7):1573–82.
21. Decaro N, Elia G, Martella V, Desario C, Campolo M, Trani LD, Tarsitano E, Tempesta M, Buonavoglia C. A real-time PCR assay for rapid detection and quantitation of canine parvovirus type 2 in the feces of dogs. Vet Microbiol. 2005;105:19–28.
22. Tamura K, Stecher G, Peterson D, Filipski A, Kumar S. MEGA6: Molecular Evolutionary Genetics Analysis version 6.0. Mol Biol Evol. 2013;30:2725–9.
23. Muhire BM, Varsani A, Martin DP. SDT: a virus classification tool based on pairwise sequence alignment and identity calculation. PLoS One. 2014;9, e108277.
24. Untergasser A, Cutcutache I, Koressaar T, Ye J, Faircloth BC, Remm M, Rozen SG. Primer3–new capabilities and interfaces. Nucleic Acids Res. 2012;40, e115.
25. The Parliament of Wallon region of Belgium. 29 mai 2013 - Arrêté royal relatif à la protection des animaux d'expérience (M.B. 10.07.2013) + erratum 26.07.2013 + erratum 21.01.2014 [http://environnement.wallonie.be/legis/bienetreanimal/bienetre008.html]

2

The impact of *Aloe vera* and licorice extracts on selected mechanisms of humoral and cell-mediated immunity in pigeons experimentally infected with PPMV-1

Daria Dziewulska*[ID], Tomasz Stenzel, Marcin Śmiałek, Bartłomiej Tykałowski and Andrzej Koncicki

Abstract

Background: The aim of the study was to evaluate the impact of herbal extracts on selected immunity mechanisms in clinically healthy pigeons and pigeons inoculated with the pigeon paramyxovirus type 1 (PPMV-1). For the first 7 days post-inoculation (dpi), an aqueous solution of *Aloe vera* or licorice extract was administered daily at 300 or 500 mg/kg body weight (BW). The birds were euthanized at 4, 7 and 14 dpi, and spleen samples were collected during necropsy. Mononuclear cells were isolated from spleen samples and divided into two parts: one part was used to determine the percentage of IgM^+ B cells in a flow cytometric analysis, and the other was used to evaluate the expression of genes encoding IFN-γ and surface receptors on $CD3^+$, $CD4^+$ and $CD8^+$ T cells.

Results: The expression of the IFN-γ gene increased in all birds inoculated with PPMV-1 and receiving both herbal extracts. The expression of the CD3 gene was lowest at 14 dpi in healthy birds and at 7 dpi in inoculated pigeons. The expression of the CD4 gene was higher in uninoculated pigeons receiving both herbal extracts than in the control group throughout nearly the entire experiment with a peak at 7 dpi. A reverse trend was observed in pigeons inoculated with PPMV-1 and receiving both herbal extracts. In uninoculated birds, increased expression of the CD8 gene was noted in the pigeons receiving a lower dose of the *Aloe vera* extract and both doses of licorice extracts. No significant differences in the expression of this gene were found between inoculated pigeons receiving both herbal extracts. The percentage of IgM^+ B cells did not differ between any of the evaluated groups.

Conclusions: This results indicate that *Aloe vera* and licorice extracts have immunomodulatory properties and can be used successfully to prevent viral diseases, enhance immunity and as supplementary treatment for viral diseases in pigeons.

Keywords: *Aloe vera*, Flow cytometry, Gene expression, Herbal extracts, Licorice, Pigeons, PPMV-1

* Correspondence: daria.pestka@uwm.edu.pl
Department of Poultry Diseases, Faculty of Veterinary Medicine, University of Warmia and Mazury in Olsztyn, ul. Oczapowskiego 13/14, 10-719 Olsztyn, Poland

Background

Immunomodulation is the stimulation or suppression of immune responses in living organisms. Numerous substances, both natural and synthetic, exert effects on immunity. Natural immunomodulators include herbal preparations whose popularity continues to increase due to the decreasing effectiveness of antibiotics and other synthetic drugs [27].

The therapeutic properties of *Aloe vera*, also known as Barbados aloe (*Aloe barbadensis Miller*), have been recognized already in ancient times. *Aloe vera* is a succulent plant of the lily family (*Liliaceae*) [6]. The part of *Aloe vera* plants that plays the most important role in natural medicine are its leaves which are a rich source of latex and gel containing 98.5% to 99.5% water and 75 biologically active compounds [9]. *Aloe vera* gel also contains polysaccharides, including acemannan which is one of the most potent plant-derived immunomodulators. Acemannan binds to macrophage receptors and stimulates the synthesis of cytokines (interleukin 1 (IL-1), interleukin 6 (IL-6)) and tumor necrosis factor-alpha (TNF-α) [8, 11]. *Aloe vera* extract could also stimulate cell-mediated immunity (CMI). Vahedi et al. (2011) reported a higher percentage of CD4$^+$ and CD8$^+$ T cells in the peripheral blood of rabbits receiving *Aloe vera* extract [43]. *Aloe vera* extracts were also found to stimulate humoral immunity in chickens experimentally infected with the Newcastle disease virus (NDV) [26]. The discussed plant delivers numerous health benefits and exerts anti-inflammatory, antibacterial, antifungal and anti-carcinogenic effects due to the presence of anthraquinones, saccharides and antioxidant vitamins (A, C and E) [38].

Licorice (*Glycyrrhiza glabra*) is also a popular medicinal plant of the legume family (*Fabaceae*). It is valued mostly for its roots which contain 1% to 9% glycyrrhizic acid (glycyrrhizin) [15]. Glycyrrhizin is a potent immunomodulator which stimulates the production of interferon [1, 42] and the proliferation of regulatory (Treg) cells in mice [16]. Licorice extracts have been found to increase the phagocytic capacity of chicken granulocytes and mononuclear cells [12]. Similarly to *Aloe vera*, licorice exerts various types of antiviral activity. Licorice inhibits viral replication not only by becoming attached to the cell membrane and compromising the cells' ability to undergo endocytosis, which prevents the virus from penetrating cells [46], but also by activating the NF-κB protein complex which plays a key role in regulating the immune response to infections and stimulates IL-8 secretion [34].

Medicinal herbs are widely used as functional additives in animal diets to improve the palatability and digestibility of feed [4, 21]. Herbal functional additives have various properties and can be used in the prevention and supplementary treatment of infectious diseases with different etiology. These properties can be directly attributed to herbal extracts' ability to stimulate immune responses, which was observed in chickens [9]. For example, Liu et al. (2010) demonstrated that the addition of four herbal extracts, *Astragalus membranaceus*, *Codonopsis pilosula*, *Epimedium spp.* and *Glycyrrhiza uralensis*, to drinking water can enhance the immune response in immunosuppressed chickens with the reticuloendotheliosis virus [23]. Further evidence of the immunomodulatory effects of herbal extracts was provided by Latheef et al. (2017) who reported that *Withania somnifera*, *Tinospora cordifolia* and *Azadirachta indica* were capable of inhibiting the replication of the chicken infectious anemia virus and increasing the cell-mediated response of chickens against this virus [22]. However, the immunomodulatory effects of herbal extracts have never been investigated in domestic pigeons.

Viral diseases, in particular infections with the pigeon circovirus (PiCV) which exerts immunosuppressive effects, pose a serious problem in pigeon breeding [36]. Since a laboratory protocol for culturing PiCV under laboratory conditions has not been developed to date [10], the pigeon paramyxovirus type 1 (PPMV-1), the pigeon variant of NDV, is successfully used for experimental inoculation of pigeons [13, 28, 37, 39].

The course of an NDV infection can differ substantially, depending on the strain's virulence [28]. Strain virulence also determines birds' immune responses to infection or inoculation with live vaccines. The early immune response to a viral infection is influenced by innate immunity, a universal mechanism that protects living organisms against infections. Innate immunity relies on pattern recognition receptors (PRRs) which identify pathogen-associated molecular patterns (PAMPs). PRRs enable an organism to discriminate between nonself and self antigens. Toll-like receptors (TLRs) are a group of PRRs which play a key role in the initiation of immune responses [40]. TLRs are found on the surface of selected immune system cells, such as lymphocytes, heterophils and macrophages, and their stimulation constitutes a signal that activates non-specific and specific immune responses. In an in vitro study, the inoculation of chicken peripheral blood heterophils and mononuclear cells with NDV stimulated the production of interferon and nitric oxide (NO) [2]. Research has also demonstrated that the expression of genes encoding interferon α (IFN-α), IFN-ß, IL-1ß and IL-6 increased in chicken splenocytes inoculated with NDV [32]. However, the observed increase in expression was determined by the NDV strain and its virulence, and it was not induced by mild viral strains [20, 24].

Cell-mediated immunity associated with T cells, including cytokine-producing CD4$^+$ lymphocytes and cytotoxic

CD8$^+$ lymphocytes (CTL), also plays an important role in the immune response to NDV. In chickens, a CMI response to NDV was observed already 2–3 days after inoculation with the vaccine virus strain [31]. Similarly to the innate immune response, the adaptive immune response is influenced by several factors, including strain virulence [30] and the breed of chickens exposed to vaccine and field isolates [7]. The humoral immune response, which involves the proliferation of B cells and the production of immunoglobulins M, Y and A (IgM, IgY, IgA), is also an important element of immunity against NDV [19]. Anti-NDV antibodies are detected in mucosal membranes of the upper respiratory tract and in blood already 6 days after infection or inoculation with an attenuated vaccine, and their concentrations peak 21–28 days after infection. The antibodies' role is to neutralize the virus by binding to it and preventing it from adhering to host cells [3].

The influence of paramyxovirus infections on the immune response in birds has been studied extensively [19, 20, 24, 32]. However, very little is known about the influence of immunomodulatory herbal extracts on viral infections and immune responses in infected birds. A few studies have been conducted to investigate the immunomodulatory effects of *Aloe vera* and licorice extracts on birds infected with the avian paramyxovirus serotype-1 (APMV-1), and their results are limited to analyses of antibodies against this virus in chickens [26].

In view of alternative immunomodulation-based strategies and the scarcity of published information relating to the applicability of immunomodulatory herbal extracts in pigeons, the main aim of this basic research was to determine the influence of *Aloe vera* and licorice extracts on selected mechanisms of cell-mediated and humoral immunity in virus-inoculated pigeons. PPMV-1 was used as an experimental model because it is easy to culture under laboratory conditions. However, it should be noted that Newcastle disease is a notifiable disease that has to be legally reported to the authorities, and treatment of PPMV-1 infections in pigeons is not allowed.

Methods
Virus
Pigeons were infected with the pigeon paramyxovirus serotype-1 (PPMV-1/pigeon/Poland/AR3/95) obtained from the National Veterinary Research Institute in Puławy. The pathogenicity of the applied isolate was classified based on biological (calculation of the Intracerebral Pathogenicity Index (ICPI) for one-day-old SPF chickens) and molecular analyses (analysis of the amino acid sequence at the cleavage site in the fusion protein). The ICPI was 1.4, and the amino acid sequence at the cleavage site in the fusion protein was [112]R-R-Q-K-R-

F[117]. Based on those results, the virus was classified as a mesogenic pathotype.

Plant extracts
Aloe vera
The *Aloe vera* extract was obtained by freeze/spray drying of aloe leaf juice. Five grams of the extract with maximum moisture content of 8% and bulk density of 0.3–0. 6 g/1 ml were obtained from 1000 g of fresh *Aloe vera* juice.

Licorice
Dry licorice extract was obtained by spray drying an aqueous solution of licorice root, a registered feed additive (European Union Register of Feed Additives, group 2b: natural products – botanically defined: CAS 68916–91-6 FEMA 2629, CoE 218, pursuant to Regulation (EC) No 1831/2003). The extract contained 20% glycyrrhizic acid, and it was characterized by maximum moisture content of 3.6% and bulk density of 0.5 g/1 ml.

Aloe vera and licorice extracts were free of pathogenic bacteria such as *Escherichia coli*, *Staphylococcus aureus* and *Pseudomonas aeruginosa* in 10 g of the product. The contamination of the extract with selected pathogenic bacteria (*Escherichia coli*, *Staphylococcus aureus* and *Pseudomonas aeruginosa*) was determined in accordance with PN-EN ISO 6887–1 [29]. First, a 10% solution of the extract was prepared in the amount of 100 mL (conc. 10^{-1}), and it was used as the initial suspension that was diluted ten-fold to obtain a concentration of 10^{-5}. Using a sterile pipette, 1 mL of the sample from each dilution was transferred to the following culture media: MacConkey Agar No. 3, Columbia Agar with sheep blood plus and Mannitol salt agar. All culture media were obtained from the same manufacturer (Oxoid, UK), and all analyses were conducted in duplicate. The suspension was spread evenly with a sterile cell spreader, and the plates were incubated at a temperature of 37 °C for 24 h. Beginning with the first dilution, an increase in the CFU of the tested pathogenic bacteria was not observed on any of the plates after incubation.

Pigeons
One hundred twenty 8-week-old fantail pigeons were obtained from a private breeder. The flock in the breeding facility had not been vaccinated against PPMV-1 since 2008, and it was free of the infection. Before the experiment, cloacal swabs and blood samples were collected from all birds to rule out PPMV-1 infection with the use of the real-time PCR method described by Wise et al. (2004) and modified by Cattoli et al. (2009) and to determine the presence of antibodies against PPMV-1 with the use of the commercial ELISA test kit (IDEXX, USA) according to the method

described by Stenzel et al. (2011) [5, 35, 45]. The birds were housed in isolated units in a PCL3 biosafety facility of the Department of Poultry Diseases, Faculty of Veterinary Medicine of the University of Warmia and Mazury in Olsztyn. The biosafety facility is equipped with a HEPA filtering system and an automated system for pressure control in corridors, bird units and hygiene stations to prevent contamination of experimental premises. Every group of pigeons was housed in a separate unit. The birds were administered seed mixtures and water ad libitum throughout the experiment.

Experimental design

Pigeons were divided into 10 groups of 12 birds each. Pigeons from groups A1, B1, C1, D1 and K1 were inoculated oculonasally with 10^6 EID_{50} of PPMV-1 at 100 μL per bird (applied to the nostril and the eye at 50 μL each). For the first 7 days post-inoculation (dpi), an aqueous solution of *Aloe vera* extract was administered daily per os at 300 mg/kg body weight (BW) (groups A and A1) or 500 mg/kg BW (groups B and B1), and an aqueous solution of licorice extract was administered at 300 mg/kg BW (group C, C1) or 500 mg/kg BW (group D, D1). Control group (K and K1) birds were orally administered 0.9% NaCl. At 4, 7 and 14 dpi, the birds were euthanized by intravenous administration of pentobarbital sodium at 70 mg/1 kg BW (Morbital, Biowet Puławy, Poland) after premedication by intramuscular injection of butorphanol tartrate at 4 mg/1 kg BW (Torbugesic, Zoetis, USA), and spleen samples were collected during an anatomopathological examination (Table 1). Mononuclear cells were isolated from spleen samples and divided into two parts: one part was used to determine the percentage of IgM^+ B cells in a flow cytometric analysis, and the other was used in RNA extraction to evaluate the expression of genes encoding IFN-γ and surface receptors on $CD3^+$, $CD4^+$ and $CD8^+$ T cells.

Isolation of mononuclear cells

Mononuclear cells were isolated from whole spleens using the manual Dounce tissue grinder (Kimble, USA) in 9 ml of a complete growth medium (RPMI – 1640, 10% fetal bovine serum (FBS), 1% MEM non-essential amino acids solution, 1% penicillin – streptomycin, 1% HEPES, 1% sodium pyruvate) (Sigma Aldrich, USA) and were filtered (70 μm mesh). A homogenous suspension was obtained, and centrifuged cell pellets (450 g for 10 min at 25 °C) were resuspended in 2.3 mL of a complete growth medium and gently layered on 2.5 mL of Histopaque-1077 (Sigma Aldrich, USA). After centrifugation (30 min, 400 g, at room temperature), the upper layer of the opaque interface containing mononuclear cells was carefully aspirated. Finally, the obtained mononuclear cells were washed twice and resuspended in 1 mL of PBS (phosphate-buffered saline) (Sigma Aldrich, USA). Cell concentrations and the percentage of viable cells were determined in the Vi-cell XR analyzer (Beckman Coulter, USA).

Flow cytometry

Before the experiment, the cross-reactivity of Goat anti-Chicken IgM-FITC polyclonal antibodies (AbD Serotec, UK) was checked in pigeon lymphocytes. For this purpose, mononuclear cells were isolated from the thymus, bursa of Fabricius and peripheral blood. The antibodies' cross-reactivity was characterized by 3.99%, 44.46% and 9.5% of IgM^+ B cells isolated from the thymus, bursa of Fabricius and peripheral blood, respectively. The quantity of the tested antibodies (3 μg antibodies per one million cells) was determined experimentally in serial dilutions (1 to 5 μg antibodies per one million cells).

Thereafter, half a million mononuclear cells isolated from spleen samples were stained with 1.5 μg of the IgM^+ polyclonal antibody for B cells. The samples were incubated in darkness on ice for 30 min. Next, the cells were twice rinsed in PBS, centrifuged at 400 g for 10 min,

Table 1 Experimental design

Group	Day of experiment					
	1–7	8	9–15	12	15	22
		Experimental inoculation with PPMV-1, 10^6 EID_{50}	Once daily administration of:			
A	Adaptation to new conditions	–	*Aloe vera*, 300 mg/kg BW	Collection of spleen samples for molecular biology and flow cytometry analyses		
A1		+				
B		–	*Aloe vera*, 500 mg/kg BW			
B1		+				
C		–	licorice, 300 mg/kg BW			
C1		+				
D		–	licorice, 500 mg/kg BW			
D1		+				
K		–	0.9% NaCL			
K1		+				

and the resulting pellets were suspended in 400 µL of PBS and analyzed with the use of the FACS Canto II (BD, USA) flow cytometer. Data were acquired in FACS Diva Software 6.1.3. (BD, USA). Cells were analyzed and immunophenotyped in FloJo 7.5.5 (Tree Star, USA).

Real-time PCR
The number of mononuclear cells isolated from spleen samples was standardized to 5×10^6 and used for RNA isolation with the use of the RNeasy Mini Kit (Qiagen, Germany) according to the manufacturer's protocol. Genomic DNA remaining in the samples after RNA isolation was digested with deoxyribonuclease I (Sigma Aldrich, USA). RNA quality was evaluated in the 2100 Bioanalyzer (Agilent, USA). The concentrations of eluted RNA were measured with the NanoDrop 2000 spectrophotometer (Thermo Fisher Scientific, USA), and the samples were stored at – 80 °C until further analysis.

Reverse transcription was carried out with the High-Capacity cDNA Reverse Transcription Kit (Life Technologies, USA) according to the manufacturer's recommendations. The concentration of RNA for the synthesis of complementary DNA (cDNA) was standardized to 0. 5 µg per sample. The expression of the gene encoding IFN-ɣ and the genes encoding receptors on the surface of T cells (CD3, CD4 and CD8) was determined by real-time PCR. The reaction mixture for all analyzed genes had the following composition: 10 µL of the Power SYBR® Green PCR Master Mix (Life Technologies, USA), 1.8 µL of each 10 µM primer, 4.4 µL of RNase-free water, and 2 µL of cDNA. The primer sequences and the accession numbers of gene sequences used for designing the primers are presented in Table 2. The reaction was carried out under the following conditions: polymerase activation at 95 °C for 10 min, followed by 40 two-stage cycles: denaturation at 95 °C for 30 min, primer annealing and chain elongation at 60 °C for 60 s. The relative expression of each gene was calculated using the $2^{-\Delta\Delta Ct}$ method [25] normalized to efficiency corrections,

expression levels of reference gene coding glyceraldehyde 3-phosphate dehydrogenase (GAPDH) and reference groups (K and K1) in GenEx 6.1.0.757 data analysis software (MultiD, Sweden).

Statistical analysis
The significance of differences between the relative expression of IFN-ɣ, CD3, CD4 and CD8 genes and the percentage of IgM⁺ B cells were analyzed using the Kruskal-Wallis non-parametric test for independent samples. The analyzed factors were the experimental group and the day of the experiment. Differences were considered significant at a confidence level of 95% ($P < 0.05$).

Results
Expression of the IFN-ɣ gene
No significant differences in the expression of the gene encoding IFN-ɣ in mononuclear cells isolated from spleen samples were observed between the experimental groups during the experiment. The expression of the above gene was higher in groups A and C at 4 and 14 dpi, and in groups B and D at 4 dpi than in control group K (expression level > 1). In birds uninfected with PPMV-1, the lowest levels of IFN-ɣ gene expression were noted at 7 dpi (Fig. 1). In all inoculated birds receiving herbal extracts, the expression of the IFN-ɣ gene was higher than in control group K1 at 4, 7 and 14 dpi (Fig. 2).

Expression of the CD3 gene
The expression of the gene encoding surface receptor CD3 in the mononuclear cells of group A and D birds at 4 and 7 dpi and group C birds at 7 dpi was higher than in the control group. In group B, the expression of the analyzed gene did not increase relative to group K throughout the experiment, and it was significantly lower than in group D at 4 and 7 dpi ($P = 0.023$ and $P = 0.045$, respectively) (Fig. 1). The expression of the CD3 gene was higher

Table 2 Primers used for real time PCR

Primer	Sequence 5′ -> 3′	Fragment size (bp)	Accession number
CD3 F	GCAATTTACGATGATCCCAGAG	112	XM_005500716.2
CD3 R	GCGTCCACTTCAATGCAATTC		
CD4 F	GAACGTGTGAATGGGACTCAGA	116	MG214789
CD4 R	GTCATTGTCTTCTATGAGGTGACA		
CD8 F	TTCATCTGGGTTCCCTTGGCA	97	MG214790
CD8 R	CTGCATCTTCGGCTCCTGGT		
IFNɣ F	CTGACAAGTCAAAGCCGCAC	125	DQ479967.1
IFNɣ R	AGTCATTCATCT GAAGCTTGGC		
GAPDH F	CCCTGAGCTCAATGGGAAGC	137	NM_001282835.1
GAPDH R	TCAGCAGCAGCCTTCACTAC		

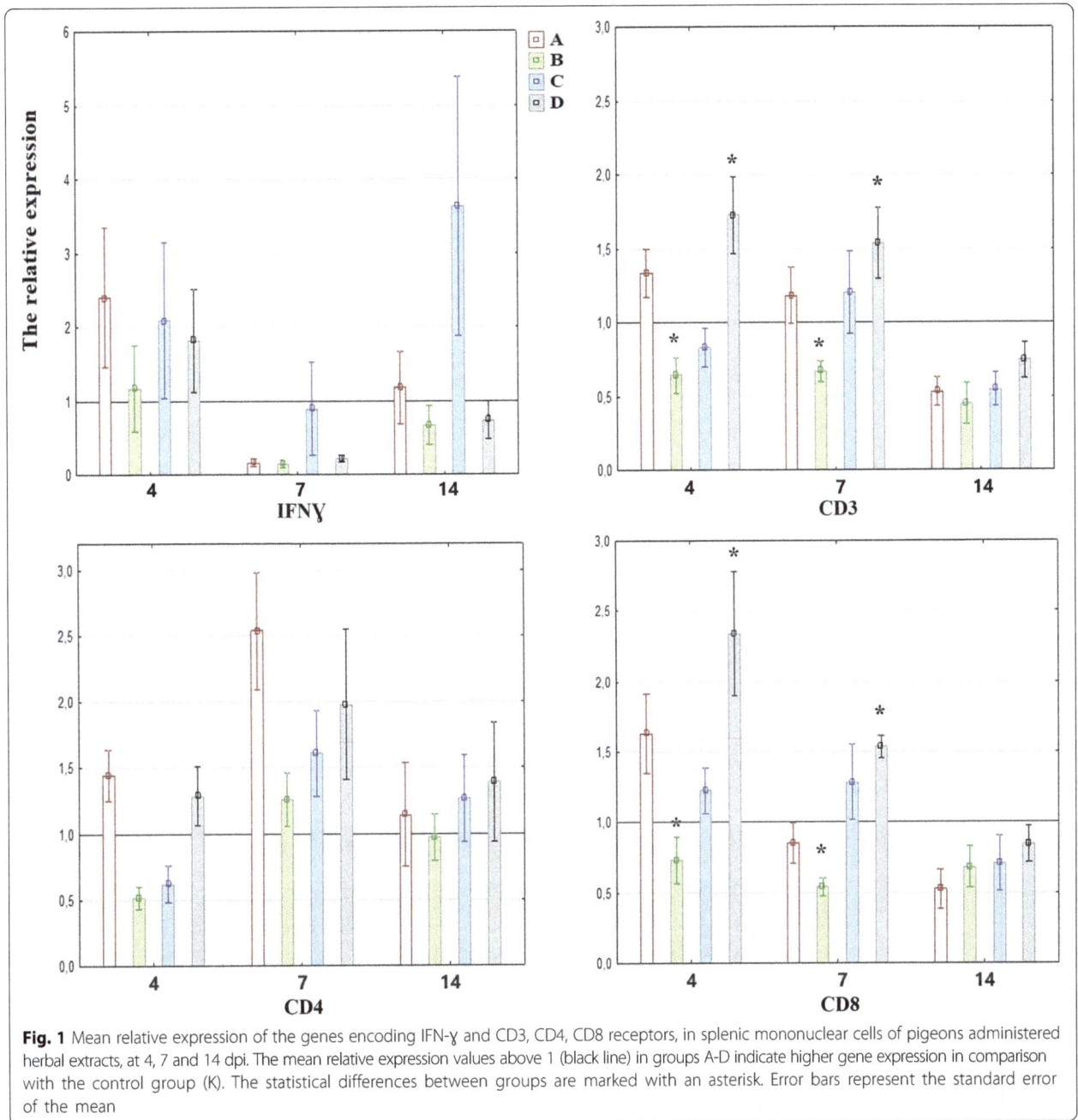

Fig. 1 Mean relative expression of the genes encoding IFN-ɣ and CD3, CD4, CD8 receptors, in splenic mononuclear cells of pigeons administered herbal extracts, at 4, 7 and 14 dpi. The mean relative expression values above 1 (black line) in groups A-D indicate higher gene expression in comparison with the control group (K). The statistical differences between groups are marked with an asterisk. Error bars represent the standard error of the mean

in groups A1 and B1 throughout the entire experiment, and in group D1 at 4 and 14 dpi than in the control group (K1). Group C1 pigeons were characterized by lower expression of the CD3 gene than group K1 birds throughout the experiment, and significantly lower expression of the CD3 gene than group B1 pigeons at 14 dpi (P = 0.006) (Fig. 2).

Expression of the CD4 gene
No significant differences in the expression of the gene encoding surface receptor CD4 in mononuclear cells isolated

from spleen samples were observed between pigeons from groups A-D. However, the expression of the above gene in groups A-D was higher than in the control group throughout the experiment. The only exceptions were pigeons from group B at 4 and 14 dpi and pigeons from group C at 4 dpi. The highest number of copies of the CD4 gene in groups A-D were detected at 7 dpi (Fig. 1). A reverse trend was noted in groups A1-D1 where CD4 gene expression was lowest at 7 dpi. At 14 dpi, the expression of the CD4 gene was significantly higher in group B1 than in group C1 (P = 0.011) (Fig. 2).

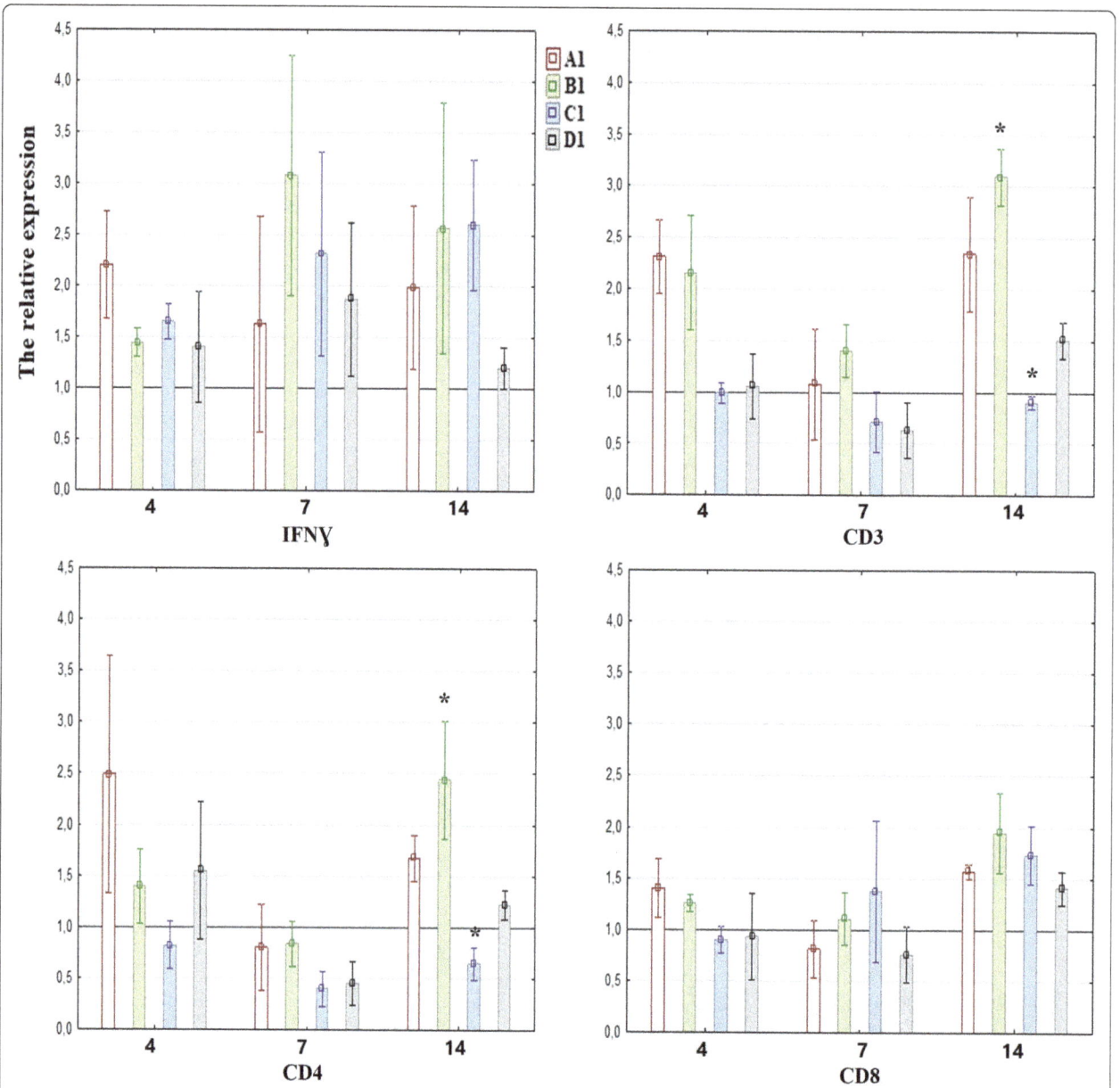

Fig. 2 Mean relative expression of the genes encoding IFN-γ and CD3, CD4, CD8 receptors, in splenic mononuclear cells of pigeons inoculated with PPMV-1 and administered herbal extracts, at 4, 7 and 14 dpi. The mean relative expression values above 1 (black line) in groups A1-D1 indicate higher gene expression in comparison with the control group (K1). The statistical differences between groups are marked with an asterisk. Error bars represent the standard error of the mean

Expression of the CD8 gene

In comparison with group K pigeons, the expression of the CD8 gene was higher only in groups A, C and D at 4 dpi, and in groups C and D at 7 dpi. The expression of the CD8 gene was significantly higher in group D at 4 and 7 dpi than in group B (P = 0.011 and $P = 0.045$, respectively) (Fig. 1). No significant differences in the number of copies of the CD8 gene were found between infected birds receiving herbal extracts and control group birds, but CD8 gene expression in the above experimental groups peaked at 14 dpi (Fig. 2).

Flow cytometric analysis

Flow cytometry data are presented in Fig. 3. No significant differences in the percentage of IgM+ B cells were found between experimental groups or between sampling dates. The highest percentage of IgM+ B cells was noted at 7 dpi in groups A, B and D (34.4%, 34.67% and 37.72%, respectively). In comparison, in control group birds at 7 dpi, the percentage of IgM+ B cells was determined at 28.14% (group K) and 27.23% (group K1). The percentage of IgM+ B cells was lowest in groups A1, B1 and C1 at 7 dpi (22.30%, 24.33% and 26.05%, respectively) and in groups

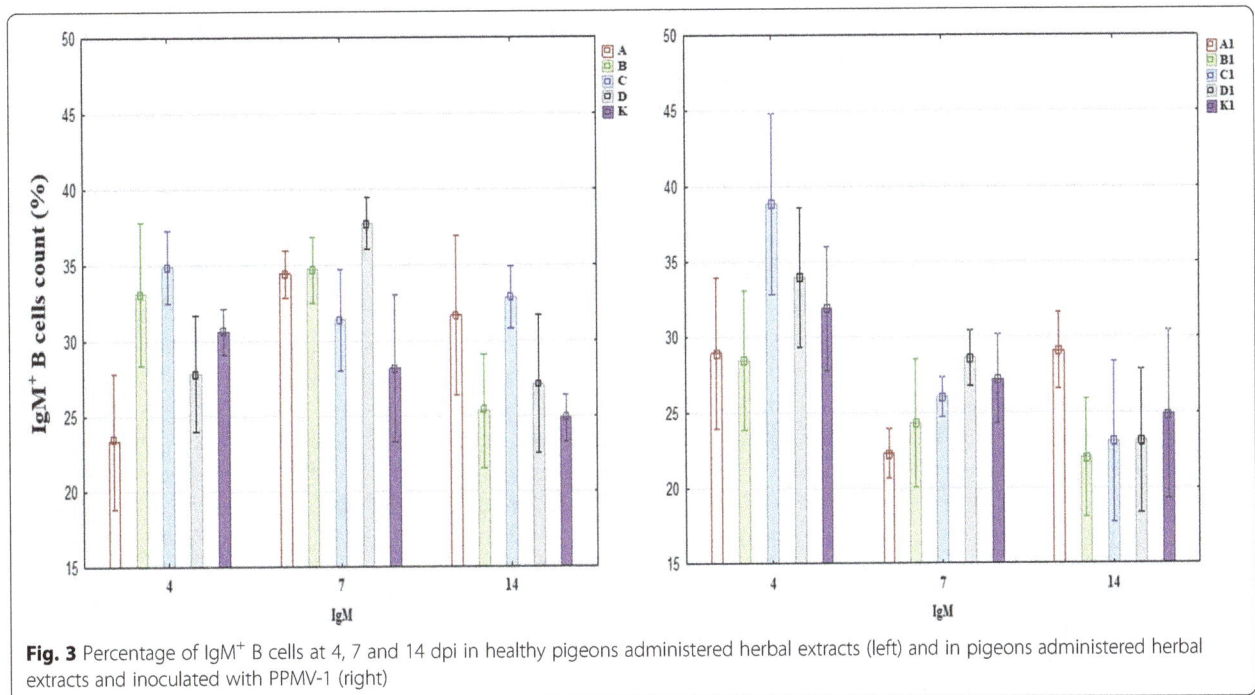

Fig. 3 Percentage of IgM$^+$ B cells at 4, 7 and 14 dpi in healthy pigeons administered herbal extracts (left) and in pigeons administered herbal extracts and inoculated with PPMV-1 (right)

B1, C1 and D1 at 14 dpi (22.04%, 23.10% and 23.14%, respectively).

Discussion

Immunomodulators are biological and synthetic substances which stimulate or suppress humoral and cell-mediated immune responses [18]. Antibiotics and synthetic drugs are increasingly replaced by plant-based immunomodulators in the treatment and prevention of animal diseases. Natural materials such as herbs, spices, essential oils and oleoresins are classified as phytogenic feed additives or phytobiotics [4, 44]. Phytobiotics are a rich source of biologically active compounds with a wide range of anti-carcinogenic, anti-inflammatory, antibacterial, antifungal and antiviral properties [17].

Diseases with a viral etiology pose a growing problem in pigeon breeding. This can be attributed mainly to pigeon rearing systems where infectious diseases spread rapidly due to the absence of biosecurity procedures. In view of the above, the present experiment was carried out to demonstrate the immunomodulatory properties of *Aloe vera* and licorice in the supplementary treatment of viral infections using PPMV-1 as an experimental model. Molecular and flow cytometry analyses were performed on mononuclear cells isolated from the spleen which is the first lymphoid organ to be colonized by pathogenic paramyxovirus strains during infection. [32]. The percentages of T cell subpopulations were not determined by flow cytometry due to the absence of monoclonal antibodies reacting with pigeon lymphocytes. The investigation conducted previously by Stenzel et al. (2011)

with the use of flow cytometry was burdened with high methodological error because commercially available antibodies against chicken lymphocytes are characterized by minimal cross-reactivity with pigeon cells [35]. For this reason, a method for evaluating the expression of genes encoding surface receptors on CD3, CD4 and CD8 T cells was developed in this study as a reliable alternative to flow cytometry.

The results of the conducted research revealed that both herbal extracts had immunomodulatory properties which differed radically depending on the dose and the presence or absence of inoculation with PPMV-1. In uninfected birds, the analyzed herbal extracts stimulated both cell-mediated and humoral immunity, as demonstrated by the higher expression of genes encoding CD4 and CD8 surface receptors in comparison with the control group. The expression of the gene encoding the CD4 receptor peaked at 7 dpi in all birds receiving herbal extracts, in particular in the group administered *Aloe vera* extract at 300 mg/kg BW. Similar results were reported by Vahedi et al. (2011) who observed increased proliferation of CD4$^+$ and CD8$^+$ lymphocytes in rabbits receiving *Aloe vera* extract [43]. Interestingly, the gene encoding receptor CD4 was least expressed in pigeons receiving the *Aloe vera* extract at 500 mg/kg BW (below control group levels at 4 and 14 dpi), which indicates that higher doses of the *Aloe vera* extract deliver immunosuppressive effects. The above phenomenon was also noted in analyses of CMI because the expression of the gene encoding the CD8 receptor was lower in the above group than in the remaining experimental groups and the control group. However, the observed differences were significant only relative to group D (licorice extract dose of 500 mg/kg BW) (Fig. 1). The expression of the gene encoding the CD3 receptor was similar to the expression of the gene encoding the CD8 receptor because the CD3 receptor is present on all T cells and, together with the T-cell receptor (TCR), it forms the TCR-CD3 complex responsible for antigen recognition and the transmission of the T-cell activation signal [14]. In view of the above, the increase in the expression of the gene encoding the CD4 receptor or the CD8 receptor should be correlated with a similar increase in the expression of the gene encoding the CD3 receptor. Such a correlation was observed in this study. The decrease in the expression of all analyzed genes at 14 dpi can be attributed to the weakening of immune responses after the elimination of immunomodulatory additives from pigeon diets.

Somewhat different results were noted in pigeons that received immunomodulatory feed additives and were then experimentally infected with PPMV-1. A clear increase in the expression of the gene encoding the CD4 receptor was observed, and it was correlated with an increase in the expression of the gene encoding the CD3 receptor in birds receiving both doses of the Aloe vera extract relative to control group birds and birds receiving licorice extracts. The highest correlation ($P = 0.011$) was noted between groups B1 (Aloe vera dose of 500 mg/kg BW) and C1 (licorice extract dose of 300 mg/kg BW) (Fig. 2). No such correlations were observed in an analysis of cytotoxic lymphocytes.

The immune response to a viral infection also leads to an increase in the synthesis of interferons which target viruses directly by inhibiting their protein synthesis or indirectly by activating defense mechanisms [41]. In the current study, special attention was paid to the expression of the gene encoding IFN-γ because this protein plays a significant role in both immediate and long-term immune responses to a viral infection. IFN-γ is produced mainly by CD4$^+$ and CD8$^+$ T cells. This cytokine is also secreted by B cells, NK cells, NKT cells and professional antigen-presenting cells [33]. In our study, the expression of the gene encoding IFN-γ increased in all birds from the inoculated groups, which can be directly linked with experimental inoculation. Higher expression of the above gene in the groups infected with PPMV-1 and administered herbal extracts than in the control group (K1) could also be attributed to the immunomodulatory properties of Aloe vera and licorice extracts. Similar results were reported by Utsunomiya et al. (1997) in whose study, glycyrrhizin stimulated T cells to produce IFN-γ in mice infected with the influenza A virus [42]. In our study, despite an absence of significant differences between groups, the gene encoding IFN-γ was most highly expressed in pigeons administered Aloe vera extract at 500 mg/kg BW, which is partially consistent with the expression of the genes encoding CD3 and CD8 receptors (Fig. 2). The highest expression of the gene encoding IFN-γ in group B1 should be associated with the immune response to the viral infection, which was exacerbated by the immunomodulatory effect of Aloe vera extract. The PPMV-1 strain used in the study was pathogenic, and pathogenic paramyxoviruses are most potent in stimulating interferon synthesis [20, 24].

The humoral immune response is the last stage of the immune response to an infection, where M class antibodies are produced first and constitute the largest group of immunoglobulins [19]. In the present study, the percentage of IgM$^+$ B cells was higher in the groups administered both Aloe vera and licorice extracts than in the control groups at 7 dpi. However, the noted differences were not significant, probably due to high values of standard deviation (Fig. 3). Therefore, the influence of the tested herbal extracts on humoral immunity could not be confirmed despite the fact an increase in the percentage of anti-NDV antibodies after the administration of Aloe vera extract has been reported by other authors [26]. However, in a study of rabbits, the Aloe vera extract did not influence serum immunoglobulin levels [43], which is consistent with our findings.

Conclusion

It can be concluded that both Aloe vera and licorice extracts exerted immunomodulatory effects, but their efficacy was clearly correlated with dose and the health status of the analyzed birds. In healthy pigeons, herbal extracts influenced both humoral and cell-mediated immune responses, but the immunostimulatory effects of Aloe vera were observed only in birds receiving a lower dose of the extract. A higher dose of Aloe vera extract exerted immunosuppressive effects on pigeons not infected with PPMV-1, and it exerted immunostimulatory effects on infected pigeons. The above correlations should be taken into consideration during the administration of Aloe vera extracts to birds.

Abbreviation

BW: Body weight; cDNA: Complementary DNA; CFU: Colony-forming unit; CMI: Cell-mediated immunity; DPI: Days post-inoculation; GAPDH: Glyceraldehyde 3-phosphate dehydrogenase; IFN-γ: Interferon γ; IgM: Immunoglobulin M; IL: Interleukin; NDV: Newcastle disease virus; PBS: Phosphate-buffered saline; PCR: Polymerase chain reaction; PPMV-1: Pigeon paramyxovirus type 1; PRR: Pattern recognition receptors; SPF: Specific pathogen free; TLR: Toll-like receptors

Funding

This publication was supported by the "Healthy Animal - Safe Food" Scientific Consortium of the Leading National Research Centre (KNOW) pursuant to a decision of the Ministry of Science and Higher Education No. 05–1/KNOW2/2015.

Authors' contributions
DD designed the study, inoculated birds, collected samples, performed PCR, real-time PCR and serological examination, analyzed the data and wrote the manuscript; TS inoculated birds, collected samples and helped to write the manuscript; MŚ and BT participated in sample collection and performed flow cytometry; AK supervised the study and helped to write the manuscript. All authors read and approved the final manuscript.

Competing interests
The authors declare that they have no competing interests.

References
1. Abe N, Ebina T, Ishida N. Interferon induction by glycyrrhizin and glycyrrhetinic acid in mice. Microbiol Immunol. 1982;26:535–9.
2. Ahmed KA, Saxena VK, Ara A, Singh KB, Sundaresan NR, Saxena M, et al. Immune response to Newcastle disease virus in chicken lines divergently selected for cutaneous hypersensitivity. Int J Immunogenet. 2007;34:445–55.
3. Al-Garib SO, Gielkens AL, Gruys DE, Hartog L, Koch G. Immunoglobulin class distribution of systemic and mucosal antibody responses to Newcastle disease in chickens. Avian Dis. 2003;47:32–40.
4. Bampidis VA, Christodoulou V, Florou-Paneri P, Christaki E, Chatzopoulou PS, Tsiligianni T, Spais AB. Effect of dietary dried oregano leaves on growth performance, carcase characteristics and serum cholesterol of female early maturing turkeys. Br Poult Sci. 2005;46:595–601.
5. Cattoli G, De Battisti C, Marciano S, Ormelli S, Monne I, Terregino C, et al. False-negative result of a validated real-time PCR protocol for diagnosis of Newcastle disease due to genetic variability of the matrix gene. J Clin Microbiol. 2009;47:3791–2.
6. Choi S, Chung M-H. A review on the relationship between aloe vera components and their biologic effects. Semin Integr Med. 2003;1:53–62.
7. Dalgaard TS, Norup LR, Pedersen AR, Handberg KJ, Jørgensen PH, Juul-Madsen HR. Flow cytometric assessment of chicken T cell-mediated immune responses after Newcastle disease virus vaccination and challenge. Vaccine. 2010;28:4506–14.
8. Darabighane B, Zarei A, Shahneh AZ. The effects of different levels of Aloe vera gel on ileum microflora population and immune response in broilers: a comparison to antibiotic effects. J Appl Anim Res. 2012;40:31–6.
9. Darabighane B, Nahashon SN. A review on effects of Aloe vera as a feed additive in broiler chicken diets. Ann Anim Sci. 2014;14:491–500.
10. Daum I, Finsterbusch T, Härtle S, Göbel TW, Mankertz A, Korbel R, Grund C. Cloning and expression of a truncated pigeon circovirus capsid protein suitable for antibody detection in infected pigeons. Avian Pathol. 2009;38:135–41.
11. Djeraba A, Quere P. In vivo macrophage activation in chickens with Acemannan, a complex carbohydrate extracted from Aloe vera. Int J Immunopharmacol. 2000;22:365–72.
12. Dorhoi A, Dobrean V, Zăhan M, Virag P. Modulatory effects of several herbal extracts on avian peripheral blood cell immune responses. Phytother Res. 2006;20:352–8.
13. Dortmans JC, Koch G, Rottier PJ, Peeters BP. A comparative infection study of pigeon and avian paramyxovirus type 1 viruses in pigeons: evaluation of clinical signs, virus shedding and seroconversion. Avian Pathol. 2011;40:125–34.
14. Erf GF. Immune system function and development in broilers. Poultry Sci. 1997;5:109–23.
15. Fiore C, Eisenhut M, Krausse R, Ragazzi E, Pellati D, Armanini D, et al. Antiviral effects of Glycyrrhiza species. Phytother Res. 2008;22:141–8.
16. Guo A, Dongming H, Hong-Bo X, Chang-An G, Zhao J. Promotion of regulatory T cell induction by immunomodulatory herbal medicine licorice and its two constituents. Sci Rep. 2015;5:14046.
17. Hashemi SR, Davoodi I I. Herbal plants and their derivatives as growth and health promoters in animal nutrition. Vet Res Commun. 2011;35:169–80.
18. Jantan I, Ahmad W, Bukhari SN. Plant-derived immunomodulators: an insight on their preclinical evaluation and clinical trials. Front Plant Sci. 2015;6:655.
19. Jeurissen SH, Boonstra-Blom AG, Al-Garib SO, Hartog L, Koch G. Defence mechanisms against viral infection in poultry: a review. Vet Q. 2000;22:204–8.
20. Kapczynski DR, Afonso CL, Miller PJ. Immune responses of poultry to Newcastle disease virus. Dev Comp Immunol. 2013;41:447–53.
21. Khattak F, Ronchi A, Castelli P, Sparks N. Effects of natural blend of essential oil on growth performance, blood biochemistry, cecal morphology, and carcass quality of broiler chickens. Poult Sci. 2014;93:132–7.
22. Latheef SK, Dhama K, Samad HA, Wani MY, Kumar MA, Palanivelu M, et al. Immunomodulatory and prophylactic efficacy of herbal extracts against experimentally induced chicken infectious anaemia in chicks: assessing the viral load and cell mediated immunity. Virusdisease. 2017;28:115–20.
23. Liu FX, Sun S, Cui ZZ. Analysis of immunological enhancement of immunosuppressed chickens by Chinese herbal extracts. J Ethnopharmacol. 2010;127:251–6.
24. Liu WQ, Tian MX, Wang YP, Zhao Y, Zou NL, Zhao FF, et al. The different expression of immune-related cytokine genes in response to velogenic and lentogenic Newcastle disease viruses infection in chicken peripheral blood. Mol Biol Rep. 2012;39:3611–8.
25. Livak KJ, Schmittgen TD. Analysis of relative gene expression data using real-time quantitative PCR and the 2 (–Delta Delta C(T)) method. Methods. 2001;25:402–8.
26. Ojiezeh TI, Eghafona N. Humoral responses of broiler chickens challenged with NDV following supplemental treatment with extracts of Aloe vera, Alma millsoni, Ganoderma lucidum and Archachatina marginata. Cent Eur J Immunol. 2015;40:300–6.
27. Perera C, Efferth T. Antiviral medicinal herbs and phytochemicals. J Pharmacogn. 2012;3:45–8.
28. Pestka D, Stenzel T, Koncicki A. Occurrence, characteristics and control of pigeon paramyxovirus type 1 in pigeons. Pol J Vet Sci. 2014;17:379–84.
29. PN-EN ISO 6887-1:2017–1: Microbiology of food and animal feeding stuffs – Preparation of test samples, initial suspension and decimal dilutions for microbiological examination – Part 1: General rules for the preparation of the initial suspension and decimal dilutions. International Organization for Standardization. https://www.iso.org/standard/63335.html.
30. Rauw F, Gardin Y, Palya V, van Borm S, Gonze M, Lemaire S, et al. Humoral, cell-mediated and mucosal immunity induced by oculo-nasal vaccination of one-day-old SPF and conventional layer chicks with two different live Newcastle disease vaccines. Vaccine. 2009;27:3631–42.
31. Reynolds DL, Maraqa AD. Protective immunity against Newcastle disease: the role of cell-mediated immunity. Avian Dis. 2000;44:145–54.
32. Rue CA, Susta L, Cornax I, Brown CC, Kopczynski DR, Suarez DL, et al. Virulent Newcastle disease virus elicits a strong innate immune response in chickens. J Gen Virol. 2011;92:931–9.
33. Schroder K, Hertzog PJ, Ravasi T, Hume DA. Interferon-gamma: an overview of signals, mechanisms and functions. J Leukoc Biol. 2004;75:163–89.
34. Shaneyfelt ME, Burke AD, Graff JW, Jutila MA, Hardy ME. Natural products that reduce rotavirus infectivity identified by a cell-based moderate-throughput screening assay. Virol J. 2006;3:68.
35. Stenzel T, Tykałowski B, Śmiałek M, Koncicki A, Kwiatkowska-Stenzel A. The effect of different doses of methisoprinol on the percentage of CD4+ and CD8+ T lymphocyte subpopulation and the antibody titers in pigeons immunised against PPMV-1. Pol J Vet Sci. 2011;14:367–71.
36. Stenzel T, Pestka D. Occurrence and genetic diversity of pigeon circovirus strains in Poland. Acta Vet Hung. 2014;62:274–83.
37. Stenzel T, Tykałowski B, Śmiałek M, Pestka D, Koncicki A. Influence of methisoprinol on the course of an experimental infection with PPMV-1 in pigeons. Med Weter. 2014;70:219–23.
38. Surjushe A, Vasani R, Saple DG. Aloe vera: a short review. Indian J Dermatol. 2008;53:163–6.
39. Śmietanka K, Olszewska M, Domańska-Blicharz K, Bocian Ł, Minta Z. Experimental infection of different species of birds with pigeon paramyxovirus type 1 virus - evaluation of clinical outcomes, viral shedding, and distribution in tissues. Avian Dis. 2014;58:523–30.
40. Thompson MR, Kaminski JJ, Kurt-Jones EA, Fitzgerald KA. Pattern recognition receptors and the innate immune response to viral infection. Viruses. 2011;3:920–40.
41. Interferons TSK. Biochemistry and mechanisms of action. Am J Obstet Gynecol. 1995;172:1350–3.
42. Utsunomiya T, Kobayashi M, Pollard RB, Glycyrrhizin SF. An active component of licorice roots, reduces morbidity and mortality of mice infected with lethal doses of influenza virus. Antimicrob Agents Chemother. 1997;41:551–6.

43. Vahedi G, Taghavi M, Maleki AK, Habibian R. The effect of Aloe vera extract on humoral and cellular immune response in rabbit. Afr J Biotechnol. 2011; 10:5225–8.

44. Windisch W, Schedle K, Plitzner C, Kroismayr A. Use of phytogenic products as feed additives for swine and poultry. J Anim Sci 2008; doi: https://doi.org/10.2527/jas.2007-0459.

45. Wise MG, Suarez DL, Seal BS, Pedersen JC, Senne DA, King DJ, et al. Development of a real-time reverse-transcription PCR for detection of Newcastle disease virus RNA in clinical samples. J Clin Microbiol. 2004;42: 329–38.

46. Wolkerstorfer A, Kurz H, Bachhofner N, Szolar OH. Glycyrrhizin inhibits influenza a virus uptake into the cell. Antivir Res. 2009;83:171–8.

3

Genetic detection of peste des petits ruminants virus under field conditions: a step forward towards disease eradication

Waqas Ashraf[1,2*], Hermann Unger[3], Sunaina Haris[1,2], Ameena Mobeen[1,2], Muhammad Farooq[1,2], Muhammad Asif[1,2] and Qaiser Mahmood Khan[1,2*]

Abstract

Background: The devastating viral disease of small ruminants namely Peste des petits ruminants (PPR) declared as target for "Global Eradication" in 2015 by the Food and Agriculture Organization (FAO) and the World Organization for Animal Health (OIE). For a successful eradication campaign, molecular diagnostic tools are preferred for their specificity, efficacy and robustness to compliment prophylactic measures and surveillance methods. However, molecular tools have a few limitations including, costly equipment, multi-step template preparation protocols, target amplification and analysis that restrict their use to the sophisticated laboratory settings. As reverse transcription-loop mediated isothermal amplification assay (RT-LAMP) has such an intrinsic potential for point of care diagnosis, this study focused on the genetic detection of causative PPR virus (PPRV) in field conditions. It involves the use of a sample buffer that can precipitate out virus envelope and capsid proteins through ammonium sulphate precipitation and exposes viral RNA, present in the clinical sample, to the LAMP reaction mixture.

Results: The test was evaluated using 11 PPRV cultures, and a total of 46 nasal swabs ($n = 32$ collected in the field outbreaks, $n = 14$ collected from experimentally inoculated animals). The RT-LAMP was compared with the reverse transcription-PCR (RT-PCR) and real-time quantitative RT-PCR (RT-qPCR) for its relative specificity, sensitivity and robustness. RT-LAMP detected PPRV in all PPRV cultures in or less than 30 min. Its detection limit was of 0. 0001TCID$_{50}$ (tissue culture infective dose-50) per ml with 10-fold higher sensitivity than that of RT-PCR. In 59.4% of the field samples, RT-LAMP detected PPRV within 35–55 min. The analytical sensitivity and specificity of the RT-LAMP were equivalent to that of the RT-qPCR. The time of detection of PPRV decreased by at least forty minutes or 3–4 h in case of in the RT-LAMP as compared with the RT-qPCR and the RT-PCR, respectively.

Conclusions: The sensitive and specific RT-LAMP test developed in this study targeting a small fragment of the N gene of PPRV is a rapid, reliable and applicable molecular diagnostic test of choice under the field conditions. RT-LAMP requiring minimal training offers a very useful tool for PPR diagnosis especially during the "Global PPR Eradication Campaign".

Keywords: Peste des petits ruminants virus (PPRV), Reverse-transcription loop mediated isothermal amplification (RT-LAMP), Nucleocapsid gene (N), Robust diagnosis

* Correspondence: waqasasahrafnibge@gmail.com; qk_5@yahoo.com
[1]National Institute for Biotechnology and Genetic Engineering (NIBGE), Faisalabad, Pakistan
Full list of author information is available at the end of the article

Background

Peste des petits ruminants (PPR) is an acute, highly contagious viral disease of small ruminants that is considered one of the major constraints for efficient small ruminant production in the developing countries. It is a notifiable, List-A disease to the OIE caused by peste des petits ruminants virus (PPRV) [1]. Since its emergence in West Africa in the 1940s, PPR has spread across vast regions of the Africa, Middle East, Arabian Peninsula, and Southern Asia [2]. PPR is endemic in Pakistan where approximately half (48.3–48.5%) of its small ruminant population is seropositive [3, 4]. The disease mainly exists in three pathogenic, per-acute, acute and sub-acute, forms. However, acute form is the most common form that causes 80–90% mortality in individual flocks [5]. Once the host is infected with PPRV, general viraemia develops within 4–6 days followed by high fever (104–106 °F), marked salivation, shallow erosions in the oral mucosa, serous to purulent oculonasal discharge, dyspnea, coughing, pneumonia and diarrhea [6, 7]. In the later stages, a sub-normal temperature of 101–102 °F and dehydration due to diarrhea can lead to hypovolumic shock and death of the affected animals [8].

PPRV; the causative agent of the disease, is a member of the genus *Morbillivirus*, in the family *"Paramyxoviridea"*. It is a negative-sense, single-stranded RNA virus. The genome of PPRV, viz. 3'-N-P-M-F-H-L-5' encodes six structural proteins namely nucleocapsid, phosphoprotein, matrix, fusion, haemagglutinin and large polymerase [9–11]. Among these, nucleocapsid (N) protein gene is the most abundantly transcribed gene in the host cells and is therefore preferred target site for genomic detection of PPRV [12]. Conventionally, diagnosis of PPRV relied on serological techniques and virus isolation from clinical samples through its propagation in adaptable cell lines [13–15]. However, these techniques are labor-intensive and insensitive for PPRV detection especially during latent phase of the infection [16, 17]. In contrast, reverse transcription polymerase chain reaction (RT-PCR) is considered as standard diagnostic test for PPR worldwide [14, 18]. However, limitations associated with RT-PCR including its high cost and pre-requisite for scientific manpower render it unsuitable for low or middle-income country settings [19]. Accordingly, the World Health Organization (WHO) recommends that an ideal diagnostic test for such countries should meet the ASSURED (Affordable, sensitive, specific, user-friendly, robust, equipment free, deliverable to the end user) guidelines [20, 21]. The invention of isothermal technologies, like loop mediated isothermal amplification (LAMP) assay, for DNA amplification is a step-forward towards the development of ASSURED diagnostic tests [21]. LAMP based amplification of target nucleic acid is based on isothermal amplification of template DNA utilizing the strand displacement

activity of Bst or Bsm DNA polymerase enzyme originated from *Bacillus stearothermophilus* or *Bacillus smithii*, respectively [22]. The stem-loop structures generated, at an initial phase, during LAMP initiate exponential amplification process that results in rapid accumulation of DNA amplicons of varying lengths [22, 23]. The LAMP products can simply be visualized with naked eye after the addition of fluorescent DNA-intercalating dyes such as SYBR Green I, propidium iodide and calcein to the reaction mixture. In addition, the generation of LAMP products can also be monitored on a real-time basis by measuring the change in fluorescence over a specified interval with an ESE-Quant tube scanner [23]. Here, we report the development of a one-step, single tube, N gene based RT-LAMP assay for direct detection of PPRV in clinical samples (swab extracts), cell culture supernatants, obviating the need for RNA extraction and cDNA synthesis step that can potentially make it a suitable diagnostic test for on-site detection of PPRV particularly during eradication campaign in low income country settings.

Results

Optimization of RT-LAMP Assay for the detection of PPRV in culture supernatants

The success of an RT-LAMP assay mainly depends on three factors including primer concentration, optimal reaction temperature and template amount. Optimal primer concentration was found to be 0.25 μM for each of the outer primers (F3 and B3) and 1.25 μM for each of the inner primers (FIP and BIP) to achieve threshold increase of >30 mV for three consecutive readings. Once the primer concentration was reduced to half of the above-mentioned values, it compromised the efficiency of the reaction thus resulting in the drop in signal below the threshold value (Fig. 1). The optimal reaction temperature for each primer pair was determined using a gradient PCR (IQ5™, BioRad, USA) prior to its use in RT-LAMP assay. Accordingly, the assay was carried out at an increasing gradient of temperature that ranged from 50 °C to 62 °C. Gel electrophoresis of reaction products revealed amplification at three different temperatures including 53 °C, 55 °C and 58 °C. Among these, best amplification curve with high specificity was achieved at 58 °C (Fig. 2). Optimal template concentration for RT-LAMP assay was determined using a standard curve method. For this, tenfold serial dilutions of PPRV cell culture supernatant were prepared and subsequently subjected to RT-LAMP assay along with positive and no template controls (NTCs). The amplification curves for the positive control and the dilution factors viz; 10^{-1} to 10^{-4} showed typical four phases of amplification including the baseline, exponential phase, linear phase and a plateau within 60 min from the start of the assay. Amplification curve for 5th dilution crossed

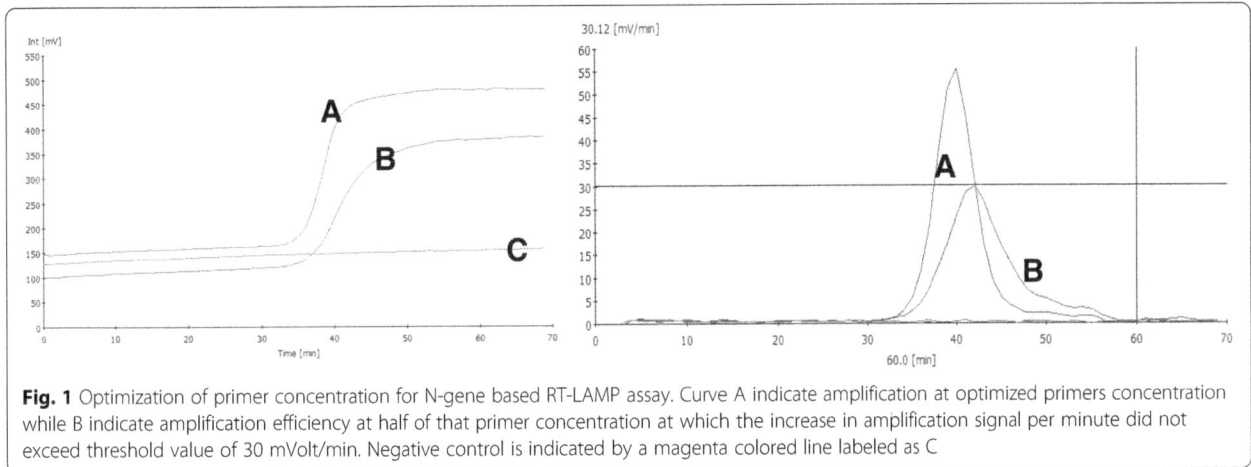

Fig. 1 Optimization of primer concentration for N-gene based RT-LAMP assay. Curve A indicate amplification at optimized primers concentration while B indicate amplification efficiency at half of that primer concentration at which the increase in amplification signal per minute did not exceed threshold value of 30 mVolt/min. Negative control is indicated by a magenta colored line labeled as C

threshold limit (30 mV/min) after 65 min (Fig. 3, Additional files 1 and 2) while NTC did not show any amplification. A standard curve was plotted with the log template concentration as the x value and the time of positivity (detection time, Td) as the y value. The y-intercept value (line representing the best fit) calculated using the least square method of linear regression was found to be 69.32 min. Accordingly, a maximum time limit of 60 min was set for the test to decide for positivity of the tested sample to ensure specific detection and rule out false positives. The detection limit of the assay (10^{-4} $TCID_{50}$/ml) was 10 fold higher as compared with conventional RT-PCR (Fig. 4). Subsequent to the optimization of primer concentration, reaction temperature and template amount, culture soups of 11 PPRVs isolates, grown in CHS-20 cells, were detected by RT-LAMP. These cultures yielded amplification curves that climbed maximally up to

60-75 mV/min well above the 30 mV/min threshold during the LAMP reaction, indicating positive detection of PPRVs with a Td value of 30 or less than 30 min (Table 1, Fig. 5).

Colorimetric and fluorimetric detection of RT-LAMP products

SYBR green staining of RT-LAMP products resulted in a colour change, observable by naked eye, from orange to light green in case of positive reaction mixtures while the negative reaction mixture remains orange (Fig. 6). Upon UV light excitation, positive reaction products produced strong bright green fluorescence while negative control depicted slight background fluorescence (Fig. 6). These observations were in accordance with those of the gel electrophoresis as a ladder-like pattern of amplified DNA product was visualized in case of positive reactions while it was absent in case of negative control reactions (Fig. 6, Additional file 3).

Robustness and specificity of RT-LAMP assay for detection of PPRV in experimentally inoculated animals

An equivalent of each swab extract collected, from experimentally inoculated goat and sheep, at 2, 4, 8, 10, 12, 14 day post-inoculation (dpi) was analyzed with RT-LAMP and RT-PCR to determine the relative robustness and sensitivity of RT-LAMP assay. During the course of experimental infection, RT-LAMP detected PPRV genome in the nasal secretions of goat from 2-dpi to 14-dpi (Table 2). It was, however, detected only on day-10 and 12 post-inoculation in sheep. The remaining samples did not produce amplification curves reaching beyond threshold limit of >30 mV increase in signal strength per minute and therefore, were considered as negative. In comparison to RT-LAMP, RT-PCR detected PPRV in the nasal discharges on 4-dpi to 14-dpi in the goat, whilst only on 10-dpi in sheep (Table 2, Additional file 4). Depending on the viral load, it was observed that RT-LAMP assay was able to detect the virus in the secretions of goat with Td

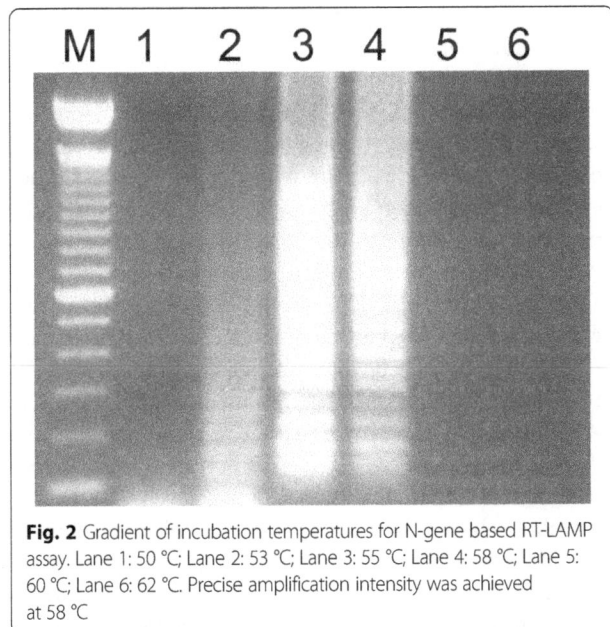

Fig. 2 Gradient of incubation temperatures for N-gene based RT-LAMP assay. Lane 1: 50 °C; Lane 2: 53 °C; Lane 3: 55 °C; Lane 4: 58 °C; Lane 5: 60 °C; Lane 6: 62 °C. Precise amplification intensity was achieved at 58 °C

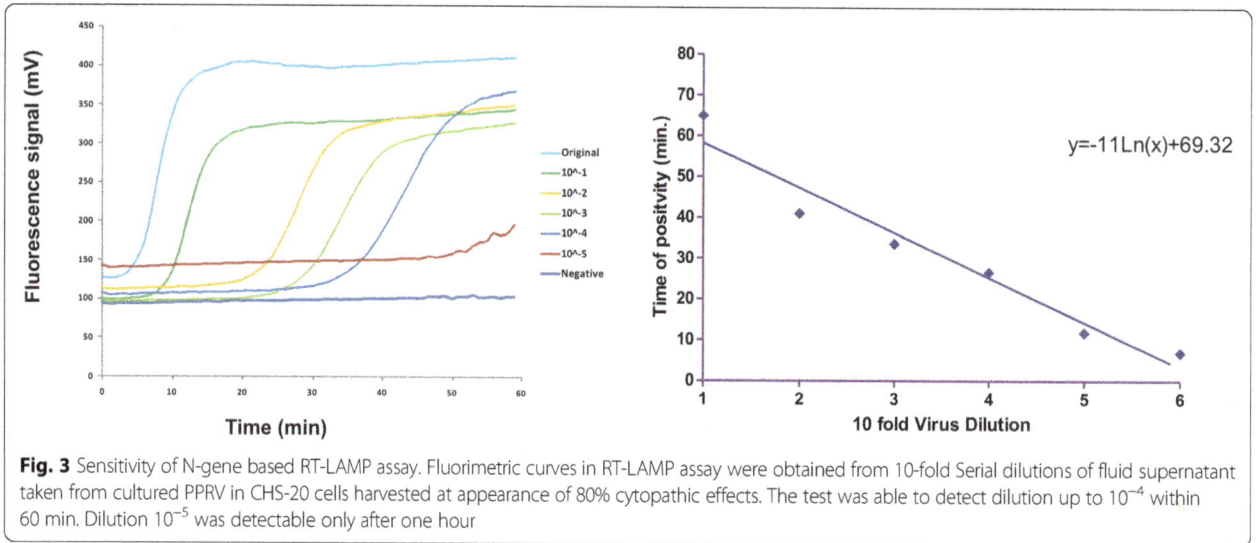

$$y = -11\mathrm{Ln}(x) + 69.32$$

Fig. 3 Sensitivity of N-gene based RT-LAMP assay. Fluorimetric curves in RT-LAMP assay were obtained from 10-fold Serial dilutions of fluid supernatant taken from cultured PPRV in CHS-20 cells harvested at appearance of 80% cytopathic effects. The test was able to detect dilution up to 10^{-4} within 60 min. Dilution 10^{-5} was detectable only after one hour

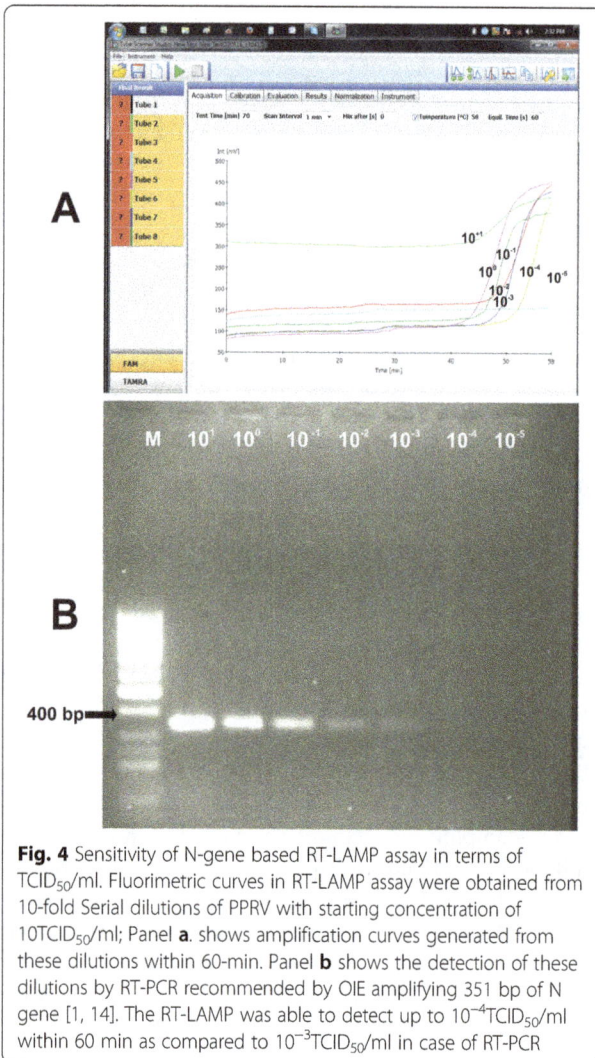

Fig. 4 Sensitivity of N-gene based RT-LAMP assay in terms of $TCID_{50}$/ml. Fluorimetric curves in RT-LAMP assay were obtained from 10-fold Serial dilutions of PPRV with starting concentration of $10TCID_{50}$/ml; Panel **a**. shows amplification curves generated from these dilutions within 60-min. Panel **b** shows the detection of these dilutions by RT-PCR recommended by OIE amplifying 351 bp of N gene [1, 14]. The RT-LAMP was able to detect up to $10^{-4}TCID_{50}$/ml within 60 min as compared to $10^{-3}TCID_{50}$/ml in case of RT-PCR

value of 23 to 40 min which were equivalent to range of ct value of 31.5-36.4 as observed in RT-qPCR (Table 3). However, the pre-requisites of RNA extraction and cDNA synthesis in case of real-time RT-PCR made it take 76.5-80.3 min for the detection of these samples (Table 3, Additional files 5 and 6). RT-LAMP was found to be more robust than RT-qPCR by taking at least forty fewer minutes for detecting PPRV in these samples.

Detection of PPRV in the animals affected in suspected outbreaks

A total of 32 nasal swabs collected from goats in field outbreaks at three districts of Punjab were subjected to the optimized RT-LAMP protocol along with positive and negative controls (Table 4). For threshold validation, only those samples were considered positive that produced amplification curve beyond 30 mV detection limit. Among the clinical samples, nineteen yielded an amplification curve with a Td value of 25-50 min (Table 4). However, thirteen clinical samples along with negative controls that consisted of measles and canine distemper viruses did not yield any amplification signal or a signal below threshold limit. In order to evaluate the sensitivity and robustness of RT-LAMP assay, equivalent set of the swabs was subjected to the conventional RT-PCR using N protein gene specific primers that generated a PCR fragment of ~450 bp in case of positive samples. Conventional RT-PCR detected PPRV in seventeen samples. These results indicated that RT-LAMP assay is somewhat as specific as the conventional RT-PCR assay; however, RT-LAMP detected PPRV in two more samples than RT-PCR. Comparatively, RT-LAMP was more robust in nature as it allowed quick detection of PPRV infection within 60 min as compared to 4–5 h in case of conventional RT-PCR that requires separate RNA extraction, cDNA synthesis, PCR and gel electrophoresis

Table 1 PPRVs isolated from clinical samples by cell culture method from four district of Punjab Province of Pakistan and detected by RT-LAMP. Time points of detection (Ct values) of cultured PPRVs in RT-LAMP assay ranged from 12 to 30 min

No.	Type of sample	Origin	Place/year of outbreak	Cell culture (CHS-20 cells)	Td value (time point of first rise in fluorescence signal above threshold)
1	[a]NS	Sheep	DGK/2012	+	30
2	NS	goat	DGK/2012	+	12
3	NS	goat	MNW/2012	+	30
4	NS	goat	MNW/2012	+	10
5	NS	goat	MNW/2012	+	13
6	NS	goat	MNW/2012	+	14
7	NS	goat	SAH/2012	+	25
8	NS	goat	SAH/2012	+	20
9	NS	goat	SAH/2012	+	15
10	NS	goat	SAH/2012	+	12
11	NS	goat	CHK/2012	+	21

+, Each isolate name includes sample ID, acronym of the district of sampling and year, *DGK* DG Khan, *MNW* Mianwali, *CHK* Chakwal, *SAH* Sahiwal. *nd* not determined. [a]*NS* nasal swab

steps. RT-qPCR detected PPRV in $n = 19$ clinical samples by with Ct values of 22.2 to 33.5 (Fig. 7, Table 4, Additional files 7, 8, 9, 10, 11, 12 and 13). As all the samples that were positive in RT-qPCR were also positive in RT-LAMP, both tests were equally sensitive in the detection of PPRV.

Discussion

The overall aim of this research was to develop a reliable and cost-effective diagnostic test for on-site detection of peste des petits ruminants virus (PPRV) infection in sheep and goats based on LAMP technology that might serve as a foundation for the development of point of care diagnostic tests for different infectious diseases of

livestock prevalent in Pakistan. We selected PPR disease to start with on the basis of two reasons viz. i) economic significance, ii) close relevance to its prototype, Riderpest (cattle plague) that has been effectively eradicated from the face of the earth. Current diagnostic tests for PPR involve the use of immunoassays to detect the antibodies against PPRV, virus isolation using adaptable cell lines and PPRV genome detection using RT-PCR or Real Time RT-PCR [13–18]. However, these techniques require costly equipment and highly trained manpower that render these tests unsuitable for on-site application.

Recently, LAMP based diagnostic tests have been developed for various viral diseases (PPR, Avian Influenza, FMDV) [24–26]. Although, these tests are relatively inexpensive and robust in nature yet challenges associated with LAMP technology including template preparation protocols (RNA extraction and cDNA synthesis) need consideration to take it to the point of care diagnostics. Our protocol involves the use of a sample buffer that can precipitate out virus envelope and capsid proteins through ammonium sulphate precipitation and exposes viral RNA, present in the clinical sample, for its subsequent amplification through LAMP reaction. Hitherto, N gene mRNAs are preferred target for molecular detection of PPRV because it is the most abundantly transcribed viral gene in all morbilliviruses [14, 27]. Hence, in this test, highly conserved region at the 5′ end of nucleocapsid gene specific to PPRV was targeted for primer designing to ensure specificity and sensitivity (Fig. 8). For an unknown reason, we found this protocol more effective for PPRV detection in fresh samples (cell culture supernatants, swab extracts) as compared to previously stored samples that had undergone many freeze-thaw cycles.

The monitoring system of ESE Quant tube scanner provided the ease of spatio-temporal curve analysis

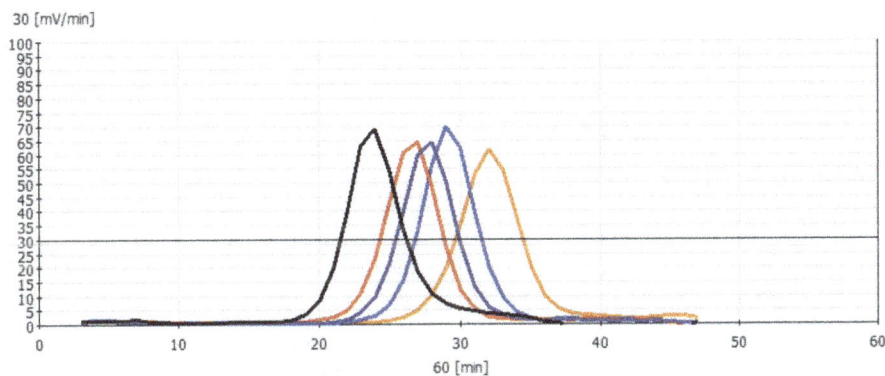

Fig. 5 Threshold validation of positive reaction in RT-LAMP assay. The amplification curves shown in the figure were produced in the assay from culture supernatants of PPRVs. The curves were generated in response to change in fluorescence signal per minute during RT-LAMP assay by ESE Quant tube scanner software. According to threshold validation, amplification curves in positive reactions increased above threshold value of 30 mV per minute (indicated by horizontal line)

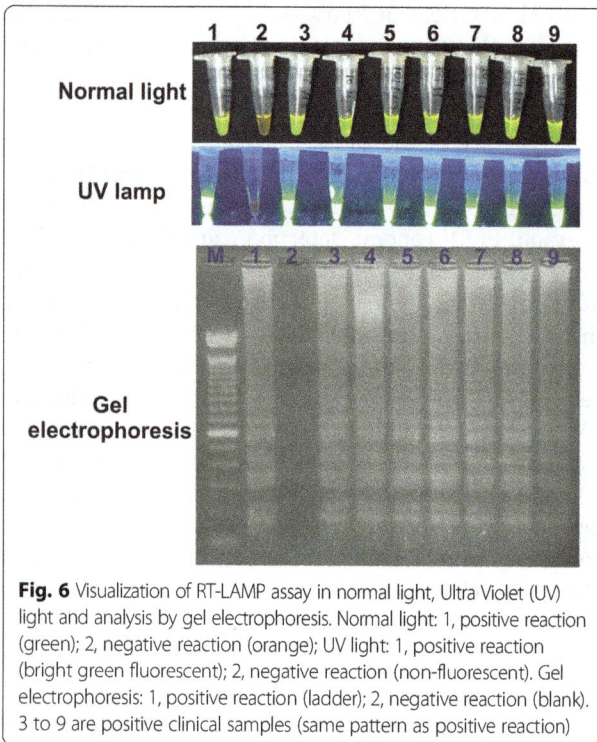

Fig. 6 Visualization of RT-LAMP assay in normal light, Ultra Violet (UV) light and analysis by gel electrophoresis. Normal light: 1, positive reaction (green); 2, negative reaction (orange); UV light: 1, positive reaction (bright green fluorescent); 2, negative reaction (non-fluorescent). Gel electrophoresis: 1, positive reaction (ladder); 2, negative reaction (blank). 3 to 9 are positive clinical samples (same pattern as positive reaction)

Table 3 Comparison of robustness of RT-LAMP assay with real-time RT-PCR for detection of PPRV in nasal swab samples collected on alternative days from experimentally infected goat

Day post inoculation (dpi)	Detection time (Td) in RT-LAMP	Real-time RT-PCR	
		Ct value	Total time to reach Ct (minutes)[a]
2	39.5	36.4	80.33
4	23	31.5	76.5
6	39	34.5	78.7
8	37	34.7	78.9
12	40	34.2	78.5

[a]Total time includes RNA extraction + cDNA synthesis (50 min) + time to reach ct value in real-time PCR steps (50 s per cycle)

during the RT-LAMP reaction. Accordingly, the test can be used even in worse field conditions where electricity is not available and transportation facilities are inadequate. To ensure specificity while maintaining high sensitivity, an algorithm is incorporated into the tube scanner software that allows threshold validation of the amplification curve to consider it as positive only if the increase of signal (slope) above the mean of the baseline is > 30 mV for at least 3 consecutive measurements. This method of curve validation adds to the confidence of RT-LAMP result and, therefore, is more reliable than gel-electrophoresis and colorimetric analysis. By doing so, all seven samples of experimentally infected goat and only two out of total of seven samples of sheep, collected

on alternative days for a period of two weeks, were considered as positive. Because all the samples that were positive by RT-PCR was also positive in RT-LAMP, RT-LAMP was rather as specific as conventional RT-PCR. Nonetheless, RT-LAMP detected PPRV in these samples 4–5 h earlier than RT-PCR. Comparative sensitivity (100%) of the RT-LAMP with RT-qPCR was assured by the fact that both detected nineteen clinical samples, collected in field outbreaks; while conventional RT-PCR (89.5% sensitive) could not detect PPRV in two samples that were positive both in RT-LAMP and RT-qPCR. The specific detection of RT-LAMP can be attributed to its intrinsic stringency by single small fragment ($\approx$200–250 bp) simultaneously by utilizing six primers, while the large amount of DNA produced during the reaction ensures high sensitivity [22–26]. LAMP can multiply the target DNA 10^9 to 10^{10} times the starting amount and detect 0.01 to 10 plaque-forming units (pfu) of viral particles [28]. The sensitivity of the RT-LAMP developed in this study is comparable to previously reported LAMP assays targeting either M or N gene for the detection of PPRV [24, 29]. In our study, RT-LAMP assay was 10-fold more sensitive (detection limit of $0.0001 TCID_{50}$/ml) and 3–4 times more robust than conventional-RT-PCR with a detection limit of $0.001 TCID_{50}$/ml as reported by Couacy-Hymann et al. [14]. Previous protocols of LAMP, however, have used extracted RNA for subsequent synthesis of cDNA prior to its amplification by LAMP. The essential steps of RNA extraction and its reverse transcription require a sophisticated laboratory environment and increase the time of detection. The leading benefit of the RT-LAMP developed in this study for the detection of genome of PPRV directly by using the virus culture supernatant or clinical samples as such without a need for separate RNA extraction and/or cDNA synthesis ensure its application in the field. Due to the added benefits of higher sensitivity, robustness and simplicity, N gene based RT-LAMP can be a proficient alternative to the

Table 2 Comparison of conventional RT-PCR and RT-LAMP in terms of detection of PPRV in nasal secretions

Species	Type of test	Days post inoculation							Total no. of positive samples
		2	4	6	8	10	12	14	
Goat	RT-PCR[a]	-	+	+	+	+	+	+	6
	RT-LAMP	+	+	+	+	+	+	+	7
Sheep	RT-PCR	-	-	-	-	+	-	-	1
	RT-LAMP	-	-	-	-	+	+	-	2

[a]These samples were also confirmed by RT-PCR mentioned in OIE terrestrial Manual [1]

Table 4 PPRV detection by RT-LAMP in clinical samples collected in outbreaks

District	Year of outbreak	Samples tested	RT-LAMP		No. positive in RT-PCR	No. positive in real-time RT-PCR (Ct value)
			Positive	Negative		
Mianwali	2012	08	04	04	04	04 (22.3–26.5)
Chakwal	2012	04	03	01	03	03 (31.7–33.1)
Sahiwal	2014	20	12	08	10	12 (26.2–33.3)
Total (percentage)	-	32	19 (59.4%)	13 (40.6%)	17 (59.4%)	19 (59.4%)

conventional RT-PCR especially in open field conditions where cold chain of sample transportation and highly equipped laboratories are not available.

Under UV light, a bright green fluorescence is observed in case of positive reaction (on addition of SYBR-green). This ability of the RT-LAMP assay to differentiate between positive and negative samples can be very useful especially in the field conditions even if the tube scanner is not available. Due to the high sensitivity of the test, it can be assumed that if a test is negative by threshold validation, colorimetric analysis and gel-electrophoresis, it would be essentially negative [21–26].

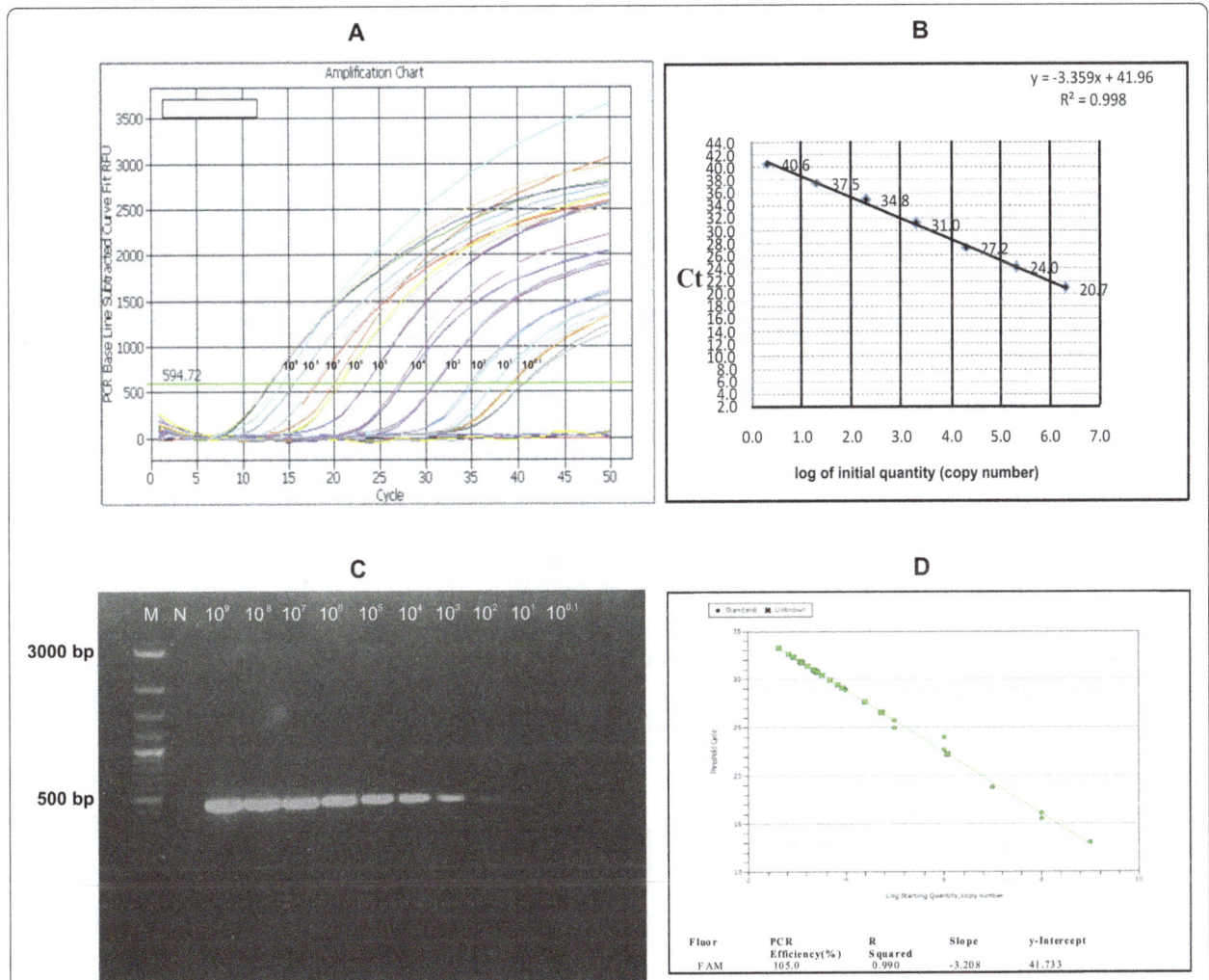

Fig. 7 Limit of detection of RT-qPCR and its application for the detection of PPRV in clinical samples; Panel **a**. In the figure are shown amplification curves generated by serial dilution of cloned standard of PPRV N gene (463 bp fragment); Panel **b**. For the estimation of limit of detection, threshold cycle (Ct) values of standard dilutions from 2.06×10^6 to 2.6×10^0 were drawn against log values of copy number. Limit of detection was estimated to be ≈ 20 copies at 37.5 ± 0.9 cycle although Ct value of 40th cycle was also detectable; Panel **c**. Detection limit of RT-PCR used in this study was up to 2.06×10^2 copies of the plasmid; Panel **d**. The threshold cycle (Ct) values obtained by amplification of PPRV in clinical samples along with standard

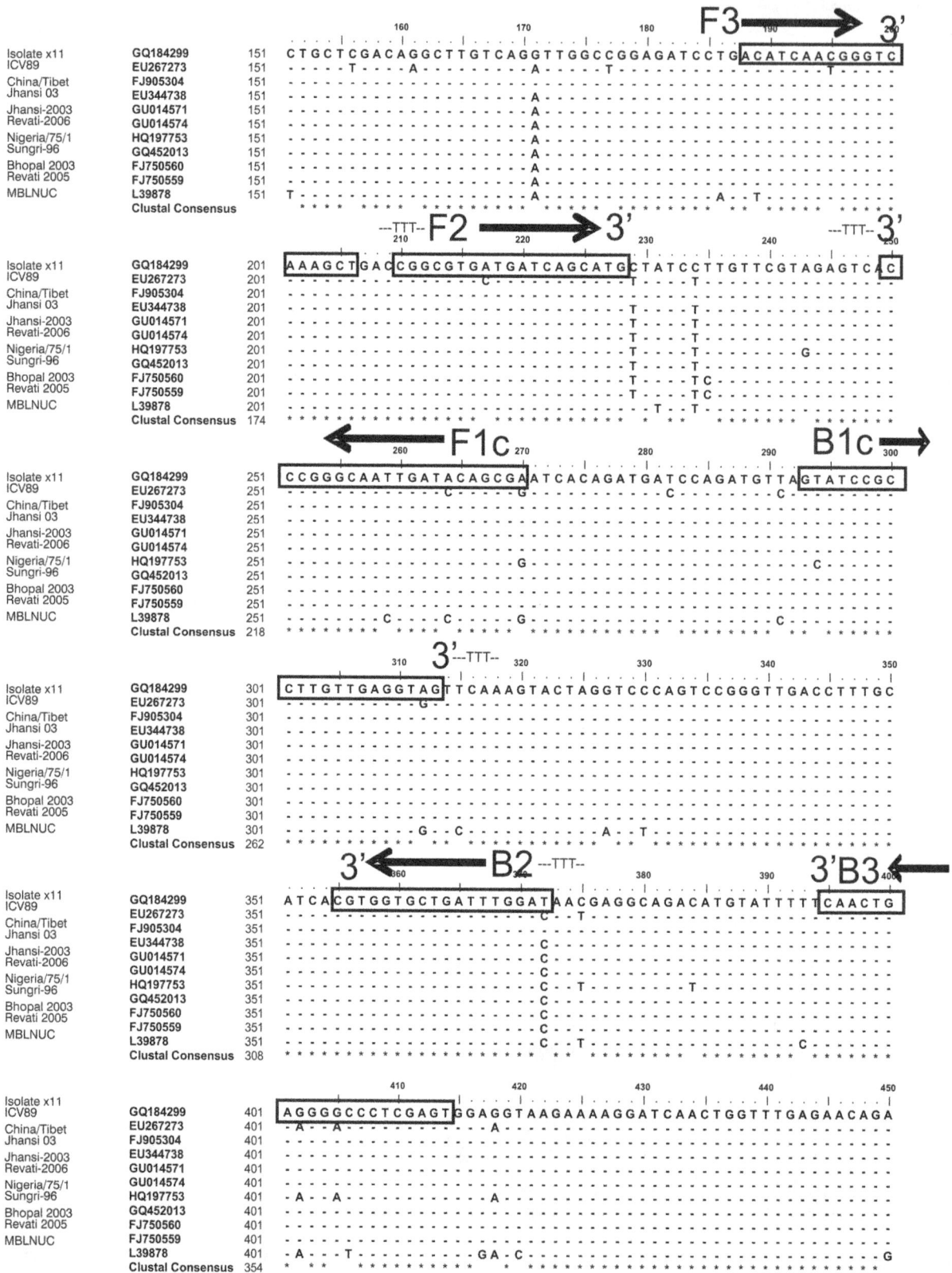

Fig. 8 Locations of RT-LAMP primers along the nucleoprotein gene sequence of PPRV. Primers are indicated by solid boxes and solid line arrows in the right (→) and left orientation (←) indicate forward (F3, F2 and B1c) and reverse primers (F1c, B2 and B3), respectively. The length of product was 187 as limited by Forward (F3) and backward outer (B3) primers

For that reason, RT-LAMP can be used for the detection of PPRV antigen in eradication campaigns to confirm the absence of disease.

The world population is experiencing continuous growth at an annual rate of 1.1% and estimated to reach 9.1 billion by 2050 [30]. Keeping it in the view, The Food and Agriculture Organization (FAO) of the United Nations estimates that the overall food production will need to increase by 70% in order to feed this projected world population thereby increasing pressure on the livestock sector to meet the growing demand for high value animal protein. Along with developed countries, where maximum livestock potential is already being utilized, developing countries like Pakistan, which have a lot of unexplored potential for livestock, need to contribute towards global food security. In these countries, there are certain challenges faced by livestock sector to cope up with these needs, among those, infectious diseases remain a key constraint and must be controlled for efficient livestock production. Accordingly, successful disease control or eradication program should be launched that would mainly rely on the prophylactic measures, surveillance & monitoring methods as well as the efficacy & robustness of the diagnostic tests. LAMP based diagnostic methods have taken the lead over other costly ones for being robust, economic and sensitive. RT-LAMP, developed in this study, for robust, specific on-site diagnosis of PPRV nucleic acid under field conditions is very practicable in the disease diagnosis during eradication of PPR. It offers a unique utility in the "Global PPR Eradication Campaign" launched recently, in the year 2015, by the FAO and OIE [31].

Conclusions

These laboratory, experimental and field evaluation of the LAMP method developed in this study for the on-site diagnosis of PPR showed that it is sensitive, robust, and easy diagnostic method and can prove to be very useful to the field practitioners. These features make it an inimitably advantageous nucleic acid detection system for the confirmation of PPRV infection under the field conditions. It is a sensitive as well as stringent assay of choice for quick detection of the virus in epidemics or during regular abattoir inspections or screening programs targeting PPRV eradication.

Methods
Cell lines, viruses and clinical samples
Vaccine strain of PPRV, Nigeria 75/1 was used as a reference strain in this study. In addition to 11 cultured PPRV strains a total of 32 nasal swabs collected from goats in three suspected PPR outbreaks during 2012–2014 were evaluated for virus detection. Those suspected goats were owned by herdsmen who were well informed about the purpose of sample collection and their consent was taken prior to collection of swab samples. The virus isolation from individual samples was carried out using CHS-20 cell line (kindly provided by Dr. Adama Diallo, IAEA Laboratories, Austria) that stably express signaling lymphocyte activation molecule (SLAM), a cell surface receptor for PPRV [8].

PPRV inoculation of sheep and goat
Two animals, one goat and one sheep bought from local livestock market, at the age of 5–6 months were tested negative for the presence of antibodies against PPRV using competitive enzyme linked immunosorbent assay (cELISA) [32] and acclimatized to controlled environment for two weeks. These were also confirmed as negative for PPRV antigen by RT-PCR [33] prior to their inoculation intravenously with 1 ml of 10^2 (TCID$_{50}$) PPRV strain 17BLK, which was isolated from a clinically infected goat in 2012 [34]. These animals were bought from local market and kept under controlled conditions at NIBGE animal house and were examined on a daily basis for the appearance of any clinical signs and symptoms up to 14 days post infection (dpi). Experimental samples including blood and nasal swabs were collected from each animal on alternate days up to 14dpi. The samples were stored at −80 °C until further use.

Primer designing
Full-length nucleotide sequences of N protein gene of lineage-IV PPRVs along with vaccine strain (Nigeria/75/1) were retrieved from the GenBank database and aligned using MegAlign 5.00, DNA star Inc., software to identify conserved regions. Analysis of the prospective target regions was performed with the LAMP primer design software (www.http://primerexplorer.jp/elamp4.0.0/) for the automated selection of the primers. A set of primers was finally selected that included an outer pair consisting of forward outer (F3) and backward outer (B3) primer, an inner pair consisting of forward inner (FIP = FIc + F2) and backward inner (BIP = B1 + B2c) primer as mentioned in Table 5. The location of these primers on the conserved target region is shown in Fig. 8.

Sample preparation and RT-LAMP reaction
Two microliter (2 µl) of swab extract or cell-culture supernatant was added to 30 µl of RT-LAMP buffer (66 mM Tris-HCl pH8.8, 32 mM KCl, 32 mM (NH4)$_2$SO$_4$, 16 mM MgCl$_2$, 0.3% (v/v) Tween 20) and incubated at room temperature for five minutes. Fifteen microliter (15 µl) of this sample mixture was added to 10 µl of reaction mixture providing final concentrations of 0.4 mM dNTP, 460 mM trehalose and 0.4X EvaGreen® dye. A ready to use primer mix was added providing a final concentration of 0.25 µM of each of the outer primers (F3 and B3) and 1.25 µM of each of the inner

Table 5 N gene based primers set used in detection of PPRV by RT-LAMP assay

Primer ID	Type	Sequence (5′ to 3′)	Position*	Length (bases)	Predicted length of amplified segment
PPRV F3	Forward outer	ACATCAACGGGTCAAAGCT	295–313	19	187 bp
PPRV B3	Backward outer	ACTCGAGGGTCCTTCAGTTG	521–502	20	
PPRV/FIP	Forward inner (F1c + F2)	CCGCTGTATCAATTGCCCGGGTTTTCGGCGTGATGATCAGCATG	376–357/317–335	44	
PPRV/BIP	Backward inner (B1 + B2c)	GCATCCGCCTTGTTGAGGTAGTTTTTTTGTCCAAATCAGCACCACG	400–421/481–462	46	

*The position of primers is indicated as per sequence position of complete genome of PPRV (Accession #hq197753)

primers (FIP and BIP). Amplification of the target genome segment of PPRV based on strand displacement activity was achieved using 8 units per reaction of Bsm polymerase (FermentasThermo Scientific, St. Leon-Rot, Germany). The assay was carried out at 58 °C in an ESE-Quant Tube Scanner TS95 (Qiagen, Hilden, Germany) and the increase in fluorescence signal was recorded once per minute for 60 min.

Fluorometric and gel electrophoresis based analysis of RT-LAMP products

ESE-Quant Tube Scanner was set at 6- carboxyfluorescein (FAM) channel (excitation = 487 nm, detection = 525 nm) for the acquisition of fluorescence data. Baseline threshold value of the amplification curve was calculated as 10 times standard deviation of the fluorescence signal during initial 5 min. Accordingly, for evaluation of the generation of "quasi" exponential amplification phase which is an indication of a positive reaction, each sigmoidal curve was analysed for a time-point corresponding to increase in florescence signal above threshold i.e. > 30 mV/min for a minimum of 3 consecutive measurements (Fig. 5). The time of first rise in signal above threshold (>30 mV/min) was given by the detection-time (Td) value. For unaided eye evaluation, 0.2 µl SYBR green fluorescent dye was added to the reaction tubes at the end of the assay. The positive reaction developed green colour while negative reaction remained orange. Furthermore, these reactions were analyzed for bright green fluorescence under UV light, which is an indication for a positive test. For agarose gel analysis, RT-LAMP products were incubated at 80 °C for 02 min to stop any residual enzymatic activity and subsequently analysed on 2% agarose gel stained with ethidium-bromide and the image was captured using a UV light transilluminator.

Comparative evaluation of RT-LAMP assay with RT-PCR and/or Quantitative RT-PCR (RT-qPCR) using clinical samples

For comparative evaluation of LAMP with reverse transcription PCR (RT-PCR), PPRV nucleic acid was detected in the clinical samples that were collected from suspected goats in field outbreaks, after taking consent from their owners, and experimentally infected animals, bought from local market and kept under control conditions, from days 1–14 post-inoculation (dpi), by these tests. The progression of disease in these infected animals was monitored by measuring the cumulative clinical score and changes in rectal temperature. After 14 days, the animals were slaughtered and observed for any pathological lesions in the organs. RT-LAMP was applied directly on swab extracts as mentioned above. For RT-PCR and RT-qPCR based detection of PPRV in these samples, total RNA was extracted from 200 µl suspension of each swab-extract using TRIzol method following the manufacturer's instructions (Invitrogen, Carlsbad, USA). RNA pellet was resuspended in 20 µl of RNase-free water, quantified using Nanodrop 1000 spectrophotometer, and stored at –80 °C until further use. Total RNA (1 µg) was reverse transcribed to cDNA by random hexamer primers using RevertAid™ First Strand cDNA Synthesis Kit following the manufacturer's instructions (Invitrogen, Paisley, UK). Subsequently, 2.5 µl of first strand cDNA product was used as a template to amplify 463 bp fragment of nucleoprotein gene of PPRV by RT-PCR [33]. To test the sensitivity of the assay, serial dilutions of virus culture supernatants were subjected to the LAMP. For the comparative detection of PPRV genome, RNA extracted from tenfold serial dilutions ranging from 10^0 to 10^{-5} dilution of PPRV starting from $10 TCID_{50}$/ml was subjected to RT-LAMP and conventional RT-PCR recommended by OIE that amplifies a 351 bp fragment of the N gene of PPRV [1, 14]. One-step RT-qPCR was carried out, in parallel with the RT-LAMP assay, using the iScript One-Step RT-qPCR Kit (BioRad, Hercules, CA, USA). PPRV N-gene specific forward (5′- ggactgggcctcgacagg-3′) and reverse (5′-ggatcgcagctttgacttcttc-3′) primers were used in combination with Taqman probe (FAM-5′tccttcctccagcataa3′-BHQ1) [8]. For standard curve generation, seven 10-fold dilutions (10^6-10^0 copies) of a defined template (pTZ57R/T that contained partial CDS of PPRV N protein gene) were subjected to a 40-cycle qPCR assay along with clinical samples and no template control (NTC). The reaction was carried out in a total volume of 20 μl (400nM of each primer, 200nM of the probe and 0.4 μl of the iScript reverse transcriptase) using IQ5™ Real-Time PCR Detection System (BioRad, USA) under the following reaction conditions: 50 °C for 10 min; 95 °C for 5 min and 40 cycles of 95 °C for 10s and 60 °C for 30s.

Additional files

Additional file 1: Detection limit of N-gene based RT-LAMP assay. Sheet 1; The excel file includes data as values of millivolts (mV) against specified time obtained from 10-fold Serial dilutions of fluid supernatant taken from cultured PPRV in CHS-20 cells harvested at appearance of 80% cytopathic effects. Sheet 2; The test was able to detect dilution up to 10^{-4} within 60 min. Sheet 3; includes data generated from real-time RT-PCR by amplifying 10-fold serial dilutions of known plasmid concentrations. Sheet 4; data generated in real-time RT-PCR based detection of PPRV in clinical samples collected in field outbreaks. (XLSX 51 kb)

Additional file 2: Data file generated by ESE software during the detection of PPRV in the field conditions. It is only readable format by software. (DAT 29 kb)

Additional file 3: Amplification curves obtained by ESE quant tube scanner software during the RT-LAMP based detection of PPRV in clinical samples. The slope of positive amplification curve becomes steeper during the exponential phase and becomes horizontal platue again as the reactions components are consumed during the amplification process. (PNG 153 kb)

Additional file 4: ESE Software file in notepad format generated from LAMP based PPRV detection in experimentally infected goat from 2-dpi to 12-dpi. In positive samples, the value of amplification signal in the form of mVolts increased with the passage of time until a platue phase is reached with no more increase in signal. (TXT 5 kb)

Additional file 5: RT-qPCR in goat. File includes information on amplification of PPRV from experimentally infected goat from 2 to 14-dpi. (RTF 3678 kb)

Additional file 6: ESE software file in notepad format generated from Detection of PPRV in clinical samples collected in outbreaks. Again in positive samples, the value of amplification signal in the form of mVolts increased with the passage of time during exponential phase until a platue phase is reached with no more increase in signal. (TXT 5 kb)

Additional file 7: RT-LAMP based detection of serial dilutions of virus supernatant. The graph was generated in "GraphPad Prism software" expressing increasing serial dilution of PPRV culture supernatant along x-axis and corresponding time of detection along y-axis. The RT-LAMP was able to detect up to 10^{-4}TCID50/ml within 60 min. (TXT 143 bytes)

Additional file 8: Threshold validation in clinical samples. In the field outbreaks, only those samples were taken as positive which crossed the threshold limit of 30mVolts in their amplification signals. (PNG 164 kb)

Additional file 9: End-point UV visualization of RT-LAMP reaction. The positive samples produced bright green fluorescence in contrast to negative control. (JPG 981 kb)

Additional file 10: Data file of End point report of RT-qPCR. It includes the data of end-point analysis for threshold validation of amplification curves generated during the test. (RTF 261 kb)

Additional file 11: Data file of RT-qPCR based detection of serial dilutions of PPR N gene standard cloned plasmid, at FAM channel. Accuracy was equivalent to $R^2 = 0.999$, 3.03 as value of slope. (RTF 3659 kb)

Additional file 12: Data file of RT-qPCR based detection of PPRV in clinical samples collected in field outbreaks. In these samples, ct values ranged from 22 to 33. (RTF 3690 kb)

Additional file 13: RT-PCR based detection of PPRV in blood of experimentally infected sheep and goat. Corresponding wells for Goat (G) and Sheep (S) are indicated in the figure along with respective number of day post inoculation (dpi). In goat virus detectable from 4-dpi to 14-dpi. (JPG 821 kb)

Abbreviations
ASSURED: Affordable sensitive specific user-friendly robust equipment free deliverable to the end user; BIP: Backward inner primer; Bsm: *Bacillus smithii*; Bst: *Bacillus stearothermophilus*; cDNA: Complimentary DNA; cELISA: Competitive enzyme linked immunosorbent assay; CHS-20: Monkey kidney cells expressing sheep signaling lymphocyte activation molecule; dpi: Day post inoculation; FAM: 6- carboxyfluorescein; FAO: The Food and Agriculture Organization of the United Nations; FIP: Forward inner primer; FMDV: Foot and mouth disease virus; mV: Milli Volts; N: Nucleocapsid gene; NTCs: No template controls; OIE: The World Organization for Animal Health; PPR: Peste des petits ruminants; PPRV: Peste des petits ruminants virus; RT-LAMP: Reverse transcription-loop mediated isothermal amplification assay; RT-PCR: Reverse transcription polymerase chain reaction; TCID50/ml: Tissue culture infective dose-50 per ml; Td: Detection time

Acknowledgements
I would also recognize Dr Imran Amin for helping in real-time PCR experiments and Dr. Wasim Abbas (Senior Scientist, (Virology/Immunology) at Environmental Biotechnology Division, NIBGE) for reviewing the draft. I would like to acknowledge field veterinarians who informed about the suspected animals and helped during sampling (Dr Zafar Iqbal (VO, Rahim Yar khan), Dr M. Irfan Alvi (VO, District Qabola), Dr Sajjad Haider (VO, District Qabola), Dr M. Imran (VO, District Chakwal), Dr M. Sajjad (SVO, District DG Khan), Dr Agha Shafique, (VO, District DG Khan), Mr. Jamrood Khan (Mianwali).

Funding
This research was funded by International Atomic Energy Agency (IAEA) through Coordinated Research Program IAEA/CRP, Contract No. 14567 and International Research Support Initiative Program (IRSIP) of the Higher Education Commission, Pakistan.

Authors' contribution
WA Conceived and designed the experiments and primers, performed the experiment, Analyzed the data, Wrote the paper. HU Supported primer design and contributed reagents, analyzed the data, reviewed the manuscript. SH Performed the experiments and collected the data. AM Maintained the care at animal house, helped in collection of samples, arranged field samples. MF Analyzed the data, Wrote the paper. MA Maintained the care and handling of experimental animals, collection of field samples. QMK Contributed reagents/materials, evaluated the design of the experiments, data analysis and reviewed the manuscript. All authors read and approved the final manuscript.

Authors' information
Ashraf W. is PhD from Biotechnology Campus (NIBGE) of Pakistan Institute of Engineering and Applied Sciences (PIEAS) and involved in research on molecular detection and evolutionary phylogenetics of PPRV since 2009. He did part of his PhD studies on partial as well as full genome sequences and evolutionary phylogeny of PPRV from Pakistan, Ghana and Benin. He has future plans to continue his research on molecular diagnosis and vaccine development and genetic studies of host- viral pathogen interaction. Dr. Khan Q. M. (corresponding author) is involved in molecular diagnostics and evolutionary genetics of transboundary animal diseases of Rinderpest, PPR, FMD, Avian Influenza and Brucellosis since last twenty years and has actively participated in Rinderpest eradication and successfully completed Co-ordinated research projects (CRP) on Early and Sensitive Detection of PPRV in collaboration with Dr Herman Unger and Dr Adama Diallo at IAEA.

Competing interests
The authors declare that they have no competing interests.

Consent for publication
Not Applicable.

Ethics approval and consent to participate

The study design and protocols for experimental goat and sheep were approved by *"The Ethical Committee, National Institute for Biotechnology and Genetic Engineering"* and *"Animal House Committee, National Institute for Biotechnology and Genetic Engineering (NIBGE), P.O. Box: 577, Jhang road, Faisalabad, Pakistan"* prior to the start of the study (ISO#08/01/2015). Experimental animals [sheep; Kajla breed and goat; Beetle breed] were purchased from local market and kept under control conditions, provided with adequate resources for illumination, thermoregulation, ventilation and bedding of fine wood husk to maintain temperature range of 20–25 °C and 60–70% relative humidity; water, feed and fodder were offered *ad libitum*. These were acclimatized to the control conditions for two weeks prior to start of experiment. Adherence to high standard of animal handling and care for experimental animals was assured during the study. The experimental animals were euthanized at the end of the study. From the sick goats in herds suspected for PPR outbreaks, swabs were collected with the prior consent of their owners.

Author details

[1]National Institute for Biotechnology and Genetic Engineering (NIBGE), Faisalabad, Pakistan. [2]Pakistan Institute of Engineering and Applied Sciences (PIEAS), Islamabad, Pakistan. [3]Animal Production and Health Section, Joint FAO/IAEA Division of Nuclear Techniques in Food and Agriculture, Vienna, Austria.

References

1. In Manual of diagnostic tests and vaccines for terrestrial animals 2013, Peste des petits ruminants, World Organisation for Animal Health (Office International des Épizooties: OIE) Chapter 2.7.10, page#5-7. (Dowloaded on 05-01-2016). http://www.oie.int/fileadmin/Home/eng/Health_standards/tahm/2.07.10_PPR.pdf.
2. Dhar P, Sreenivasa BP, Barrett T, Corteyn M, Singh RP, Bandyopadhyay SK. Recent epidemiology of peste des petits ruminants virus (PPRV). Vet Microbiol. 2002;88:153–9. doi:10.1016/s0378-1135(02)00102-5 .
3. Abubakar M, Jamal SM, Arshed MJ, Hussain M, Ali Q. Peste des petits ruminants virus (PPRV) infection; its association with species, seasonal variations and geography. Trop Anim Health Prod. 2009;41:1197–202. doi:10.1007/s11250-008-9300-9.
4. Zahur AB, Ullah A, Hussain M, Irshad H, Hameed A, Jahangir M, Farooq MS. Sero-epidemiology of peste des petits ruminants (PPR) in Pakistan. Prev Vet Med. 2011;102:87–92. doi:10.1016/j.prevetmed.201.06.011.
5. Abu-Elzein EME, Hassanien MM, Alfaleq AI, Abd-Elhadi MA, Housawi FMI. Isolation of PPR virus from goats in Saudi Arabia. Vet Rec. 1990;127:309–10. [PubMed]: https://www.ncbi.nlm.nih.gov/pubmed/2238415.
6. Diallo A, Minet C, Le Goff C, Berhe G, Albina E, Linbeau G, Barrett T. The threat of peste des perits ruminants: progress in vaccine development for disease control. Vaccine. 2007;26:5591–7. doi:10.1016/j.vaccine.2007.02.013.
7. Couacy-Hymann E, Bodjo SC, Danho T, Koffi MY, Libeau G, Diallo A. Early detection of viral excretion from experimentally infected goats with peste-des-petits ruminants virus. Prev Vet Med. 2007;78:85–8. doi:10.1016/j.prevetmed.2006.09.003.
8. Adombi CM, Lelenta M, Lamien CE, Shamaki D, Koffi YM, Traore A, Silber R, Couacy-Hymann E, Bodjo SC, Djaman JA, Luckins AG, Diallo A. Monkey CV1 cell line expressing the sheep-goat SLAM protein: a highly sensitive cell line for the isolation of peste des petits ruminants virus from pathological specimens. J Virol Methods. 2011;173:306–13. doi:10.1016/j.jviromet.2011.02.024.
9. Bailey D, Banyard A, Dash P, Ozkul A, Barrett T. Full genome sequence of peste des petits ruminants virus, a member of the Morbillivirus genus. Virus Res. 2005;110:119–24. doi:10.1016/j.virusres.2005.01.013.
10. Nanda SK, Baron MD. Rinderpest virus blocks type I and type II interferon action: role of structural and nonstructural proteins. J Virol. 2006;80:7555–68. doi:10.1128/jvi.02720-05.
11. Chard LS, Bailey DS, Dash P, Banyard AC, Barrett T. Full genome sequences of two virulent strains of peste-des-petits ruminants virus, the Côte d'Ivoire1989 and Nigeria 1976 strains. Virus Res. 2008;136:192–7. doi:10.1016/j.virusres.2008.04.018.
12. Muthuchelvan D, Sanyal A, Balamurugan V, Dhar P, Bandyopadhyay SK. Sequence analysis of the nucleoprotein gene of Asian lineage peste des petits ruminants vaccine virus. Vet Res Commun. 2006;30:957–63. doi:10.1007/s11259-006-3407-0.
13. Libeau G, Diallo A, Colas F, Guerre L. Rapid differential diagnosis of rinderpest and peste des petits ruminants using an immunocapture ELISA. Vet Rec. 1994;134:300–4. doi:10.1136/vr.134.12.300.
14. Couacy-Hymann E, Roger F, Hurard C, Guillou JP, Libeau G, Diallo A. Rapid and sensitive detection of peste des petits ruminants virus by a polymerase chain reaction assay. J Virol Methods. 2002;100:17–25. doi:10.1016/s0166-0934(01)00386-x.
15. Parida S, Muniraju M, Mahapatra M, Muthuchelvan D, Buczkowski H, Banyard AC. Peste des petits ruminants. Vet Microbiol. 2015;181:90–106. doi:10.1016/j.vetmic.2015.08.009.
16. Barrett T. Morbillivirus infections, with special emphasis on morbilliviruses of carnivores. Vet Microbiol. 1999;69:3–13. doi:10.1016/s0378-1135(99)00080-2.
17. Singh RP, Sreenivasa BP, Dhar P, Shah LC, Bandyopadhyay SK. Development of a monoclonal antibody based competitive-ELISA for detection and titration of antibodies to peste des petits ruminants (PPR) virus. Vet Microbiol. 2004;98:3–15. doi:10.1016/j.vetmic.2003.07.007.
18. Toplu N, Oguzoglu TC, Albayrak H. Dual infection of fetal and neonatal small ruminants with border disease virus and peste des petits ruminants virus (PPRV): neuronal tropism of PPRV as a novel finding. J Comp Pathol. 2012;146:289–97. doi:10.1016/j.jcpa.2011.07.004.
19. Morshed MG LM, Jorgensen D, Isaac-Renton JL. Molecular methods used in clinical laboratory: prospects and pitfalls. FEMS Immunol Med Microbiol. 2007;49:184–91. doi:10.1111/j.1574-695x.2006.00191.x.
20. Mabey D, Peeling RW, Ustianowski A, Perkins MD. Diagnostics for the developing world. Nat Rev Microbiol. 2004;2:231–40. doi:10.1038/nrmicro841.
21. Peeling RW, Mabey D. Point-of-care tests for diagnosing infections in the developing world. Clin Microbiol Infect. 2010;16:1062–9. doi:10.1111/j.1469-0691.2010.03279.x.
22. Nagamine K, Hase T, Notomi T. Accelerated reaction by loop-mediated isothermal amplification using loop primers. Mol Cell Probes. 2002;16:223–9. doi:10.1006/mcpr.2002.0415.
23. Njiru ZK. Loop-Mediated Isothermal Amplification Technology: Towards Point of Care Diagnostics. PLoS Negl Trop Dis. 2012;6:1–4. doi:10.1371/journal.pntd.0001572.
24. Lin L, Bao J, Wu X, Wang Z, Wang J. Rapid detection of peste des petits ruminants virus by a reverse transcription loop-mediated isothermal amplification assay. J Virol Methods. 2010;170:37–41. doi:10.1016/j.jviromet.2010.08.016.
25. Nemoto M, Yamanaka T, Bannai H, Tsujimura K, Kondo T, Matsumura T. Development and evaluation of a reverse transcription loop-mediated isothermal amplification assay for H3N8 equine influenza virus. J Virol Methods. 2011;178:239–42. doi:10.1016/j.jviromet.2011.07.015.
26. Yamazaki W, Mioulet V, Murray L, Madi M, Haga T, Misawa N, Horii Y, King DP. Development and evaluation of multiplex RT-LAMP assays for rapid and sensitive detection of foot-and-mouth disease virus. J Virol Methods. 2013;192:18–24. doi:10.1016/j.jviromet.2013.03.018.
27. Kwiatek O, Ali YH, Saeed IK, Khalafalla AI, Mohamed OI, Obeida AA, Abdelrahman MB, Osman HM, Taha KM, Abbas Z, El Harrak M, Lhor Y, Diallo A, Lancelot R, Albina E, Libeau G. Asian lineage of peste des petits ruminants virus, Africa. Emerg Infect Dis. 2011;17:1223–31. doi:10.3201/eid1707.101216.
28. Hong TC, Mai QL, Cuong DV, Parida M, Minekawa H, Notomi T, Hasebe F, Morita K. Development and evaluation of a novel loop-mediated isothermal amplification method for rapid detection of severe acute respiratory syndrome coronavirus. J Clin Microbiol. 2004;42:1956–196. doi:10.1128/jcm.42.5.1956-1961.2004.
29. Dadas RC, Muthuchelvan D, Pandey AB, Rajak KK, Sudhakar SB, Shivchandra SB, Venkatesan G. Development of loop-mediated isothermalamplification (lamp) assay for rapid detection of peste des petits ruminants virus (PPRV) genome from clinical samples. Indian J Comp Microbiol Immunol Infect Dis. 2012;33:7–13. doi:10.1016/j.jviromet.2015.06.005.
30. Christopher Delgado RM, Steinfeld H, Ehui S, Courbois C. Livestock to 2020: The Next Food Revolution. 1999. BRIEF 61; International Food Policy Research Institute 2033 K Street, NW Washington, DC 20006-1002 USA. doi:10.5367/000000001101293427.
31. Semedo MH. Towards the eradication of peste des petits ruminants/sheep and goat plague. Empress 360, (Animal Health) FAO. 2015;45:3. http://www.fao.org/3/fa6c6714-abb4-4bb7-8164-2c4975e6b329/i4484e.pdf.

32. Libeau G, Prehaud C, Lancelot R, Colas F, Guerre L, Bishop DH, Diallo A. Development of a competitive ELISA for detecting antibodies to the peste des petits ruminants virus using a recombinant nucleoprotein. Res Vet Sci. 1995;58:50–5. doi:10.1016/0034-5288(95)90088-8.

33. Kerur N, Jhala MK, Joshi CG. Genetic characterization of Indian peste des petits ruminants virus (PPRV) by sequencing and phylogenetic analysis of fusion protein and nucleoprotein gene segments. Res Vet Sci.2008;85:176–83. doi:10. 1016/j.rvsc.2007.07.007.

34. Ashraf W, Khan QM, Mobeen A, Kamal H, Farooq M, Jalees MM, Diallo A, Unger H. Molecular detection and genotyping characterization of PPRV in various areas of Pakistan, 2007 to 2012. Empress 360, (Animal Health), FAO. 2015;45:14–6. http://www.fao.org/3/fa6c6714-abb4-4bb7-8164-2c4975e6b329/i4484e.pdf.

4

Topical use of 5% acyclovir cream for the treatment of occult and verrucous equine sarcoids

Maarten Haspeslagh* ⓘ, Mireia Jordana Garcia, Lieven E. M. Vlaminck and Ann M. Martens

Abstract

Background: Previous studies mention the use of topical acyclovir for the treatment of equine sarcoids. Success rates vary and since the bovine papillomavirus (BPV) lacks the presence of a kinase necessary to activate acyclovir, there is no proof of its activity against equine sarcoids.

Results: Twenty-four equine sarcoids were topically treated with acyclovir cream and 25 with a placebo. Both creams were applied twice daily during 6 months. Before the start of the treatment and further on a monthly basis, photographs and swabs were obtained. On the photographs, sarcoid diameter and surface area were measured and verrucosity of the tumours was quantified using a visual analog scale (VAS). The swabs were analysed by PCR for the presence of BPV DNA and positivity rates were calculated as the number of positive swabs divided by the total number of swabs for each treatment group at each time point. Success rates were not significantly different between both treatment groups. There was also no significant effect of treatment on sarcoid diameter, surface area or VAS score. For the swabs, a significantly higher BPV positivity rate was found for acyclovir treated tumours compared to placebo treated sarcoids only after 1 month of treatment and not at other time points.

Conclusions: None of the results indicate that treatment with acyclovir yields any better results compared to placebo treatment.

Keywords: Acyclovir, Bovine Papillomavirus, Equine Sarcoid, Topical treatment

Background

Acyclovir (acycloguanosine) is an antiviral drug developed for the treatment of herpes simplex virus (HSV) infections in humans [1]. The drug relies on competitive inhibition of the viral DNA polymerase, but needs to be phosphorylated by a HSV specific thymidine kinase and then further by cellular enzymes to a triphosphate form to exert its action [2]. Nevertheless, inhibition of viral replication also occurs for other viral species then HSV [3], suggesting phosphorylation of the drug may also occur in cells where the HSV thymidine kinase is not present, albeit to a lesser extent.

Equine sarcoids are tumours originating in the dermal layers of equine skin. The pathogenesis is not entirely clear yet, but there is agreement in literature that the bovine papillomavirus (BPV) (mainly type 1 and type 2) most likely plays an important role in the development of these tumours [4]. Many treatments have been reported, but no universal treatment has been found to cure all sarcoids on all body locations.

Topical treatment with acyclovir has been described to result in complete regression in 68% of occult, verrucous, nodular or mixed equine sarcoids [5]. In addition, a recent ex vivo study has shown that acyclovir concentrations reached in the dermal layers after topical administration on sarcoid-affected equine skin are high enough to possibly achieve an antiviral effect [6]. However, the thymidine kinase necessary for initial phosphorylation of the drug is

---

* Correspondence: maarten.haspeslagh@ugent.be
Department of Surgery and Anaesthesiology of Domestic Animals, Faculty of Veterinary Medicine, Ghent University, Salisburylaan 133, 9820 Merelbeke, Belgium

missing in BPV DNA [7] and the susceptibility of the BPV DNA polymerase for acyclovir is unknown. A retrospective study further described complete regression after topical acyclovir treatment in only 53% of the cases [8].

The goal of the present study was to establish if topical treatment of occult equine sarcoids with a 5% acyclovir cream is more effective compared to the application of a placebo cream following the same administration protocol.

Methods

Subjects

A power analysis revealed that at least 15 unrelated sarcoids were needed in each treatment group to be able to show a difference between acyclovir and placebo treatments (power = 0.8; type-1 error = 0.05). Because multiple sarcoids on the same horse would be treated the same way, a minimum of 15 horses was necessary in each treatment group.

All horses that were presented to the Department of Surgery and Anaesthesiology of the Faculty of Veterinary Medicine of Ghent University for the treatment of previously untreated occult or partly verrucous equine sarcoids were considered for inclusion in the study. Horses that had fibroblastic or nodular sarcoids in addition to the occult and/or partly verrucous tumours and horses that received concurrent medical treatment for other indications were excluded. The diagnosis of equine sarcoid was made by clinical examination by an experienced veterinarian. As a compensation for taking part in the study, the treatments and consultations were offered free of charge and sarcoids that would have been treated with placebo during the study would be treated afterwards without additional costs. When owners were willing to participate, an informed consent was signed in which the owners also committed to apply the topical treatment as instructed.

Treatment and sampling

Sarcoids were topically treated with either a generic 5% acyclovir cetomacrogol cream or a placebo consisting of the same cetomacrogol cream without active component. The choice between both treatments was made at random and owners were blinded to the treatment. Packaging of the creams was identical and the labels were coded. When multiple occult or partly verrucous sarcoids were present on the same horse, all tumours were treated with the same product.

The sarcoids of both treatment groups were completely covered with cream twice daily by the owner. If cream remnants were still present, the lesion was cleaned with water and dried with a paper towel before applying more. Owners were instructed and demonstrated to beforehand how to apply the cream. Treatment continued for 6 months or until the sarcoid had

disappeared completely. The experiment was stopped if sudden aggressive growth of the sarcoid would occur.

Before the start of the treatment (T0) and further at monthly intervals (T1 until T6), a close-up photograph of the sarcoid was taken with a ruler next to, but not covering the tumour. To gain more insight in the antiviral effect of acyclovir on BPV in equine sarcoids, a swab sample for BPV DNA analysis was taken at the same time as the photographs by rubbing a sterile cottontip swab soaked in sterile distilled water over the surface of the sarcoid [9]. Swabs were stored in separate containers at $-20\ °C$ until further processing. When the horses were stabled too far away from the clinic, private practitioners were responsible for obtaining the swab samples and photographs and provide them to the clinic. When this was the case, the private practitioners received clear instructions on how to do this to ensure a high sample quality.

All pictures and samples were processed together at the end of the experiment. Pictures were given a random coded file name and put in random order to ensure blinded processing. On all pictures, sarcoid maximal diameter and surface area were measured twice using image measuring software (ImageJ). The mean of two measurements was used for further analysis. Severity of the sarcoid was determined by three diplomates of the European College of Veterinary Surgeons. A visual analog scale (VAS) was used ranging from no visible abnormalities of the skin (score 0) to distinct skin verrucosity (score 1000) (Fig. 1). The mean of three scores was used for further analysis.

Swabs were examined for the presence of BPV DNA. DNA was extracted using a commercial kit (DNeasy Blood & Tissue Kit, Qiagen). Swabs were first incubated for 12 h in 180 µl buffer ATL and 20 µl proteinase K at 56 °C. The samples were then vortexed and 200 µl buffer AL was added. After vortexing again, swabs were removed from the vials and discarded. Further DNA extraction was continued as described in the manual of the kit, following the tissue protocol. After DNA extraction, real-time PCR analysis was performed using general BPV primers and BPV-1 and -2 specific TaqMan probes as described by Bogaert et Al. [10]. All samples were processed in duplicate and when quantification cycle values between repeats differed by more than one, the PCR was repeated. Positive controls consisting of a mixture of known BPV-1 and BPV-2 samples and negative controls consisting of distilled water were included in each run. All samples were also tested for the presence of equine interferon beta (IFNb) DNA to confirm successful DNA extraction [11]. Samples were considered positive when either BPV-1 or BPV-2 DNA was detected. Samples were only considered negative when IFNb DNA could be detected, but no BPV DNA. When

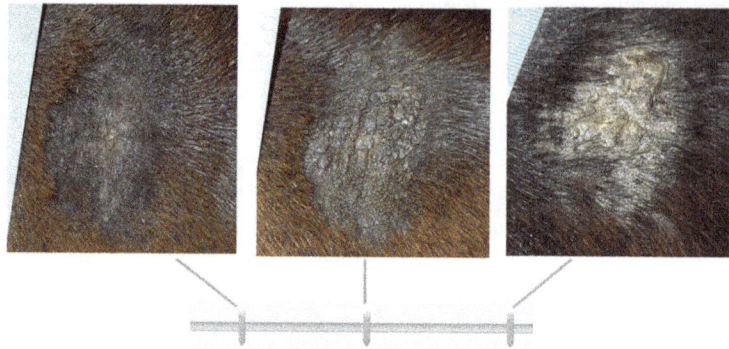

Fig. 1 Example of how a slider bar was used to determine severity of a sarcoid on a visual analog scale (VAS), based on verrucosity

no IFNb DNA was detected, samples were marked as missing data for further analysis.

Statistical analysis

All data analysis was performed using statistical software (SPSS 20, IBM). Statistical significance was set at $p \leq 0.05$. To estimate the effect of treatment on the number of fully regressed sarcoids, a Fisher exact test was used. For all continuous data (sarcoid diameter, sarcoid surface area and VAS score), the effect of time and treatment on the dependent variables was determined by repeated measures ANOVAs with horse as a blocking factor. When sphericity could not be assumed, a Greenhouse-Geisser correction was applied. Additionally, a "change parameter" was calculated for continuous data from all sarcoids as the difference between the measurement at T0 and the last measurement. The effect of treatment on this "change parameter" was estimated by a generalized linear model, corrected for follow-up time and with horse as a blocking factor. For each time point, the effect of treatment on the number of samples positive for BPV DNA was tested using a binary logistic regression with horse as a blocking factor. To test the effect of time on the number of samples positive for BPV DNA, a binary logistic regression with horse as a blocking factor was used for both treatments with T0 as the reference category.

Results

Twenty-eight horses and three ponies were included in the study. In total, 24 sarcoids on 15 individuals were treated topically with 5% acyclovir cream and 25 sarcoids on 16 individuals were treated topically with placebo cream. Multiple sarcoids were present in four horses in the acyclovir group, and in five horses in the placebo group. For both groups, median treatment time was 6 months (min: 1 month, max: 6 months). The study was stopped early because of sudden aggressive tumoural growth in one horse, which was part of the acyclovir treatment group. For three placebo treated

horses and one acyclovir treated horse, the study was stopped early because of complete sarcoid regression. No side effects were observed in any of the treated horses.

Complete regression during treatment occurred in two of the acyclovir treated sarcoids (8.3%; 95% CI: 1.0% - 27.0%), while this was the case in four (16%; 95% CI: 4.5% - 36.1%) of the placebo treated sarcoid. This difference was not significant ($p = 0.67$).

Figure 2 shows the mean measurements along with the 95% confidence interval of sarcoid diameter (A), sarcoid surface (B) and VAS score (C) at each time point for both treatment groups. The intraclass correlation coefficient of the raters for average measures of VAS was 77.0%. There was no significant effect of treatment or time on any of these variables. The mean calculated "change parameters" are listed in Table 1 along with the p-values for the effect of treatment on them. A positive "change parameter" indicates a decrease in measurement between T0 and the last sample point whereas a negative change indicates an increase. The mean largest sarcoid diameter of acyclovir treated tumours increased during treatment, while it decreased for the placebo treated sarcoids (Table 1). Mean surface area increased during the treatment for both groups (Table 1). Mean VAS score decreased, indicating that sarcoids were found to be less verrucous towards the end of the treatment (Table 1). Differences in "change parameters" between both treatment groups were never significant.

No genomic DNA was present on the swab in 15.9% of the samples in the acyclovir treated group and in 22.8% of the samples in the placebo group. Figure 3 shows the percentage of positive PCR samples in each treatment group at each time point. Only at T1, a significantly higher percentage of samples was positive for the presence of BPV DNA in the acyclovir group compared to the placebo group ($p = 0.005$). At all other time points, there were no significant differences between groups. In the acyclovir group, the percentage of positive samples was significantly higher at T1 compared to T0

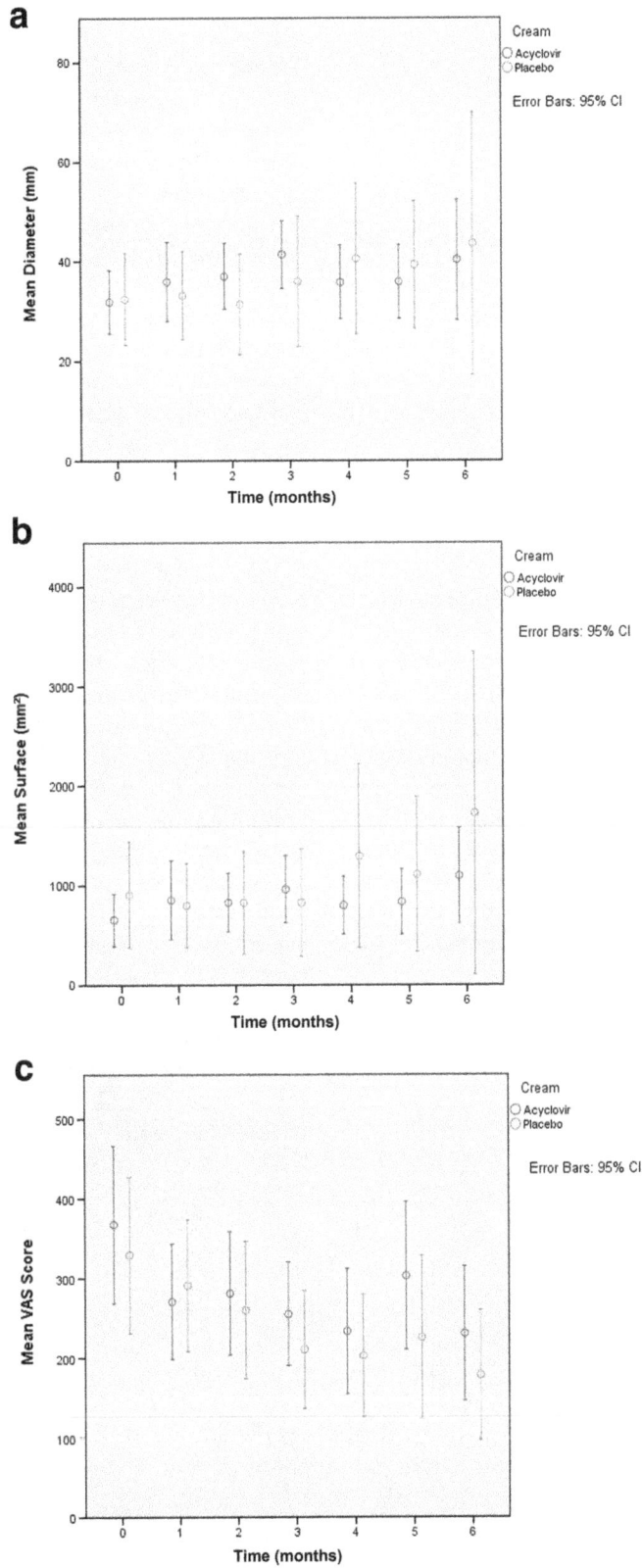

Fig. 2 Mean measurements along with the 95% confidence interval of sarcoid diameter (**a**), sarcoid surface (**b**) and visual analog scale (VAS) score (**c**) at each time point for both treatment groups

Table 1 Mean "change parameters" and associated *p*-values for the difference between both treatment groups (SE = Standard Error; VAS = visual analog scale)

Change	Acyclovir (± SE)	Placebo (± SE)	*p*-value
Diameter (mm)	−6.10 (± 4.63)	2.02 (± 5.67)	0.52
Surface area (mm^2)	−314.26 (± 203.44)	−60.25 (± 247.67)	0.87
VAS score	146.43 (± 53.75)	166.58 (± 39.41)	0.65

($p = 0.004$). At all other time points, the percentage of positive samples was not significantly different from T0. In the placebo group, the percentage of positive samples was significantly lower at all time points except for T1, compared to the percentage of positive samples at T0 (T2: $p = 0.06$; T3: $p = 0.03$; T4: $p = 0.04$; T5: p = 0.04; T6: p = 0.04).

Discussion

This is the first study on equine sarcoid treatment which is placebo controlled and double blinded. Results of this study are therefore valuable for treatment selection in practice. Because of the long study duration, the horses were cared for by their owners at home and this implicated that it was hard to check for treatment compliance. However, all owners signed a commitment to the experiment beforehand and were contacted regularly by phone to monitor the course of the study. Because the horses were stabled at home and this was in many cases too far away to allow for visits to the clinic, monthly visits were often performed by trusted private practitioners. While this implicates a possible variation in sample quality, practitioners received clear instructions on how to take pictures and swabs to maximize uniformity. A benefit of this was that the persons evaluating the pictures and analyzing the samples never saw the patients in real life, which enabled an unprejudiced evaluation.

The success rate for acyclovir treatment was lower compared to previously reported success rates [5, 8]. The sample population of the present study was smaller, which could explain the lower success rates. Moreover, acyclovir treatment was stopped after 6 months of treatment in the present study, while under normal clinical circumstances it is often continued longer when the tumour has not fully regressed yet, but a beneficial effect is observed. In both previous studies, a certain number of sarcoids have indeed been treated for over 6 months, which could have increased the success rate. While the concentration of acyclovir in the cream used in the present study was the same as in all earlier studies [5, 8], the constitution of the cream vehicle used in this and earlier studies by Haspeslagh et al. [6, 8] differed from the one used by Stadler et al. [5].

Time and treatment type did not have a significant influence on mean sarcoid dimensions or mean VAS score. Mean "change parameters" were also not significantly different between treatment groups. Nevertheless, the mean VAS score decreased for both treatment groups during the study, indicating that equine sarcoids were found less verrucous towards the end of the treatment. Perhaps the previously observed benign effect is therefore not due to the application of acyclovir, but merely to the effect of a cream that keeps the skin hydrated, preventing the formation of a thick verrucous layer. Anti-keratotic creams have analogously been used to lessen the verrucosity of equine sarcoids prior to other treatments [12]. This hypothesis can be tested by including a third "no treatment" group, which was not done here due to ethical considerations towards the owners. The presence of a keratinous layer in equine sarcoids with a high degree of verrucosity could interfere with acyclovir penetration and the presence of highly verrucous tumours could have influenced the results. Nevertheless, the mean VAS score, based on verrucosity, was

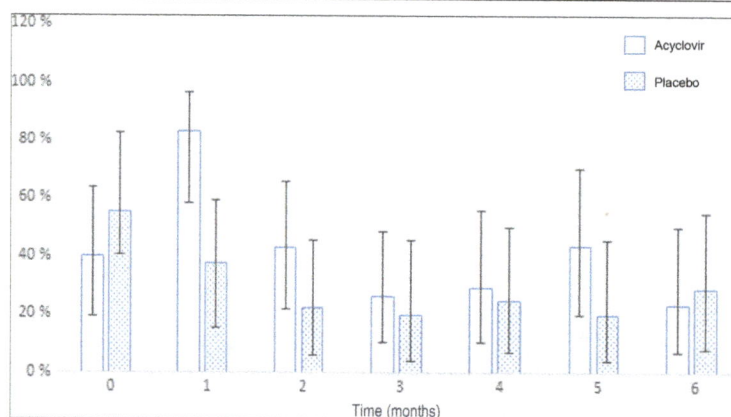

Fig. 3 The percentage of samples positive for the presence of BPV DNA at each time point (months after start of the treatment) and corresponding 95% confidence intervals. No fill: acyclovir; dotted fill: placebo

not very high at T0 and decreased during treatment. VAS scores at T0 did also not differ significantly between the acyclovir and placebo group, indicating that the comparison between both groups is valid. To evaluate the effect of a thick verrucous layer on acyclovir treatment, a similar experiment could be performed comparing treatment of strictly occult versus strictly verrucous tumours.

In order to obtain a good indication of the BPV load in non-ulcerated equine sarcoids, a quantitative real-time PCR on tissue from a biopsy probably yields the most reliable results. However, this would have required the sarcoids to be biopsied prior to the study and further at each time point of evaluation, which could have influenced the outcome, as equine sarcoids often become more aggressive after being damaged [13, 14]. For this reason, swabs were obtained instead of biopsies. This also implies that histological examination to confirm the diagnosis of equine sarcoid could not be performed and that there is a chance that some lesions which regressed spontaneously were not actual equine sarcoids. Nevertheless, the clinical appearance of equine sarcoids is so typical that clinical examination by an experienced veterinarian should be sufficient to make a correct diagnosis [14, 15]. While the presence of BPV DNA can be shown in up to 100% of swabs from equine sarcoids, this is only the case in ulcerated tumours where the dermis is exposed [9]. As this was never the case in this study, the percentage of positive samples was lower and in range with the earlier reported positivity rate originating from occult sarcoids [9]. Nevertheless, in the placebo group, a clear and significant decrease in the positivity rate could be seen over time, which was not the case for acyclovir treated tumours. No plausible explanation could be given for this observation.

Conclusions

None of the results presented in this study indicate that topical treatment of occult or partly verrucous equine sarcoids with acyclovir yields any better results compared to treatment with placebo cream.

Abbreviations

BPV: Bovine papillomavirus; IFNb: Interferon beta; VAS: Visual analog scale

Acknowledgements

The results of this research were presented at the 25th ECVS annual scientific meeting, July 7-9, 2016 in Lisbon, Portugal, and were published in abstract form in the proceedings.
The authors thank Cindy De Baere for her help with analyzing the samples and all owners and veterinarians who participated in this study.

Funding

The research was entirely funded by the department of Surgery and Anaesthesiology of Domestic Animals, Faculty of Veterinary Medicine, Ghent University, Belgium.

Authors' contributions

MH, LV and AM contributed to the study concept and design; MH, MJ, LV and AM contributed to the acquisition of data; MH contributed to data analysis; MH, MJ, LV and AM contributed to the writing and revising of the manuscript. All authors read and approved the final manuscript.

Consent for publication

Not applicable.

Competing interests

The authors declare that they have no competing interests.

References

1. Elion GB, Furman PA, Fyfe JA, de Miranda P, Beauchamp L, Schaeffer HJ. Selectivity of action of an antiherpetic agent, 9-(2-hydroxyethoxymethyl) guanine. Proc Natl Acad Sci U S A. 1977;74:5716–20.
2. Miller WH, Miller RL. Phosphorylation of acyclovir (acycloguanosine) monophosphate by GMP kinase. J Biol Chem. 1980;255:7204–7.
3. Datta AK, Colbyt BM, Shawt JE, Paganotf JS. Acyclovir inhibition of Epstein-Barr virus replication. Biochemistry. 1980;77:5163–6.
4. Nasir L, Campo MS. Bovine papillomaviruses: their role in the aetiology of cutaneous tumours of bovids and equids. Vet Dermatol. 2008;19:243–54.
5. Stadler S, Kainzbauer C, Haralambus R, Brehm W, Hainisch E, Brandt S. Successful treatment of equine sarcoids by topical aciclovir application. Vet Rec. 2011;168:187–90.
6. Haspeslagh M, Taevernier L, Maes A, Vlaminck L, De Spiegeleer B, Croubels S, et al. Topical distribution of acyclovir in normal equine skin and equine Sarcoids: an in vitro study. Res. Vet. Sci. Elsevier Ltd. 2016;106:107–11.
7. Baker CC, Howley PM. Differential promoter utilization by the bovine papillomavirus in transformed cells and productively infected wart tissues. EMBO J. 1987;6:1027–35.
8. Haspeslagh M, Vlaminck L, Martens A. Treatment of sarcoids in equids: 230 cases (2008-2013). J Am Vet Med Assoc 2016;249:311–318.
9. Martens A, De Moor A, Ducatelle RPCR. Detection of bovine papilloma virus DNA in superficial swabs and scrapings from equine sarcoids. Vet J. 2001; 161:280–6.
10. Bogaert L, Van Poucke M, De Baere C, Dewulf J, Peelman L, Ducatelle R, et al. Bovine papillomavirus load and mRNA expression, cell proliferation and p53 expression in four clinical types of equine sarcoid. J Gen Virol. 2007;88: 2155–61.
11. Haralambus R, Burgstaller J, Klukowska-Rötzler J, Steinborn R, Buchinger S, Gerber V, et al. Intralesional bovine papillomavirus DNA loads reflect severity of equine sarcoid disease. Equine Vet J. 2010;42:327–31.
12. Quinn G. Skin tumours in the horse : clinical presentation and management. In Pract. 2003;25:476–83.
13. Bergvall KE. Sarcoids. Vet Clin North Am Equine Pract. 2013;29:657–71.
14. Knottenbelt D, Edwards S, Daniel E. Diagnosis and treatment of the equine sarcoid. In Pract. 1995;17:123–9.
15. Bogaert L, Martens A, Depoorter P, Gasthuys F. equine sarcoids - part 1 : clinical presentation and epidemiology. Vlaams Diergeneeskd. Tijdschr 2008; 77:2–9.

5

Syndromic surveillance for West Nile virus using raptors in rehabilitation

Alba Ana[1,3]* ⓘ, M. Perez Andrés[1], Ponder Julia[1], Puig Pedro[2], Wünschmann Arno[1], Vander Waal Kimberly[1], Alvarez Julio[1] and Willette Michelle[1]

Abstract

Background: Wildlife rehabilitation centers routinely gather health-related data from diverse species. Their capability to signal the occurrence of emerging pathogens and improve traditional surveillance remains largely unexplored. This paper assessed the utility for syndromic surveillance of raptors admitted to The Raptor Center (TRC) to signal circulation of West Nile Virus (WNV) in Minnesota between 1990 and 2014. An exhaustive descriptive analysis using grouping time series structures and models of interrupted times series was conducted for indicator subsets.

Results: A total of 13,080 raptors were monitored. The most representative species were red-tailed hawks, great horned owls, Cooper's hawks, American kestrels and bald eagles. Results indicated that temporal patterns of accessions at the TRC changed distinctively after the incursion of WNV in 2002. The frequency of hawks showing WNV-like signs increased almost 3 times during July and August, suggesting that monitoring of hawks admitted to TRC with WNV-like signs could serve as an indicator of WNV circulation. These findings were also supported by the results of laboratory diagnosis.

Conclusions: This study demonstrates that monitoring of data routinely collected by wildlife rehabilitation centers has the potential to signal the spread of pathogens that may affect wild, domestic animals and humans, thus supporting the early detection of disease incursions in a region and monitoring of disease trends. Ultimately, data collected in rehabilitation centers may provide insights to efficiently allocate financial and human resources on disease prevention and surveillance.

Keywords: Wildlife rehabilitation, Syndromic surveillance, Raptors, Big data, Time series, West Nile

Background

Wild animals play a key role in the transmission of many infectious diseases into humans by serving as reservoirs for important pathogens such as West Nile virus (WNV), avian influenza virus (AIV), and Lyme disease. Animal health surveillance may contribute to the early detection and prevention of disease outbreaks in human populations [1–3]. In addition to human health, the industries involved in livestock, poultry and fishery production are also vulnerable to infectious disease transmitted by wildlife, which may result in economic losses from disease outbreaks and the potential imposition of stringent trade restrictions [4].

Although there are systems in place for monitoring infectious diseases in humans and some domestic animals, as well as some programs for specific diseases in free-ranging wildlife, there is currently no comprehensive, integrated strategy for monitoring wildlife health issues in the United States [5]. Furthermore, although the need for such a monitoring system has been identified, challenges such as cost, time, case acquisition, and practicality of sampling strategies remain difficult to overcome. Sample collection is currently limited to expensive active surveillance activities, often requiring trapping of animals, or convenience or passive sampling of hunted animals or wildlife submitted to public health or wildlife agencies [6].

Novel approaches, such as syndromic surveillance based on the monitoring of non-specific digital data, may help to enhance current surveillance systems. The Centers for Disease Control have defined syndromic

* Correspondence: aalbacas@umn.edu
[1]University of Minnesota, St. Paul, MN, USA
[3]Univ of Minnesota College of Veterinary Medicine, 1920 Fitch Avenue, St. Paul, MN 55108, USA
Full list of author information is available at the end of the article

surveillance as an investigational approach based on using real-time health-related data that precedes diagnosis to detect an outbreak of a disease and decrease morbidity and mortality [7]. Besides enhancing the early detection of infectious disease, syndromic surveillance may also contribute to monitoring endemic diseases and/or be used to accumulate proof of absence of a disease [8].

It is estimated that more than 500,000 amphibians, reptiles, birds, and marine and terrestrial mammals are admitted into wildlife rehabilitators across the United States annually [5], representing a diverse array of animal species from disparate geographic regions and a range of ecosystems. Use of wildlife rehabilitation centers as an alternative means of monitoring wildlife and environmental health has been proposed [5, 6, 9–13]. However, there are a number of challenges to the routine use of rehabilitation center data for syndromic surveillance, including: absence of specific surveillance goals and objectives; lack of comprehensive, integrated database systems; limited infrastructure for wildlife rehabilitators; and data quality-, integrity-, and timeline-related issues. Prerequisite for overcoming these challenges and, ultimately, facilitating early detection and prevention of disease incursions, is the implementation of novel surveillance strategies and analytical tools into data routinely collected by wildlife rehabilitation centers. Indeed, it is critical to develop and rigorously validate analytical approaches for monitoring rehabilitation data, including for the detection of aberrations in the data that may indicate a health event. Here, we use the West Nile Virus (WNV) in the state of Minnesota (MN), USA, as a proof-of-concept as to whether the initial incursion of WNV into the state would result in aberrations in raptor rehabilitation data that would be detectable via syndromic surveillance.

West Nile Virus (family *Flaviviridae*, genus *Flavivirus*) is primarily maintained through a bird-mosquito-bird transmission cycle, which sporadically results in epidemics affecting humans and horses. This pathogen has caused significant morbidity and mortality in humans, horses, and wildlife since its introduction into North America in 1999 [14]. The first report of WNV in MN was in two dead crows (*Corvus brachyrhynchos*) in the Minneapolis/St. Paul metropolitan area in July, 2002 [15]; WNV is now endemic/enzootic in MN, manifesting a seasonal pattern during the period of adult mosquito activity. Many species of raptors are susceptible to WNV, which results in a broad range of clinical signs and/or death. It has been suggested that birds represent the first wave of infections during the transmission season, meaning that they may be infected earlier as compared to mosquito pools, humans, or equine cases [6, 16]. This suggests that surveillance of raptors may

provide indicators for early detection of WNV incursions into free regions. The study was aimed at assessing the utility of wildlife rehabilitation data to support early detection and monitoring of wildlife pathogen activity as it relates to public, food animal, and environmental health. Specifically, retrospective data from raptors admitted to The Raptor Center (TRC) from Minnesota (MN) were assessed to evaluate its potential for detecting and monitoring of WNV circulation.

Methods
Data
This study included retrospective data from raptors admitted to TRC in MN between 1990 and 2014. As a veterinary facility admitting over 800 sick and injured raptors each year, TRC has extensive medical records with the majority of admission data available in a digital format, including clinical signs at admission. Data were collected in Microsoft Access (Microsoft Corp, Redmond, WA) using a relational database. The final data set contained the attributes of "case number", "species", "avian group", "date of admission", "state" where the bird was found, "age" and "clinical signs".

Exploratory descriptive analysis
A descriptive analysis was performed to check data quality with the veterinary clinicians and determine basic traits of the raptor admissions received from MN between 1990 and 2014. Raptor admissions were aggregated by year and month of admission and stratified by avian group, species, age, sex, clinical signs, and state where the birds were found. Raptors could be categorized by age since most species were accurately aged as either a first year (hatch year) or as an adults due to the change in plumage. Frequency of clinical signs was also described for each avian group. Thirty standardized clinical signs recorded upon admission during the entire span of the study were used. Clinical signs were grouped by organ systems into 10 categories, namely: integumentary system (damaged feather/cere/ft, bumblefoot, soft tissue injury); musculoskeletal system (fracture, luxation/subluxation, posterior paralysis); nervous system (convulsions, head tremors, head tilt/disorientation, postural problems/imbalance); gastrointestinal system (anorexia, diarrhea, lesions in mouth, regurgitation); respiratory system (respiratory distress, swollen sinus); urinary system (polydipsia, polyuria, urates in cloaca); special senses (ear injury and ocular disease); non-specific (assymetrical wing beats, injured but alert/feisty, unable to fly or stand, moderate weight loss, not emaciated); systemic (dehydration, weight loss, emaciation, depression/weakness); no problems observed.

Time series analysis

Descriptive analysis of time series for underlying patterns of raptor admissions

The purpose of analysing grouped time series was to identify subsets of raptor admissions that may suggest WNV circulation over time. Frequencies of admissions aggregated by month were assessed by avian taxonomic group, species, clinical signs, and age using grouping time series structures [17–19]. Series were described and compared to assess differences before and after the incursion of WNV in terms of their seasonality and trend. Potential indicators of WNV were selected based on: 1) visual evidence, 2) testing significant changes between the frequencies observed before and after 2002 using non-parametric Mann-Whitney tests [20], and 3) manifestation of WNV-like clinical signs according to previous scientific findings [21, 22].

Interrupted time series analysis in the subset of raptors identified as indicator for West Nile virus before and after 2002

Changes over time that could signal WNV circulation were assessed with an interrupted time series (ITS) analysis using the subset of raptors selected as indicator for its circulation [23, 24]. The time series Y_t, that represented the number of raptor admissions by month, was segmented in two parts, namely, data collected between January 1990 and December 2001, and data collected between January 2002 and December 2014. The time series was fitted using a linear regression model including trigonometric covariants as seasonal and cyclical components, such as $\alpha_i \cos(\omega_i t)$ or/and $\beta_i \sin(\omega_i t)$, where $\omega_i = 2\pi/T_i$, and T_i corresponded to the periods. Linear-trend components, which slopes are denoted as δ and δ', were also included, and the errors of the linear model (ϵ_t) were fitted as autoregressive–moving-average (ARMA) time series with different variances for the two parts considered. The regression coefficients before and after WNV epidemics were estimated introducing a dummy variable (I_t), being 1 if t ≤ December 2001 and 0 if t > January 2002. Therefore, the model was expressed as:

$$Y_t = [(\mu + \delta t + \alpha \cos(\omega t) + \beta \sin(\omega t) + \cdots +)I_t] +$$

$$+ \left[\left(\mu' + \delta' t + \alpha' \cos(\omega t) + \beta' \sin(\omega t) + \cdots + (1 - I_t) \right) \right] + \epsilon_t$$

with $\epsilon_t = \phi_1 \epsilon_{t-1} + \cdots + \phi_p \epsilon_{t-p} + \cdots + Z_t + \theta_1 Z_{t-1} + \cdots + \theta_q Z_{t-q}$,

where p corresponded to the order of the autoregressive part of the model errors and q indicated the moving average order.

The orders of the components of the ARMA process and the regression coefficients were selected using the Bayesian Information Criterion (BIC) [25]. The standardized residuals of the final model were analysed to verify absence of autocorrelation and partial autocorrelation.

Positive WNV cases in raptors confirmed by laboratory tests

To support the plausibility of our evidence, raptors admitted to TRC between 2007 and 2014 that were suspected of West Nile (WN) disease based on clinical signs were assessed by the Veterinary Diagnostic Laboratory of the University of MN (VDL). A raptor admitted to TRC was considered to be suspicious for WN infection if it presented clinical signs typical of the disease in combination with either an elevated white blood cell count (absolute heterophilia) or splenomegaly as demonstrated on radiology. Typical clinical manifestations include neurological signs (including head tilt, nystagmus, and tremors), blindness or visual impairment in Cooper's hawks, northern goshawks, red-tailed hawks and bald eagles, repetitive head movements ("bobble head"), ataxia, circling, and dysphagia in great horned owls. In hawks (buteos and accipiters), exudative chorioretinal lesions and chorioretinal scarring in a linear pattern were highly suggestive of WN disease. Additionally, other raptors with non-specific signs such as emaciation, dehydration, and dull mentation could be considered as clinical suspicion of WN infection.

During this period different specific studies underwent necropsies in suspicious raptors and performed histopathological analysis. A raptor was considered positive for WNV when individual or pooled tissues (brain, heart, and kidney) were positive for WNV RNA by PCR [26] and/or if WNV antigen was detected in tissue sections of at least one organ by immunohistochemistry using a monoclonal antibody specific for WNV antigen [22].

Software

Analyses were implemented in the R software [27], with the "base" and "hts" [19], "forecast" [28], and "nlme" packages [29].

Results

Exploratory descriptive analysis

A total of 16,595 raptors from 37 different states were admitted to TRC between 1990 and 2014, although our analysis focused only on admissions from the state of MN (n = 13,080; 78.8%). Five of the 28 raptor species admitted accounted for 66% of the admissions: 2360 red-tailed hawks (*Buteo jamaicensis*) (18%); 2148 great horned owls (*Bubo virginianus*) (16.4%); 1538 Cooper's hawks (*Accipiter cooperii*) (11.8%); 1347 American kestrels (*Falco sparverius*) (10.3%); and 1193 bald eagles (*Haliaeetus leucocephalus*) (9.1%). Most raptors fell into 4 taxonomic groups: 5278 hawks (including buteos and accipiters, 39%); 4272 owls (31%); 1897 falcons (14%);

and 1208 eagles (9%). Only 425 (3%) raptors belonged to other groups (Table 1: Summarized table of admitted species).

Time series analysis

The number of admitted raptors increased over time; the median number of admissions per year was 542 with a minimum of 318 in 1990 and a maximum of 818 in 2013. Admissions increased annually in July and August.

When the admissions were analyzed together, considering a unique time series, did not appear to signal the WNV circulation. To distinct underlying patterns and identify a subset of raptors as indicator of WNV circulation, the series had to be grouped by taxomonic group, species, age and clinical signs.

Exploring the admissions grouped by avian taxonomic group, the results showed that after 2002 the admissions of hawks evidenced a marked increase, whereas in the owl and eagle groups the increases were small. In contrast, the number of admissions of falcons decreased over time. Seasonal patterns were more evident in hawks and falcons than in owls and eagles (see Fig. 1a–e).

ARMA models are used for our data analysis because they are powerful tools that can be directly fitted using standard packages in R. However other families of continuous-time ARMA processes [30], or nonparametric regression models based on P-splines [31] can be explored in further research.

Examining admissions by age, hatch year birds accounted for 33.1% admissions with a marked increase in July and August over all the period. However, it is interesting to

Table 1 Summary of the raptors received at The Raptor Center from Minnesota between 1990 and 2014 detailing: avian group, species, number and percentage

Avian Group	Species	Latin name	No. admissions received at The Raptor Center	Percent
Hawk	red-tailed hawk	*Buteo jamaicensis*	2360	18.0%
	Cooper's hawk	*Accipiter cooperii*	1538	11.8%
	broad-winged hawk	*Buteo platypterus*	618	4.7%
	sharp-shinned hawk	*Accipiter striatus*	380	2.9%
	red-shouldered hawk	*Buteo lineatus*	145	1.1%
	rough-legged hawk	*Buteo lagopus*	123	0.9%
	goshawk (Northern)	*Accipiter gentilis*	93	0.7%
	Swainson's hawk	*Buteo swainson*	18	0.1%
	ferruginous hawk	*Buteo regalis*	3	0.0%
Owl	great horned owl	*Bubo virginianus*	2148	16.4%
	barred owl	*Strix varia*	797	6.1%
	saw-whet owl (Northern)	*Aegolius acadicus*	372	2.8%
	screech-owl (Eastern)	*Megascops asio*	368	2.8%
	great gray owl	*Strix nebulosa*	197	1.5%
	long-eared owl	*Asio otus*	136	1.0%
	snowy owl	*Bubo scandiacus*	111	0.8%
	short-eared owl	*Asio flammeus*	92	0.7%
	boreal owl	*Aegolius funereus*	43	0.3%
	hawk-owl (Northern)	*Surnia ulula*	8	0.1%
Falcon	kestrel (American)	*Falco sparverius*	1347	10.3%
	peregrine falcon	*Falco peregrinus*	300	2.3%
	merlin	*Falco columbarius*	244	1.9%
	prairie falcon	*Falco mexicanus*	6	0.0%
Eagle	bald eagle	*Haliaeetus leucocephalus*	1193	9.1%
	golden eagle	*Aquila chrysaetos*	15	0.1%
Others	osprey	*Pandion haliaetus*	225	1.7%
	turkey vulture	*Cathartes aura*	132	1.0%
	harrier (Northern)	*Cyrcus cyaneus*	68	0.5%
			13,080	100.0%

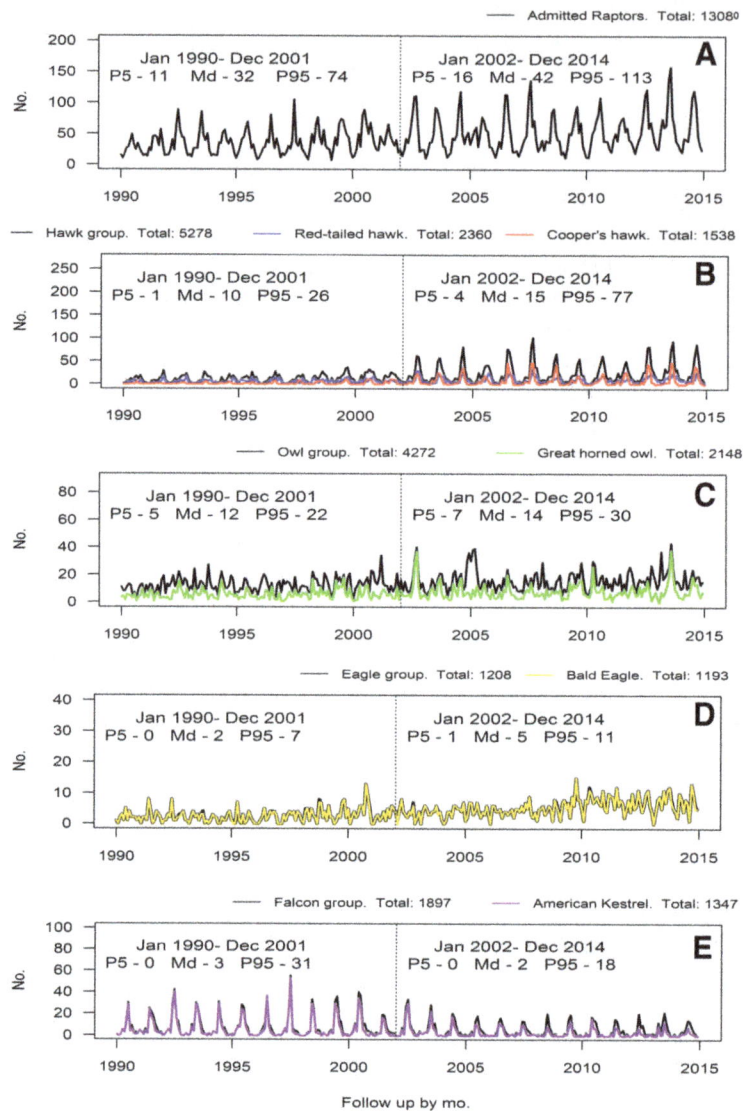

Fig. 1 a-e Time series plots of total admissions, avian groups, and most representative raptors

varied after the incursion of WNV. Indeed, after 2002 during the periods of WNV circulation (between July–October), the proportions of hatch year in hawks increased, whereas in owls increased the proportion of adults (Fig. 2a-b).

The signs compatible with WNV rose between July and October after its incursion in 2002, whereas the amount of admissions with no problems noted a decrease.

Identification of raptor subpopulations that potentially indicate WNV circulation

The cojoint analysis of time series considering taxonomic groups, age and clinical signs suggested that the admissions of hawks with compatible clinical signs could indicate the circulation of WNV. The admissions of the hawk group showed the most evident changes before

and after 2002 and the monthly medians of systemic, integumentary, ocular problems, ear injury, nervous, gastrointestinal, respiratory signs or other non-specific signs increased signicantly after 2002 (p-value < 0.005, Mann-Whitney test) (Fig. 3a–f).

The monthly patterns of clinical signs in this subset were regular before 2002. However, in the summer of 2002 and in consecutive summers, the number of admissions for the most frequent clinical signs significantly increased in comparison with the 1990–2001 period. These increases coincided with the incursion of WNV detected in humans, horses and birds in MN. These changes were more marked in hatch year hawks.

Based on these findings all hawk admissions showing WNV-like signs (namely, systemic, integumentary, ocular problems, ear injuries, nervous, gastrointestinal

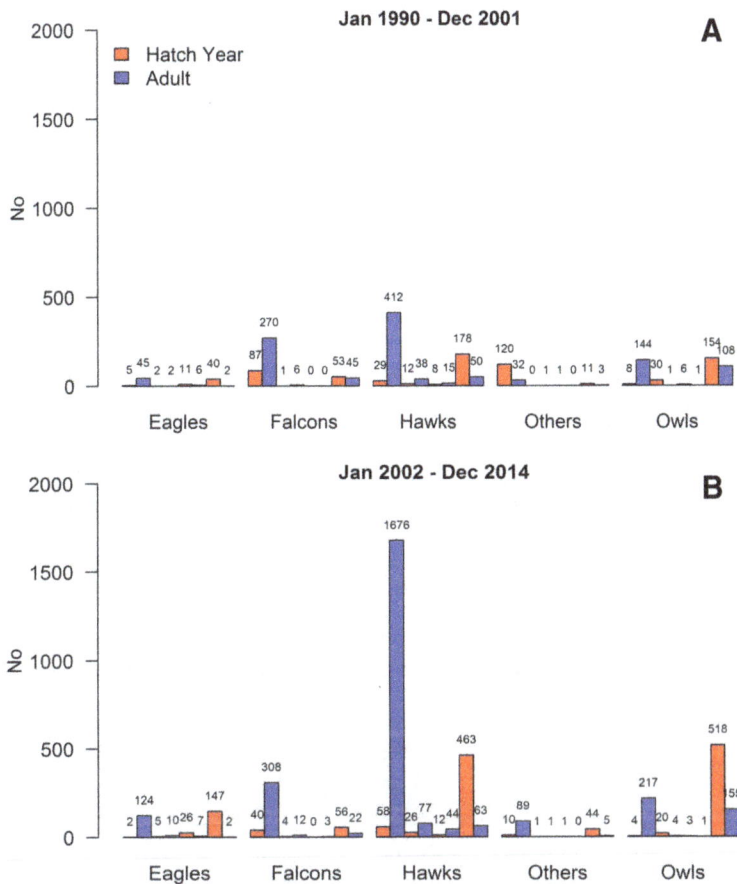

Fig. 2 a-b Frequency of admissions between July–October by avian group and age category before and after 2002

or non-specific clinical signs) were selected as indicators of WNV circulation.

The changes observed in the patterns during the incursion of WNV were not so apparent in other taxonomic groups.

Interrupted time series analysis (ITS) in hawks showing WNV-like signs before and after 2002

The results of the ITS model provided statistical evidences about the augment in the number of hawks showing WNV-like signs admittted to TRC during the periods with circulation of WNV. This effect was assessed by examining and comparing the coefficients of the regression model before and after 2002 (see Table 2).

The coefficients of the trigometric covariates (β_i and α_i for $i = 12$, 6, 4 and 3) showed that the frequency of admissions of hawks with signs compatible with WNV followed a marked annual seasonality and a less evident cyclical pattern every 6, 4 and 3 months (Fig. 4 and Table 2). The value of δ evidenced that the overall trend of admissions increased between 1990 and 2002. In contrast, the trend after 2002 represented by δ' was null over time.

The circulation of WNV appeared to increase the frequency of hawks showing WNV-like signs almost 3 times during July and August. Furthermore, the high standard deviations obtained in the errors after 2002 indicated that the circulation of WNV also increased the uncertainty in the number of admissions over time.

The diagnostic checking evidenced the lack of autocorrelation or partial autocorrelation in the residuals of our model.

WNV cases in raptors confirmed by laboratorial tests

Between 2007 and 2014 a total of 333 raptors were submitted by TRC to the Minnesota Veterinary Diagnostic Laboratory (VDL) as suspicious for WNV; 162 were WNV positive. Most cases (110, 67.9%) were hawks (i.e. red-tailed hawks, Coopcr's hawks and northern goshawks), and most positive hawks (80%) were hatch year birds. The most frequent clinical signs observed in these raptors were systemic disorders (i.e. dehydratation, depression, weakness, emaciation), neurological (i.e. head tremors, postural problems, imbalance, head tilts) and ocular signs; followed by integumentary (i.e. soft tissue injuries, damaged feathers/cere/ft or bumblefoot) and

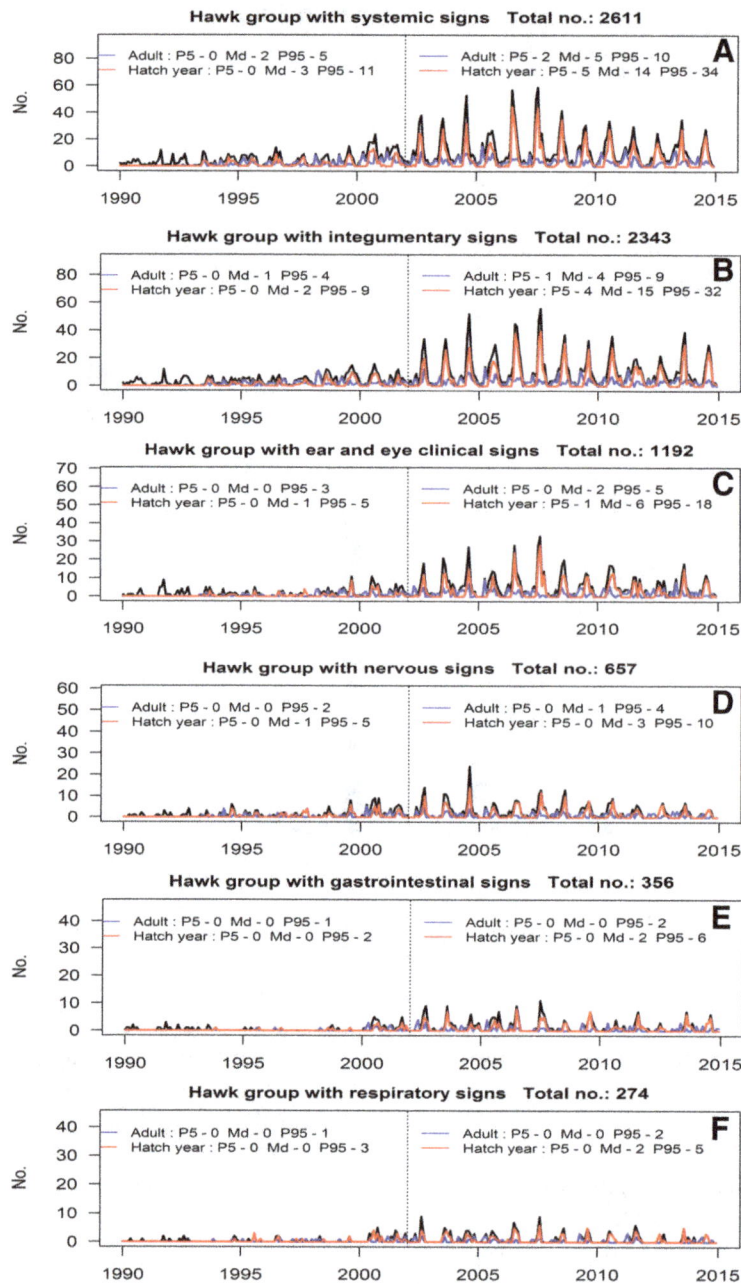

Fig. 3 a-f Time series plots of monthly admissions of hawks according to clinical signs and age

gastrointestinal signs such as anorexia, and lesions in mouth, among others.

Discussion

Our results demonstrate that raptor rehabilitation data, specifically hawks, shows marked temporal differences before and after the incursion of WNV in MN. These results coincide with previous studies conducted in different regions of USA, in which these species showed to be highly susceptible to WNV [7, 32]. The conclusion that these changes were likely due to WNV were further supported by the abrupt change in the frequency of admissions of hawks with clinical signs consistent with WNV in 2002, and the diagnostic confirmation of WNV-positive cases in hawks post-2002. This study serves as a proof-of-concept that some animal health events, such as the incursion of a new infectious disease, result in aberrations of long-term data trends collected by rehabilitation centers. Thus, near real-time analysis of rehabilitation data within a syndromic surveillance framework could potentially contribute to the detection of health anomalies and monitoring of animal health trends in wildlife populations.

Table 2 Coefficients of the regression model and errors fitted to an ARMA structure

Coefficients of the model before 2002	Period January 1990–December 2001				Coefficients of the model after 2002	Period January 2002–December 2014			
	Value Std	Error	t-value	p-value		Value Std	Error	t-value	p-value
Intercept	4.04	0.83	4.87	0.00	Intercept	21.50	3.80	5.66	0.00
δ	0.08	0.01	7.66	0.00	δ'	−0.004	0.02	−0.22	0.83
β_{12}	−5.79	0.57	−10.10	0.00	β'_{12}	−16.60	1.05	−15.85	0.00
a_{12}	−3.20	0.57	−5.60	0.00	a'_{12}	−12.96	1.04	−12.42	0.00
β_6	−0.36	0.54	−0.68	0.50	β'_6	8.55	0.98	8.69	0.00
a_6	−2.89	0.54	−5.36	0.00	a'_6	−4.96	0.98	−5.05	0.00
a_4	1.08	0.49	2.21	0.03	a'_4	6.90	0.89	7.72	0.03
a_3	−1.58	0.44	−3.62	0.00	a'_3	−4.53	0.80	−5.69	0.00
$_t$					$Z_t + 0.22_1 Z_{t-1}$				

AIC: 1913.46 BIC: 1983.83 logLik −937.73

Standardized residuals: Min: −3.47 Q1: −0.61 Med: −0.12 Q3: 0.47 Max: 4.60

Various WNV surveillance programs have been implemented in humans, horses, birds and mosquitoes throughout the United States, including MN, though many programs relying on active surveillance for WNV have been discontinued due to low sensitivity and high cost [33]. Thus, the syndromic surveillance approach here demonstrates the potential impact that an alternative and affordable source of information may have in supporting early detection and monitoring of WNV. Ultimately, the working example here shows the potential of data from wildlife rehabilitation centers to signal the introduction and circulation of emergent pathogens in wild animal populations.

Additionally, because raptors are placed at the top of the food chain and occupy broad areas, monitoring of raptor admissions may help to monitor the health status of other populations in the ecosystem [34, 35]. Routine monitoring of rehabilitation center data may help to assess the impact and evolution of a specific disease on wildlife and provide a better understanding of its transmission in the natural ecosystems. However, monitoring of raptors admissions may not be straightforward. For example, when raptor admissions were analyzed collectively, there was no evident change in patterns that could signal WNV circulation. In contrast, when frequency of admissions was assessed by taxonomic group, age and clinical signs, it was evident that certain patterns changed substantially coinciding with the occurrence of WNV outbreaks. Temporal variations were clear in hawks with WNV-like signs during the summer of 2002 and consecutive summers, especially in hatch year birds. Thus, syndromic surveillance programs based on rehabiliaton data will need to not only monitor overall trends, but also focus on identifying aberrations or trends when data is subset by taxomony, age, and suites of clinical signs. Monitoring various taxomonic or syndromic subsets for data aberrations is particularly important if syndromic surveillance is meant to detect the emergence of pathogens in a geographic region rather than monitor pathogens already circulating.

Despite the potential of wildlife rehabilitation center data for syndromic surveillance, there are certain limitations that should be considered. Most importantly, monitoring of those data is only a proxy for wildlife health status, and consequently, other complementary information might be essential to support evidence before extracting definitive conclusions. Additionally, animals admitted into the rehabilitation centers may not be representative of the status of the wildlife population because animal collection is linked to the human activity.

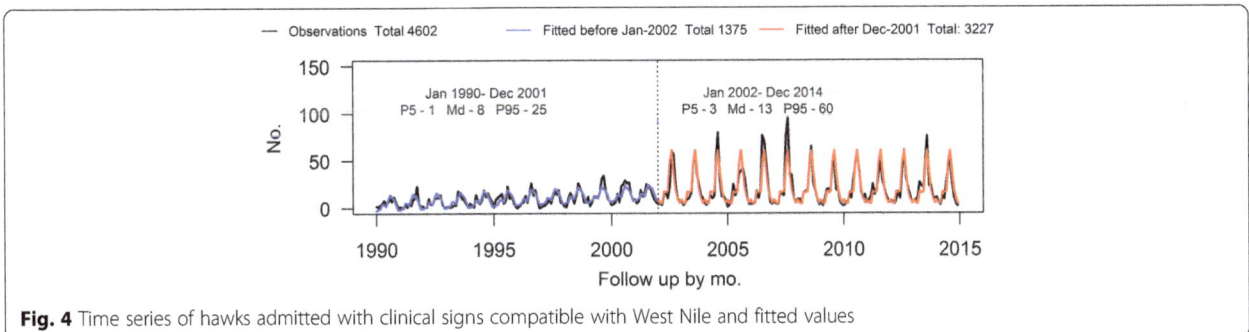

Fig. 4 Time series of hawks admitted with clinical signs compatible with West Nile and fitted values

Moreover, the timeliness of these collections can also be determined by the degree of awareness. Presence of such collection bias may have, for example, affected our results given that the number of raptors found after 2002 was higher. However, the total number of admissions did not significantly change (Fig. 1a–e) as much as the specific pattern of young hawk admissions with WNV-like signs (Fig. 3a–f).

Another potential limitation may be due to the lack of specificity of clinical signs. Different pathogens and conditions may result in similar clinical patterns. For that reason, assessment of underlying patterns by species, age or clinical signs complemented with specific diagnosis in the event of peaks in admissions might be necessary to identify indicators for specific diseases. Given that these limitations are acknowledged and, ideally, controlled, wildlife rehabilitation center data may serve as an affordable and reliable source of information for monitoring conditions at the interface of public, animal, and environmental health, thus supporting the One Health concept at local and regional levels.

Conclusions

Retrospective analysis of raptor rehabilitation data evidenced marked temporal differences before and after the incursion of West Nile Virus in MN, indicating that monitoring of data routinely collected by wildlife rehabilitation centers has the potential to capture the geographical emergence of a new pathogen in free-ranging wildlife. This study serves as a proof-of-concept that aberrations in long-term trends in datasets collected by wildlife rehabilitation centers may reflect animal health events occurring within free-ranging wildlife populations. These results demonstrate that wildlife rehabilitation centers may serve as an affordable resource to complement the routine monitoring, and ultimately, early detection and prevention of public, animal, and environmental health conditions in the country.

Abbreviations

AIC: Akaike Information Criterion; AIV: Avian Influenza Virus; ARMA: AutoRegressive-Moving Average; BIC: Bayesian Information Criterion; ITS: Interrupted Time Series; MN: Minnesota; PCR: Polymerase Chain Reaction; RNA: Ribonucleic Acid; TRC: The Raptor Center; USA: United States of America; VDL: Veterinary Diagnostic Laboratory; WNV: West Nile Virus

Acknowledgements
We thank the staff of The Raptor Center with special acknowledgement to Dr. Devin Tunseth and Dr. Kathleen MacAulay for their preparation of data.

Funding
This work was supported by the University of Minnesota Office of the Vice President for Research Grant-in-Aid Program. The contribution of Pedro Puig was partially funded by the grant MTM2015–69493-R from the Spanish Ministry of Economy and Competitiveness.

Authors' contributions
Conceptualization of the work: MW, AA, AMP, JP, PP, AW, KV, JA; data curation: AA, MW, AW; formal analysis and interpretation: AA, MW, PP; funding acquisition: MW, JP, AMP; supervision: MW, AMP, JP, PP; preparation of the manuscript: AA, MW, AMP, JP, AW, PP, KV, JA. All authors read and approved the final manuscript.

Consent for publication
"Not applicable".

Competing interests
The authors declare that they have no competing interests.

Author details
[1]University of Minnesota, St. Paul, MN, USA. [2]Universitat Autònoma de Barcelona, Cerdanyola del Vallès, Barcelona, Spain. [3]Univ of Minnesota College of Veterinary Medicine, 1920 Fitch Avenue, St. Paul, MN 55108, USA.

References
1. Butler D. Disease surveillance needs a revolution. Nature. 2006;440(7080):6–7.
2. Jebara KB. Surveillance, detection and response: managing emerging diseases at national and international levels. Rev Sci Tech Off Int Epiz. 2004;23(2):709–15.
3. Morner T, Obendorf DL, Artois M, Woodford MH. Surveillance and monitoring of wildlife diseases. Rev Sci Tech Off Int Epiz. 2002;21(1):67–76.
4. Thiermann AB. Globalization, international trade and animal health: the new roles of OIE. Prev Vet Med. 2005;67(2):101–8.
5. Willette M, Ponder J, McRuer DL, Clark EE Jr. Wildlife Health Monitoring Systems in North America: From Sentinel Species to Public Policy. In: Aguirre A, editor. Conservation Medicine: Applied Cases of Ecological Health; 2013.
6. Randall NJ, Blitvich BJ, Blanchong JA. Efficacy of wildlife rehabilitation centers in surveillance and monitoring of pathogen activity: a case study with West Nile virus. J Wildl Dis. 2012;48(3):646–53.
7. Henning KJ. What is syndromic surveillance? MMWR Morb Mortal Wkly Rep. 2004;24:7–11.
8. Katz R, May L, Baker J, Test E. Redefining syndromic surveillance. J Epidemiol Glob Health. 2011;1(1):21–31.
9. Camacho M, Hernández JM, Lima-Barbero JF, Höfle U. Use of wildlife rehabilitation centres in pathogen surveillance: a case study in white storks (Ciconia Ciconia). Prev Vet Med. 2016;130:106–11.
10. Cox-Witton K, Reiss A, Woods R, Grillo V, Baker RT, Blyde DJ, Boardman W, Cutter S, Lacasse C, McCracken H, Pyne M. Emerging infectious diseases in free-ranging wildlife–Australian zoo based wildlife hospitals contribute to national surveillance. PLoS One. 2014;9(5):e95127.
11. Doell D, Locky DA. Trends in wildlife intake at a rehabilitation center in Central Alberta: a retrospective analysis of birds, mammals, and herptiles, from 1990 through 2012. J Wildlife Rehabil. 2016;36(1):17–29.
12. Pultorak E, Nadler Y, Travis D, Glaser A, McNamara T, Mehta SD. Zoological institution participation in a West Nile virus surveillance system: implications for public health. J Public Health. 2011;125(9):592–9.
13. Stitt T, Mountifield J, Stephen C. Opportunities and obstacles to collecting wildlife disease data for public health purposes: results of a pilot study on Vancouver Island. British Columbia Can Vet J. 2007;48(1):83.
14. Beasley DW, Barrett AD, Tesh RB. Resurgence of West Nile neurologic disease in the United States in 2012: what happened? What needs to be done? Antivir Res. 2013;99(1):1–5.
15. Bell JA, Brewer CM, Mickelson NJ, Garman GW, Vaughan JA. West Nile virus Epizootiology, central Red River valley, North Dakota and Minnesota, 2002-2005. Emerg Infect Dis. 2005;12(8):1245–7.
16. Nemeth N, Kratz G, Edwards E, Scherpelz J, Bowen R, Komar N. Surveillance for West Nile virus in clinic-admitted raptors, Colorado. Emerg Infect Dis. 2007;13(2):305.
17. Athanasopoulos G, Ahmed RA, Hyndman RJ. Hierarchical forecasts for Australian domestic tourism. Int J Forecast. 2009;25(1):146–66.

18. Hyndman RJ, Ahmed RA, Athanasopoulos G, Shang HL. Optimal combination forecasts for hierarchical time series. Comput Stat Data Anal. 2011;55(9):2579–89.
19. Hyndman RJ, Lee A, Wang E, Wickramasuriya S, Wang ME. Package 'hts'. Hierarchical and Grouped Time Series. R package version 4.5. 2015. http://CRAN.R-project.org/package=hts. Accessed 21 Jun 2017.
20. Mann HB, Whitney DR. On a test of whether one of two random variables is stochastically larger than the other. Ann Math Stat. 1947:1950–60.
21. Ellis AE, Mead DG, Allison AB, Stallknecht DE, Howerth EW. Pathology and epidemiology of natural West Nile viral infection of raptors in Georgia. J Wildl Dis. 2007;43(2):214–23.
22. Wünschmann A, Shivers J, Bender J, Carroll L, Fuller S, Saggese M, van Wettere A, Redig P. Pathologic and immunohistochemical findings in goshawks (Accipiter Gentilis) and great horned owls (Bubo Virginianus) naturally infected with West Nile virus. Avian Dis. 2005;49(2):252–9.
23. Afonso ET, Minamisava R, Bierrenbach AL, Escalante JJ, Alencar AP, Domingues CM, Morais-Neto OL, Toscano CM, Andrade AL. Effect of 10-valent pneumococcal vaccine on pneumonia among children. Brazil Emerg Infect Dis. 2013;19(4):589–97.
24. McDowall D. Interrupted time series analysis, Vol. 21. Sage. 1980.
25. Schwarz G. Estimating the dimension of a model. Ann Stat. 1978;6(2):461–4.
26. Lanciotti RS, Kerst AJ, Nasci RS, Godsey MS, Mitchell CJ, Savage HM, Komar N, Panella NA, Allen BC, Volpe KE, Davis BS. Rapid detection of West Nile virus from human clinical specimens, field-collected mosquitoes, and avian samples by a TaqMan reverse transcriptase-PCR assay. J Clin Microbiol. 2000;38(11):4066–71.
27. Team RC. R: A Language and Environment for Statistical Computing. Vienna: R Foundation for Statistical Computing. 2014. http://www.R-project.org.
28. Hyndman RJ, O'Hara-Wild M, Bergmeir C, Razbash S, Wang E, Hyndman MR. Package 'forecast'. 2007. http://www.cran.r-project.org/web/packages/forecast/forecast.pdf. Accessed 23 Feb 2017.
29. Pinheiro J, Bates D, DebRoy S, Sarkar D. R Core Team. nlme: Linear and Nonlinear Mixed Effects Models. R package version 3.1–127. 2016. http://CRAN.R-project.org/package=nlme. Accessed 6 Feb 2017.
30. Arratia A, Cabana A, Cabana EM. A construction of continuous-time ARMA models by iterations of Ornstein-Uhlenbeck processes. SORT-Statistics and Operations Research Transactions. 2016;40(2):267–302.
31. Eilers PHC, Marx BD, Durban M. Twenty years of P-splines. SORT-Statistics and Operations Research Transactions. 2015;39(2):149–86.
32. Kilpatrick AM, LaDeau SL, Marra PP. Ecology of West Nile virus transmission and its impact on birds in the western hemisphere. Auk. 2007;124(4):1121–36.
33. Hadler JL, Patel D, Nasci RS, Petersen LR, Hughes JM, Bradley K, Etkind P, Kan L, Engel J. Assessment of arbovirus surveillance 13 years after introduction of west Nile virus, United States. Emerg Infect Dis. 2015;21(7):1159.
34. Bowerman WW, Roe AS, Gilbertson MJ, Best DA, Sikarskie JG, Mitchell RS, Summer CL. Using bald eagles to indicate the health of the Great Lakes' environment. Lake Reserv. Manage. 2002;7(3):183–7.
35. Giesy JP, Bowerman WW, Mora MA, Verbrugge DA, Othoudt RA, Newsted JL, Summer CL, Aulerich RJ, Bursian SJ, Ludwig JP, Dawson GA. Contaminants in fishes from Great Lakes-influenced sections and above dams of three Michigan rivers: III. Implications for health of bald eagles. Arch Environ Contam Toxicol. 1995;29(3):309–21.

6

A 2015 outbreak of Getah virus infection occurring among Japanese racehorses sequentially to an outbreak in 2014 at the same site

Hiroshi Bannai*[iD], Akihiro Ochi, Manabu Nemoto, Koji Tsujimura, Takashi Yamanaka and Takashi Kondo

Abstract

Background: As we reported previously, Getah virus infection occurred in horses at the Miho training center of the Japan Racing Association in 2014. This was the first outbreak after a 31-year absence in Japan. Here, we report a recurrent outbreak of Getah virus infection in 2015, sequential to the 2014 one at the same site, and we summarize its epizootiological aspects to estimate the risk of further outbreaks in upcoming years.

Results: The outbreak occurred from mid-August to late October 2015, affecting 30 racehorses with a prevalence of 1.5 % of the whole population (1992 horses). Twenty-seven (90.0 %) of the 30 affected horses were 2-year-olds, and the prevalence in 2-year-olds (27/613 [4.4 %]) was significantly higher than that in horses aged 3 years or older (3/1379 [0.2 %], $P < 0.01$). Therefore, the horses newly introduced from other areas at this age were susceptible, whereas most horses aged 3 years or older, which had experienced the previous outbreak in 2014, were resistant. Among the 2-year-olds, the prevalence in horses that had been vaccinated once (10/45 [22.2 %]) was significantly higher than that in horses vaccinated twice or more (17/568 [3.0 %], $P < 0.01$). Horse anti-sera raised against an isolate in 2014 neutralized both the homologous strain and a 2015 isolate at almost the same titers (256 to 512), suggesting that these viruses were antigenically similar. Among horses entering the training center from private surrounding farms in 2015, the seropositivity rate to Getah virus increased gradually (11.8 % in August, 21.7 % in September, and 34.9 % in October). Thus, increased virus exposure due to the regional epizootic probably allowed the virus to spread in the center, similarly to the outbreak in 2014.

Conclusions: The 2015 outbreak was caused by a virus which was antigenically close to the 2014 isolate, affecting mostly 2-year-old susceptible horses under epizootiological circumstances similar to those in 2014. The existence of 2-year-olds introduced from regions free from Getah virus could continue to pose a potential risk of additional outbreaks in upcoming years. Vaccination on private farms and breeding farms would help to minimize the risk of outbreaks.

Keywords: Getah virus, Japan, Racehorses, Sequential outbreak

* Correspondence: bannai@equinst.go.jp
Equine Research Institute, Japan Racing Association, 1400-4 Shiba, Shimotsuke, Tochigi 329-0412, Japan

Background

Getah virus is classified in the genus *Alphavirus* in the family *Togaviridae* [1]. It is mosquito borne and is widespread from Eurasia to Australasia. This virus causes fever, generalized rash, and edema of the legs in horses [1], and it causes fetal death and reproduction disorders in pigs [2, 3].

We previously reported that an outbreak of Getah virus infection occurred in racehorses at the Miho training center of the Japan Racing Association in autumn 2014, affecting 33 horses [4]. It was the first reported outbreak of infection with this virus among vaccinated horse populations worldwide, and the first one in Japan since 1983 [5]. The indirect causes of this outbreak included the existence of susceptible horses that did not complete the vaccination program at the training center and an increased risk of exposure because of epizootic infection around the training center [6]. However, the direct cause of the outbreak was still unclear, and the epizootic pattern of this re-emerging virus in upcoming years was unpredictable.

Following the outbreak in 2014, we took control measures to prevent a possible outbreak in the coming season. Measures included the reinforcement of pest control for vector mosquitoes at the training center and recommendations for Getah virus vaccination and pest control on the private farms surrounding the center. However, in 2015, another outbreak of Getah virus infection occurred at the Miho training center—the same site as in 2014. Here, we summarize the epizootiological aspects of the current outbreak and analyze the antigenic properties of the isolated virus to estimate the risk of possible outbreaks in upcoming years.

Methods

Study site

The Miho training center is in Ibaraki Prefecture in the Kanto region of Japan. About 2000 racehorses are trained at the center, and about 1000 racehorses are replaced with new ones every month. The horses are generally accommodated at the center for 1 to 6 months for training. After they leave the center they are usually kept on other farms for several months for rest; they then re-enter the center. The Japan Racing Association is the sporting authority which administers the horses in the training center, and the clinical samples were collected as a part of regular activities for disease prevention. The owners of racehorses have been notified that their horses might be subjected to mandatory sampling of clinical specimens for diagnostic and research purposes.

A two-dose priming course of Getah virus vaccine is given to 2-year-olds. The vaccination period generally starts in May and finishes in October to cover the mosquito season. Horses that are present at the training center in spring receive the first dose in May and the second dose in June. In the case of horses that enter after the mosquito season has started, the first dose is administered when they enter, and the second dose is given about 1 month after the first. From the second season onward, the horses are vaccinated annually before mosquito season.

Prevalence of Getah virus infection among populations stratified by age and number of vaccine doses received

We investigated the age distribution and vaccination histories of horses that were present at the Miho training center on August 15, 2015, i.e. a few days before the outbreak started ($n = 1992$). On the basis of the number of vaccination doses they had received before the outbreak, horses in populations of each age were categorized into two groups, namely 1) one dose; and 2) two doses or more. The prevalence of disease onset of Getah virus infection in each population during the period from August 15 to October 30 2015 was calculated by dividing the number of affected horses in each category by the number of horses in the corresponding population. The statistical significance of differences in prevalence was evaluated by using Fisher's exact test.

Cell culture

For virus isolation and virus-neutralizing (VN) testing, we used Vero cells (Sumitomo Dainippon Pharma, Tokyo, Japan). Cells were cultured in minimum essential medium (MEM, MP Biomedicals, Irvine, CA, USA) containing 10 % fetal calf serum (Sigma Aldrich Inc., St. Louis, MO, USA), 100 units/ml penicillin, and 100 µg/ml streptomycin (Sigma Aldrich Inc.). MEM containing 2 % fetal calf serum, 100 units/ml penicillin, and 100 µg/ml streptomycin was used as a maintenance medium for virus isolation and VN testing.

Detection of Getah virus RNA in blood samples of pyretic horses at the Miho training center in 2015

Viral gene detection was performed in blood samples of pyretic horses (≥ 38.5 °C) at the Miho training center. The test period started in June 2015 and finished in mid-November 2015. Viral RNA was extracted from EDTA-treated blood samples by using a nucleic acid isolation kit (MagNA Pure LC Total Nucleic Acid Isolation Kit, Roche Diagnostics, Mannheim, Germany), and viral gene detection was performed by an RT-PCR for the Getah virus *nsP1* gene using primer sets M_2W-S and M_3W-S [7]. For some of the positive samples, the RT-PCR products ($n = 7$) were sequenced as described previously [4].

VN test for Getah virus in paired sera collected from pyretic horses at the Miho training center in 2015

From among 95 horses that developed pyrexia between 1 August and 30 October at the Miho training center, we collected acute and convalescent sera (2- to 11-week intervals between paired sera collection) from 52. These included 14 horses that were positive and 38 that were negative on the above-mentioned Getah virus RT-PCR. The sera were subjected to a VN test for Getah virus using the 14-I-605 strain, which had been isolated from a race-horse during the 2014 outbreak, as described previously [6]. The VN test for Getah virus was performed as described previously [6]. The VN titer was defined as the reciprocal of the highest dilution that completely inhibited virus growth. Horses that showed a ≥4-fold increase between the paired sera were defined as seroconverted.

Virus isolation

EDTA-treated blood samples collected from 23 horses that had developed pyrexia and were positive on Getah virus RT-PCR during the period from 18 August to 26 September 2015 were used for virus isolation. For some of the 23 horses, buffy-coat specimens containing leukocytes instead of whole blood were used for virus isolation. Getah virus was isolated by using Vero cells as described previously [6]. Briefly, the samples were frozen and thawed three times and then centrifuged at $800\,g$ for 20 min at 4 °C. The supernatants were inoculated onto 1-day monolayer cultures of Vero cells or inoculated with the Vero cells simultaneously. The next day, the cells were washed three times with phosphate-buffered saline (pH 7.2) and cultured in maintenance medium. To identify Getah virus–specific nucleotide sequences, the supernatants of specimens that showed cytopathic effects were tested by RT-PCR for the nsP1 gene, as described above.

Antigenic comparison of the vaccine strain and Getah virus strains isolated in 2014 and 2015

Cross-neutralizing tests between the strain isolated in 2014 (14-I-605), the strain isolated in 2015, and the vaccine strain (MI-110) were performed. Horse anti-sera that were raised against the MI-110 strain ($n = 2$) and 14-I-605 strain ($n = 2$) and prepared in our previous study were used [8]. VN tests were performed as described above.

Investigation of Getah virus epizootic infection among horses on surrounding farms in Ibaraki and Chiba prefectures

Among horses that were introduced into the Miho training center between June and October 2015, those that met all of the following criteria ($n = 51$ to 81 in each month) were tested: 1) 2 years old; 2) transferred from a farm in Ibaraki Prefecture or the neighboring Chiba Prefecture; and 3) no history of vaccination with inactivated Getah virus vaccine. Sera collected on the day each horse entered the Miho training center were subjected to VN testing for Getah virus.

Results

Detection of Getah virus infection among pyretic horses at the Miho training center in 2015

The numbers of pyretic horses each week at the Miho training center are shown in Fig. 1. During the period from June to mid-August, there were 4 to 9 pyretic horses each week. These numbers increased to 10 or more in all of the weeks except one from late August to early October, and thereafter decreased to the earlier baseline. Out of 171 pyretic horses in the whole period, 162 were tested for Getah virus by RT-PCR, and 29 of them were positive. The first and last samples that were positive by RT-PCR were collected from pyretic horses on 18 August and 30 October, respectively (Fig. 1). We collected paired sera from 14 of the 29 RT-PCR-positive horses and subjected them to VN testing for Getah virus; all of them had seroconverted (≥4-fold increase).

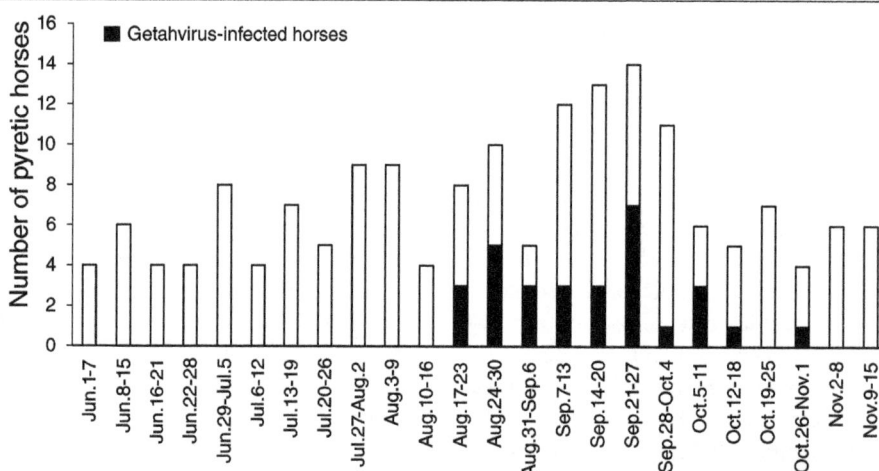

Fig. 1 Numbers of pyretic horses and Getah virus–infected horses at the Miho training center. Bars indicated the number of pyretic horses each week from 1 June to 15 November. Black, number of horses positive on RT-PCR for Getah virus or seroconverted to Getah virus on VN testing, or both

Paired sera collected from 38 RT-PCR-negative horses that had developed pyrexia between 18 August and 30 October were also tested for VN antibodies; one horse, which had developed pyrexia on 25 August, showed seroconversion. In total, from among the 95 horses that developed pyrexia between 18 August and 30 October, 30 were positive for Getah virus infection by RT-PCR or VN testing, or both (Fig. 1). Among the 30 affected horses, seven (23.3 %) had edema of their legs, and three (10.0 %) had body rashes, but all of them recovered within a few days. Two-year-olds accounted for 27 (90.0 %) of the 30 affected horses; the remainder consisted of two 3-year-olds (two horses) and one 7-year-old. Of the affected 2-year-olds, 10 had been vaccinated only once before disease onset, and 17 had been vaccinated twice. All of the affected horses aged 3 years or older had been vaccinated at least twice.

Prevalence of Getah virus infection among populations stratified by age and number of vaccine doses received

We investigated the age distribution and vaccination histories of the whole horse population at the Miho training center on August 15, 2015, i.e. a few days before the outbreak started (Table 1). All horses (n = 1992) had been vaccinated at least once before the outbreak. Among the 2-year-olds (n = 613), 568 (92.7 %) had been vaccinated twice or more, and the remaining 45 (7.3 %) had been vaccinated only once (Table 1). Among the 3-year-olds (n = 794), 4-year-olds (n =314) and 5-year-olds or older (n = 271), almost all horses had been vaccinated two times or more (788 [99.2 %], 312 [99.4 %] and 271 [100.0 %], respectively, Table 1).

The prevalence of Getah virus infection in the whole population during the outbreak was 30/1992 (1.5 %). The prevalence in 2-year-olds (27/613 [4.4 %]) was significantly higher than that in horses aged 3 years or older (3/1379 [0.2 %], $P < 0.01$). Among the 2-year-olds, the prevalence in horses that had been vaccinated once was 10/45 (22.2 %); this was significantly higher than that in horses vaccinated twice or more (17/568 [3.0 %], $P < 0.01$).

Virus isolation and analysis of nucleic acid sequence

We tried to isolate Getah virus from blood samples that were positive on RT-PCR. In testing on 23 blood samples, primary cocultivation resulted in the isolation of two strains confirmed as Getah virus by RT-PCR. We analyzed the sequences of the *nsP1* genes of the two strains and those of the RT-PCR products amplified from some of the clinical samples (n = 7). All of the samples analyzed had completely identical nucleic acid sequences, and the sequences were also identical to that of the 14-I-605 strain (381 bases, GenBank/EMBL/DDBJ accession number, LC012885) isolated during the 2014 outbreak [6]. From this result, we used one of the two 2015 isolates (15-I-752) for the further studies.

Antigenic comparison of the vaccine strain and the 2014 and 2015 Getah virus strains

To assess whether the current Getah virus vaccine was effective against the circulating virus in 2015, we performed cross-neutralization tests between the vaccine strain (MI-110) and the strains isolated in 2014 (14-I-605) and 2015 (15-I-752). The results of the cross-neutralization tests are summarized in Table 2. Our previous report revealed that horse sera (n = 2) raised against MI-110 neutralized the homologous virus at titers of 512 and neutralized the 14-I-605 strain at almost the same titers (256) [8]. In the current experiment, the same set of sera neutralized the 15-I-752 strain at titers of 256 (Table 2). The horse sera (n = 2) raised against the 14-I-605 strain—which were also used in the previous study [8]—neutralized the homologous virus at a titer of 256 or 512 and neutralized the 15-I-752 strain at titers of 512 (Table 2). These results indicated that the two strains isolated in 2014 and 2015 were antigenically close to each other, and that the current vaccine containing the MI-110 strain was likely sufficiently effective against the circulating viruses.

Investigation of Getah virus epizootic infection among horses on surrounding farms in Ibaraki and Chiba prefectures

During the 2014 outbreak, we found that epizootic Getah virus infection occurred not only at the training

Table 1 Numbers (%) of horses that had received Getah virus vaccine and were being kept at the Miho training center on August 15 in 2015

Age	Vaccination dose		Total
	One	Two or more	
2	45 (7.3)	568 (92.7)	613
3	6 (0.8)	794 (99.2)	794
4	2 (0.6)	312 (99.4)	314
5 or older	0 (0.0)	271 (100.0)	271
Total	53 (2.7)	1,939 (97.3)	1,992

Table 2 Virus-neutralizing titers of sera from horses inoculated with Getah virus MI-110 or 14-I-605 strain

Inoculated strain	Horse	Strain used in virus-neutralization test		
		MI-110	14-I-605	15-I-752
MI-110	1	512[a]	256[a]	256
	2	512[a]	256[a]	256
14-I-605	3	512[a]	256[a]	512
	4	1024[a]	512[a]	512

[a]Data quoted from our previous study [8]

center but also on private farms surrounding the center [6]. To assess whether this regional epizootic had also occurred in the 2015 outbreak, we calculated seropositivity rates among horses entering the center from farms in Ibaraki Prefecture and the neighboring Chiba Prefecture each month in 2015. The horses were 2-year-olds with no history of Getah virus vaccination. Seropositivity rates in June (5.5 %) and July (4.9 %) were comparable to those in 2014 (Fig. 2) [6]. An increase in seropositivity was observed in August (11.8 %), and those in September and October were 21.7 and 34.9 %, respectively (Fig. 2), indicating that Getah virus was epizootic in the area in autumn 2015, similarly to the 2014 season.

Discussion

The epizootic Getah virus infection in 2015 seems to have started earlier than that in 2014, which started in mid-September [4]. This trend was observed both at the Miho training center and on the private farms (Figs. 1 and 2). Despite the early start, the prevalence of Getah virus infection in whole population at the training center in 2015 (30/1992 [1.5 %]) was comparable to that in 2014 (33/1950 [1.7 %]) [4], suggesting that although the epizootic in 2015 started earlier than that in 2014 it progressed more slowly. In support of this finding, the seropositivity rate in horses entering the center was lower in October 2015 (34.9 %) than in October 2014 (42.9 %, Fig. 2). We found that the age proportions of the horses affected in 2014 and 2015 differed greatly; this might

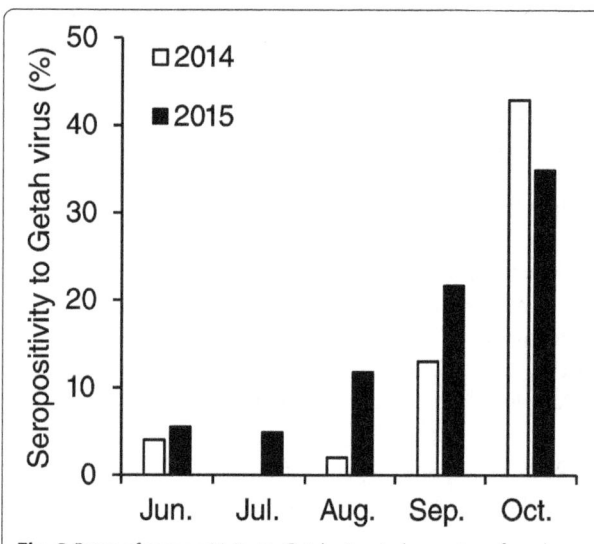

Fig. 2 Rates of seropositivity to Getah virus in horses transferred from farms surrounding the Miho training center. Sera were collected from horses introduced to the Miho training center between June to October in 2014 or 2015. The horses (n = 51 to 81 in each month) were 2-year-olds that had been transferred from Ibaraki or Chiba prefecture and had no history of Getah virus vaccination. Sera were subjected to VN testing using the 14-I-605 strain. Data for 2014 are quoted from our previous report [6]

have explained the relatively slow mild progression of the 2015 outbreak. The proportion of 2-year-olds among the horses affected in 2014 was 60.6 % [6], whereas that in 2015 was 90.0 %, suggesting that horses aged 3 years or older were relatively resistant to infection. This was probably due to the fact that, because of the 2014 outbreak, many horses aged 3 years or older in 2015 had already been exposed to Getah virus at the center or on the private farms. Therefore, the existence of these resistant horses might have delayed virus spread in the horse population to some extent.

The significantly higher prevalence in the 2-year-olds (4.4 %) than that in the 3-year-olds or older population (0.2 %) was one of the most characteristic aspects of the current outbreak, because such significant difference was not observed in the 2014 outbreak [6]. In this regard, even though older horses at the training center become resistant after natural infection, we are still concerned about the possibility of outbreaks of Getah virus infection in upcoming years. In Japan, more than 95 % of racehorses are bred in Hokkaido Prefecture, in northern Japan, and 2-year-olds are introduced to the training center every year. At the time of introduction, most have not been infected with Getah virus or vaccinated with Getah virus vaccine. Serological surveillance also suggests that there is a low prevalence of Getah virus infection in Hokkaido, with seropositivity rates of 0 % among unvaccinated horses transferred out of Hokkaido in 2013 and 2014 (Bannai et al., unpublished data). Therefore, the existence of newly introduced 2-year-olds could continue to pose a risk of additional outbreaks.

More than 30 private farms are located in Ibaraki Prefecture, where the Miho training center stands, and in neighboring Chiba Prefecture, and horses are repeatedly transferred between the farms and the center for training and rest. Unlike in the training centers, where Getah virus vaccination is mandatory, on the farms vaccination is not common. In addition, this area is one of the biggest producers of pigs. Although the exact prevalence of Getah virus in pigs and its association with the recent outbreaks in horses have not been studied, the horses in this area are considered to be at high risk of infection. As in the 2014 outbreak, in 2015 our results indicated that there was a high prevalence of Getah virus infection among horses on the private farms (Fig. 2). This regional epizootic probably resulted in an increased risk of virus exposure and allowed the virus to spread in the center. Therefore, the 2015 outbreak seems to have occurred under epizootiological circumstances similar to those in 2014. It will be helpful to increase vaccination coverage on private farms to prevent regional virus circulation. Unfortunately, despite our recommendation, coverage on the farms in 2015 seemed to be as low as in previous years, although exact data were not available. In support of this speculation, among the 2-year-olds at the Miho training center in mid-September 2015, the proportion of those that had been vaccinated only once for Getah virus was 9.0 % (75/831 head); this was almost the same as that in 2014 (8.3 % [71/858 head]) [6]. The existence of horses in this category reflects the low vaccination coverage on the private farms, because horses with no history of Getah virus vaccination are given their first, priming, dose on entry to the training center. Among the horses affected in the 2015 outbreak, 33.3 % (10/30 head) had been vaccinated only once before disease onset; this was comparable to the level in 2014 (30.3 % [10/33 head]) [6], suggesting that these horses were highly susceptible and were involved in spread of the virus at the center, similarly to the situation in 2014. The higher prevalence in 2-year-old horses vaccinated only once (22.2 %) than that in those vaccinated twice or more (3.0 %) also suggested the requirement of two-doses priming vaccination for the protection of individual horses. Therefore, further efforts are needed to increase vaccination coverage on the farms.

As described above, the 2014 and 2015 outbreaks of Getah virus infection occurred after an absence of detectable outbreaks for more than three decades. The sudden outbreaks in the two sequential years at the same site suggest the occurrence of direct factors such as mutations that might alter the features of the virus. Although we were initially concerned about antigenic mismatch between the circulating virus and the vaccine strain, our previous and current results suggest that the 2014 and 2015 isolates are antigenically similar to the vaccine strain (Table 2) [8]. Other viral features which may influence the epizootic includes the vector specificity and the efficacy of replication in mosquitoes, because Getah virus is a typical mosquito-borne virus. A previous ecological surveillance in 1979 at the Miho training center revealed that Getah virus was isolated from *Aedes vexans nipponii* and *Culex tritaeniorhynchus*, and these two species were considered to be involved in the circulation of Getah virus in Japan [9]. In our previous study, we compared the full-genome sequences of the isolate in 2014 and the vaccine strain, and found that non-structural protein 3 (nsP3) included 7 amino acid substitutions while the other non-structural proteins had only 1 or 2 substitutions [8]. The carboxyl-terminus domain of nsP3 was reported to be involved in viral replication of genus *Alphaviruses* to which Getah virus belongs [10], and also reported to be a determinant of vector specificity in O'nyoug nyong virus [11]. In this regard, further study on the mutations in the nsP3 protein might provide clues to the causes of the current outbreaks. In addition, investigation of the density of vector species in the region surrounding the training center, the prevalence of Getah virus in the

mosquitoes, and the epizootic situation in pigs, the natural host, would help us to understand the risk of future outbreaks in horses.

Conclusions

In conclusion, an outbreak of Getah virus infection occurred at the Miho training center in 30 horses from mid-August to late October 2015. It was sequential to the 2014 outbreak at the same site. The 2015 outbreak was caused by a virus closely related to the 2014 isolate and affected mostly 2-year-old susceptible horses under epizootiological circumstances similar to those in 2014. The existence of 2-year-olds introduced from non-epizootic regions could continue to pose a risk of additional outbreaks in upcoming years. Vaccination on private farms and breeding farms would help to minimize the risk of outbreaks. Continuous surveillance at the training center, as well on the farms surrounding the center, will be required.

Abbreviations
MEM, minimum essential medium; nsP3, non-structural protein 3; VN, virus-neutralizing

Acknowledgments
We are grateful to all the equine practitioners of the Japan Racing Association for collecting clinical samples. We also thank Akira Kokubun, Kazue Arakawa, Akiko Suganuma, and Kaoru Makabe of the Equine Research Institute of Japan Racing Association for their technical help.

Funding
This study was supported by the Japan Racing Association.

Authors' contributions
HB, MN, and KT performed RT-PCR of the clinical samples collected at the training center. HB drafted the manuscript and performed VN testing, virus isolation, and sequencing. AO analyzed the clinical data and vaccination histories. MN, KT, TY, and TK participated in the design of the experiment and discussed the draft of the manuscript. We all read and approved the final manuscript.

Competing interests
The authors declare that they have no competing interests.

Consent for publication
Not applicable.

References
1. Fukunaga Y, Kumanomido T, Kamada M. Getah virus as an equine pathogen. Vet Clin North Am Equine Pract. 2000;16:605–17.
2. Izumida A, Takuma H, Inagaki S, Kubota M, Hirahara T, Kodama K, et al. Experimental infection of Getah virus in swine. Jpn J Vet Sci. 1988;50:679–84.

3. Yago K, Hagiwara S, Kawamura H, Narita M. A fatal case in newborn piglets
    with Getah virus infection: Isolation of the virus. Jpn J Vet Sci. 1987;49:989–94.
4. Nemoto M, Bannai H, Tsujimura K, Kobayashi M, Kikuchi T, Yamanaka T,
    et al. Outbreak of Getah virus infection among racehorses in Japan in 2014.
    Emerg Infect Dis. 2015;21:883–5.
5. Sentsui H, Kono Y. Reappearance of Getah virus infection among horses in
    Japan. Jpn J Vet Sci. 1985;47:333–5.
6. Bannai H, Nemoto M, Ochi A, Kikuchi T, Kobayashi M, Tsujimura K, et al.
    Epizootiological investigation of Getah virus infection among racehorses in
    Japan in 2014. J Clin Microbiol. 2015;53:2286–91.
7. Wekesa SN, Inoshima Y, Murakami K, Sentsui H. Genomic analysis of some
    Japanese isolates of Getah virus. Vet Microbiol. 2001;83:137–46.
8. Nemoto M, Bannai H, Tsujimura K, Yamanaka T, Kondo T. Genomic,
    pathogenic, and antigenic comparisons of Getah virus strains isolated in
    1978 and 2014 in Japan. Arch Virol. (in press)
9. Kumanomido T, Fukunaga Y, Ando Y, Kamada M, Imagawa H, Wada R, et al.
    Getah virus isolations from mosquitoes in an enzootic area in Japan. Jpn J
    Vet Sci. 1986;48:1135–40.
10. Lasterza MW, Grakoui A, Rice CM. Deletion and duplication mutations in the
    C-terminal nonconserved region of Sindbis virus nsP3: effects on
    phosphorylation and on virus replication in vertebrate and invertebrate
    cells. Virology. 1994;202:224–32.
11. Saxton-Shaw KD, Ledermann JP, Borland EM, Stovall JL, Mossel EC, Singh AJ,
    et al. O'nyong nyong virus molecular determinants of unique vector specificity
    reside in non-structural protein 3. PLoS Negl Trop Dis. 2013;7:e1931.

7

Bat rabies surveillance and risk factors for rabies spillover in an urban area of Southern Brazil

Juliano Ribeiro[1] (ID), Claudia Staudacher[2], Camila Marinelli Martins[3], Leila Sabrina Ullmann[4], Fernando Ferreira[3], João Pessoa Araujo Jr[4] and Alexander Welker Biondo[5*]

Abstract

Background: Bat rabies surveillance data and risk factors for rabies spillover without human cases have been evaluated in Curitiba, the ninth biggest city in Brazil, during a 6-year period (2010–2015). A retrospective analysis of bat complaints, bat species identification and rabies testing of bats, dogs and cats has been performed using methodologies of seasonal decomposition, spatial distribution and kernel density analysis.

Results: Overall, a total of 1003 requests for bat removal have been attended to, and 806 bats were collected in 606 city locations. Bat species were identified among 13 genera of three families, with a higher frequency of *Nyctinomops* in the central-northern region and *Molossidae* scattered throughout city limits. Out of the bats captured alive, 419/806 (52.0%) healthy bats were released due to absence of human or animal contacts. The remaining 387/806 (48.0%) bats were sent for euthanasia and rabies testing, which resulted in 9/387 (2.32%) positives. Linear regression has shown an increase on sample numbers tested over time (regression: $y = 2.02 + 0.17x$; $p < 0.001$ and $r^2 = 0.29$), as well as significant seasonal variation, which increases in January and decreases in May, June and July. The Kernel density analysis showed the center-northern city area to be statistically important, and the southern region had no tested samples within the period. In addition, a total of 4769 random and suspicious samples were sent for rabies diagnosis including those from dogs, cats, bats and others from 2007 to 2015. While all 2676 dog brains tested negative, only 1/1136 (0.088%) cat brains tested positive for rabies.

Conclusion: Only non-hematophagous bats were collected during the study, and the highest frequency of collections occurred in the center-northern region of the city. Rabies spillover from bats to cats may be more likely due to the registered exposure associated with cats' innate hunting habits, predisposing them to even closer contact with potentially infected bats. Although associated with a very low frequency of rabies, cats should always be included in rabies surveillance and vaccination programs.

Keywords: Non-hematophagous bat, Dog, Cat, Rabies, AgV-3. Geo-referencing, Kernel, Seasonal decomposition

Background

Bats (order Chiroptera) have been considered one of the most diverse worldwide mammal groups, accounting for 20.7% of 5416 currently known mammal species, with 18 families and 1120 species [1, 2]. The presence of bats has been reported in all geographic areas of the world except the Arctic, Antarctic, extreme desert areas, and some isolated oceanic islands [3]. Brazil has been ranked as the second highest country in bat species, harboring 178 (15.9%) of the known species worldwide [4, 5].

Of the species of bats identified worldwide, only three feed exclusively on blood: *Desmodus rotundus*, *Diphylla ecaudata* and *Diaemus youngi*. *D. rotundus* is known as the common vampire bat and is the only one that feeds on mammalian blood, while the other two species feed on bird blood. Vampire bats are distributed from Mexico to South America [6]. Deforestation has drastically reduced the number of natural prey for *D. rotundus*;

* Correspondence: abiondo@ufpr.br
[5]Department of Veterinary Medicine, Federal University of Paraná, Rua dos Funcionários, 1540, Curitiba, Paraná 80035-050, Brazil
Full list of author information is available at the end of the article

faced with this change, vampire bats have found a great source of food in cattle, which were introduced by man in South America. This has given rise to the numbers of vampire bats and their contact with cattle and man, causing a direct impact on human and animal health by the transmission of the rabies virus [7].

The rabies virus (RABV) can affect all mammals; however, the orders Carnivora and Chiroptera act as reservoirs for the virus [8]. The rabies virus (RABV) has been divided into two main variants: the first is associated with carnivores, mostly dogs, on an urban cycle, and the second is associated with bats, raccoons, and skunks on a sylvatic cycle [6–9]. The rabies cycle is divided into 4 cycles in several publications in South America: urban (domestic dog and cat), rural (livestock, cattle, horses, pigs, etc.), sylvatic (fox, raccoon, opossum, etc.) and air cycles (bats). However, in this study, this context was simplified to two major cycles, urban (dog and cat) and sylvatic (which covers all free-living animals, including bats) [10].

Although human cases in developing countries have been mostly associated with dog bites, bat species may also be infected by RABV, and human fatalities in Latin America have recently been connected to spillover from hematophagous, insectivorous and frugivorous bats [10, 11]. Not surprisingly, the highest recorded rabies outbreaks in Brazil were bat-transmitted and occurred in Brazilian northern rural (21 deaths) and remote areas of the Amazon forest (16 deaths) due to rabies virus variant 3 (AgV3), which is mainly found in *Desmodus rotundus*, a vampire bat species [10, 12, 13].

Meanwhile, a switch in the habits of non-hematophagous bats has also been recently observed, with migration from rural to urban areas probably due to increased food supply in urban centers and environmental impact on their natural habitats, increasing potential contact with domestic and wild animal populations and human beings [14, 15]. As a result, 20/41 (49.1%) positive bat specimens currently reported for rabies in Brazil were from non-hematophagous species, followed by 12/41 (29.0%) hematophagous and 9/41 (21.9%) unidentified species [16]. In addition, despite a decrease in human and canine rabies in Brazil, human cases have mostly (78.0%) occurred from bat variants between 2000 and 2009 [17, 18].

Cats have been considered a high-risk species for rabies transmission to humans in some European countries mainly due to their hunting habits, particularly toward flying animals including bats, which may connect rabies from the sylvatic-aerial cycle to urban settings [19]. Such scenarios may similarly occur in major cities of Brazil such as Curitiba, the ninth biggest Brazilian city, where a cat has been diagnosed with bat variant rabies after almost 30 years of no pet rabies cases [20].

Accordingly, this study aimed to analyze the bat rabies surveillance and risk factors for rabies spillover in an area without human cases in southern Brazil during a 6-year period (2010–2015). In addition, a retrospective analysis of bat complaints, bat species identification and rabies testing of bats, dogs and cats in the same area has been performed using methodologies of seasonal decomposition, spatial distribution and kernel density analysis.

Methods

Curitiba (25°25′48″ S, 49°16′15″ W), the capital of Paraná state, southern Brazil, has been currently ranked as the ninth biggest Brazilian city with approximately 1.8 million inhabitants [21]. Although categorized as a 100% urban area, Curitiba city has been considered to be environmentally friendly and the first in sustainability and quality of life in Brazil, with a high green-area ratio distributed throughout more than 40 city parks and preservation areas [22].

Since 1984, an official central telephone system has been used in Curitiba as a communication channel between the population and public managers; this system allows the population to request government services of all areas (health, urbanism, education, etc.), and among the available services are requests for the collection of dead animals (dogs and cats), removal of fallen bats inside houses and removal and/or observation of aggressive animals. Complaints of dead animals have been used as a source of brain samples from dogs and cats, most of which are sent for rabies diagnosis at the Parana State Reference Laboratory (LACEN) and used for monitoring rabies virus circulation. In addition, complaints for bats have followed another specific protocol: local inspection by professionals from the Curitiba Zoonosis Control Center (ZCC), capture or collection of bats, an epidemiological questionnaire and bat health status. If bats were healthy and had no human or pet contact, they were released using an open box at sunset of the same day at the ZCC, which was located nearby preserved areas at the time. If bats were dead, had contact with pets or human beings, or were unhealthy (no flying, neurological signs, injuries), they were euthanized, and their brains were sent for rabies testing at the LACEN.

Official city records of bat complaints, local inspections and bat destinations were obtained from January 2010 to December 2015. Additionally, records of bats, dogs and cats sent for rabies testing were obtained from the ZCC from January 2007 to December 2015. Bats were individually identified based on two standard taxonomy references [23, 24]. All rabies tests were performed by the Central Reference Laboratory of the State of Paraná (LACEN-PR) following international guidelines for laboratory and diagnostic techniques and using the fluorescent antibody test (FAT) with a panel of monoclonal antibodies as well as intracerebral inoculation in 21-day-old mice [25, 26].

A database was constructed with a commercially available statistical package (Microsoft Excel 2007, Microsoft Company, Redmond, WA, USA) and included collection location, situation in which the animal was collected or captured, number of animals, animal genus and species, procedures at ZCC, date and rabies result. Descriptive statistics were conducted with frequencies and distributions, followed by calculation of seasonal indices and a linear regression model with significance of 5% with Minitab software (Minitab 17 Statistical Software (2010). [Computer software]. State College, PA: Minitab, Inc.) [27]. A simple linear model was performed after tests were fitted to a normal distribution of data.

A geo-referencing approach was applied on the address data, using the "RDSTK" package [28] in the R software environment [29]. A map was built in commercial software [30] and contained bat points (positives/negatives), urbanization information, and neighborhood boundaries with shape files obtained from the City Geography Services (Institute of Urban Planning and Research of Curitiba, IPPUC). Finally, a kernel density analysis was performed with the "stats" package in the R environment [29]. These spatial treatments of data were performed to visualize the points (the map build) of bats collection and to test patterns of their distribution (kernel analysis). The kernel analysis is a density analysis that estimates the contribution of each point when compared to the distance to other points. The contribution extension is dependent on the bandwidth adopted (in this study, 50 m, considering the households as reference), and this analysis provides a density evaluation in which the hot areas represent the most important areas of the study when compared to the cold areas [31].

Results

Overall, a total of 4769 samples were sent for rabies diagnosis, including dogs, cats, bats and other animal species, from 2007 to 2015 (Table 1). The highest number of brain samples were collected from dogs (2676; 56.1%), followed by cats (1136; 23.8%), bats (940; 19.7%) and other animals (17; 0.35%), which included three rabbits (*Oryctolagus* sp.), three bush dogs (*Speothos venaticus*), two ferrets (*Galictis* sp.), two horses (*Equus ferus caballus*), a non-human primate (*Cebus* sp.), a squirrel (*Sciurus ingrami*), an opossum (*Didelphis albiventris*), a deer (*Cervus* sp.), a raccoon (*Procyon* sp.), a marmoset (*Callithrix* sp.), and a gerbil (*Meriones* sp.). Out of the tested samples, only 9/4769 (0.18%) bats and 1/4769 (0.02%) cats were positive for the rabies virus.

The central phone system had registered 1003 bat removal requests from 2010 to 2015 (Table 2), resulting in a total of 806 captured or collected bats. Due to environmental preservation and no evident risk of rabies transmission, 419 healthy bats that did not have contact with other animal species or human beings were systematically

Table 1 Animal samples sent for rabies surveillance in Curitiba, Parana, Brazil from 2007 to 2015

Year	Dogs	Cats	Bats	Other	Total
2007	93	8	52		153
2008	49	3	37	1 (ferret)	90
2009	26	1	45		72
2010	38	119	54		211
2011	21	116	64	2 (non-human primate and rabbit)	203
2012	250	173	86	2 (rabbit and horse)	511
2013	911	235	66	2 (bush dog)	1214
2014	916	230	351	5 (rabbit, horse, bush dog, squirrel and opossum)	1502
2015	372	251	185	5 (deer, raccoon, ferret, marmoset, gerbil)	813
Positives[a]	0	1	9	0	10
Total	2676	1136	940	17	4769

[a]Values not added to avoid overlapping

released within city preserved areas. The remaining 387 bats were immediately submitted for euthanasia and rabies testing, resulting in 9/387 (2.32%) positive bats.

During the investigation, a total of 806 bats were captured or collected, and they were categorized in 13 genera from three families (*Molossidae*, *Vespertilionidae* and *Phyllostomidae*). The family *Molossidae* was the most frequent with 658/806 (81.5%) bats, followed by *Vespertilionidae* with 57/806 (7.1%) bats and *Phyllostomidae* with 45/806 (5.6%) bats; 46/806 (5.8%) bats were not identified (Table 3).

The case distribution map showed all the points where bats were captured or collected in Curitiba from 2010 to 2015, including the nine positive cases (Fig. 1). A seasonal decomposition was made for the same period to identify in which part of the year more captures or collections had occurred (Fig. 2). The kernel density for negative cases presented a homogeneous distribution, despite the aggregation observed in downtown Curitiba (Fig. 3a). The kernel density estimation for positive bats

Table 2 Bat complaints and proceedings for rabies surveillance in Curitiba, Parana, Brazil, 2010 to 2015

Year	Complaints	Collected	Released	Rabies test	Positive
2010	129	54	27	27	1
2011	72	64	21	43	3
2012	139	86	28	58	1
2013	140	66	22	44	0
2014	250	351	267	84	2
2015	273	185	54	131	2
Total	1003	806	419	387	9

Table 3 Family and genus of bats collected for rabies surveillance in Curitiba, Parana, Brazil from 2010 to 2015 (Additional file 1)

Family	Genus	n	tested	Positives[a]	Genus (%)	Families (%) (Positives, %)
Molossidae (Total: 658)	*Molossus*	241	136	2 (1.47%)	29.9	81.6 (7/283, 2.47%)
	Promops	61	50	2 (4.00%)	7.5	
	Tadarida	19	10	–	2.3	
	Nyctinomops	336	86	3 (3.48%)	41.6	
	Eumops	1	1	–	0.12	
Vespertilionidae (Total: 57)	*Eptesicus*	13	10	–	1.6	7.1 (1/43, 2.32%)
	Myotis	23	16	1 (6.25%)	2.8	
	Histiotus	7	5	–	0.86	
	Lasiurus	14	12	–	1.7	
Phyllostomidae (Total: 45)	*Artibeus*	27	18	–	3.3	5.6 (1/32, 3.12%)
	Sturnira	14	12	1 (8.33%)	1.7	
	Glossophaga	3	1	–	0.37	
	Pygoderma	1	1	–	0.12	
Not identified		46	29	–	5.7	5.7
Total		806	387	9 (2.32%)	100	100

[a]Values were not added to avoid overlap and show the percentage of positive test results

showed an aggregation of bat points in north Curitiba (Fig. 3a).

Discussion

Although the Brazilian National Program for Rabies Control and Prevention has historically recommended a 0.2% sampling of total estimated city dog population, consisted by dead dogs sent every year for rabies testing [32], animal sampling has increased above dog population growth, particularly between 2012 and 2014 (Table 1). Moreover, majority of samples were dogs (mostly killed by car, elderly or euthanized in shelters), which all resulted negative for rabies. Despite Curitiba has been reportedly considered a free-rabies city since 1975 [20], such "healthy" dog sampling not based on suspicious nervous clinical signs or critical bat rabies areas may have lowered the surveillance sensitivity through these years.

On the other hand, one cat tested positive for rabies virus variant 4 in 2010, compatible with isolates from insectivorous bat *Tadarida brasiliensis* [20], which may suggest that a direct contact between a bat and a cat occurred [33]. The predatory behavior of cats may include bat hunting, which can raise the risk of cat rabies infection, making cats a potential rabies source for other animal species and human beings [34]. The last human case of rabies in the nearby São Paulo state was recorded in 2001, when a woman was likely infected by a bite from her cat with variant 3, commonly found in vampire bats (*Desmodus rotundus*) [35]. In 2008, in Santander de Quilichao, Colombia, rabies transmission was recorded from a cat, leading to the death of two people; in both cases, the virus type was AgV3, which is mostly

associated with hematophagous bats [36]. Colombia recorded another human case of rabies in 2013, with the owner bitten by a cat described as a bat hunter; the rabies type was identified as variant 4, which is associated with insectivorous bats [37].

A recent study has shown the importance of rabies spillover from bats to other animal species and the likelihood of rabies transmission through the bat-cat-human chain, but it did not estimate the risk of bat-dog and bat-cat transmission [38]. Recipient hosts have been exposed to virus source in a sufficient amount to establish an infection, showing susceptibility to the virus [38, 39]. The positive cat rabies case from Curitiba reported in 2010 [20], associated with the data presented herein, may emphasize the importance of the surveillance service and monitoring to suspect bats for rabies, providing substantial information to authorities to establish strategies and actions.

The identification of bat species can be important to understanding rabies transmission. The behavior of some species can expose them more or less to the virus [17]. *Tadarida brasiliensis*, *Molossus rufus*, and *Molossus molossus* (species identified in this study) form maternal colonies, which may push males to competition and segregation and make females have more body contact [40, 41]. Spatially, the *Molossidae* family (insectivorous family in general) may be attracted to insects near urban artificial lights and may find artificial shelters in roofs, ceilings, attics, etc. [42, 43]. This is reflected in the study at hand, where the highest bat capture was at the central-northern region, the high human population density of the city, providing artificial shelters and food supply [6, 44].

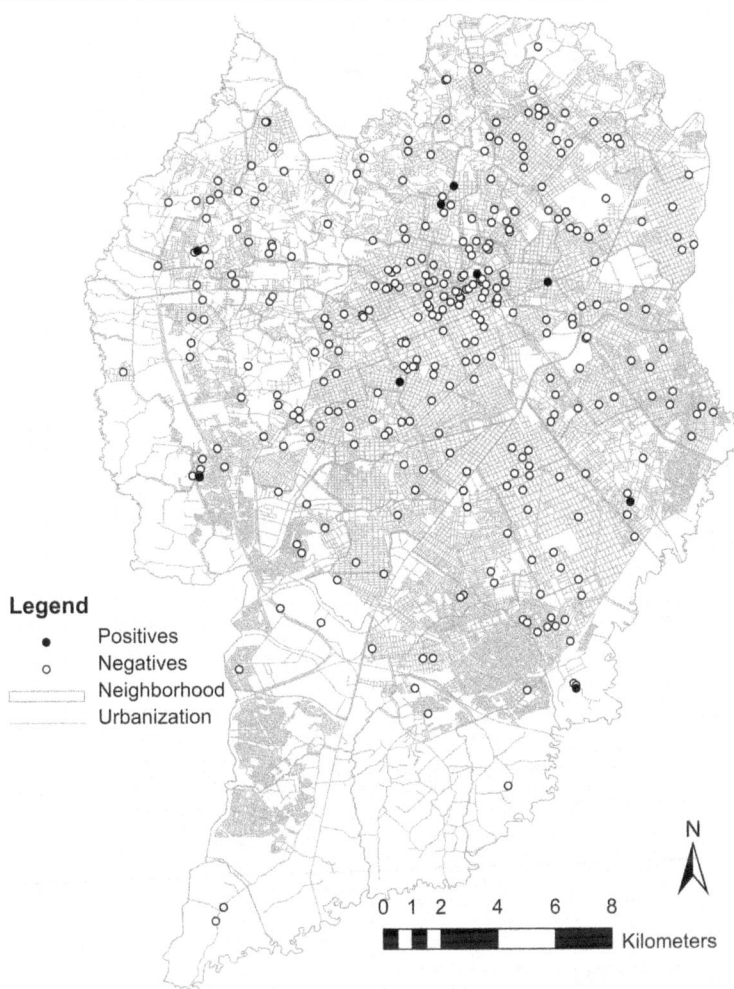

Fig. 1 Urbanization map of the city of Curitiba showing the site where bats were collected during the period 2010–2015. The dark circles indicate bats collected that were positive for the rabies virus (9 bats); the white circles indicate bats collected that were not positive for the rabies virus

Requests to remove the bats were higher than the number of animals collected since requests have occasionally involved bat colonies, which were not considered an imminent risk for rabies transmission by the Curitiba ZCC (Table 2). However, identification of bat colony genus and geo-referencing has been prioritized by the ZCC for rabies sanitary blocking, preventive informative and pet vaccination programs [20, 43].

The analysis of seasonal distribution has shown a close relationship between the warmer tropical months with the number of requests made by citizens for bat removal. Studies of *Tadarida brasiliensis* bats made in Argentina showed that weather conditions directly influence the bats' behavior; on very hot days (temperatures > 27 °C), they were more active [40]. Higher temperatures were recorded from December to March in the study area, which may have led to increasing food supply for non-hematophagous bats, mainly insectivorous bats.

Insectivorous bats were collected throughout the study area but more frequently in the central region, and the same was observed for fruit bats, probably due to the abundance of food and shelter for bats in the region. Food source may be a key factor influencing bat activity during periods of high temperatures, which may increase the activity of flying insects and consequently attract bats to areas of high concentration of insects due to food availability [6, 44]. Another important point regarding higher temperature periods has been people's habit of leaving windows opened, which may facilitate bat entry overnight, bat sightings the next day and phone system complaints, accounting for most of the requests for bat removal by the surveillance service. The area with the highest concentration observed in Fig. 3 corresponds to the Matriz sanitary district, which houses the most populous region of the city. In this sector, the ZCC technical staff identified several artificial shelters, such

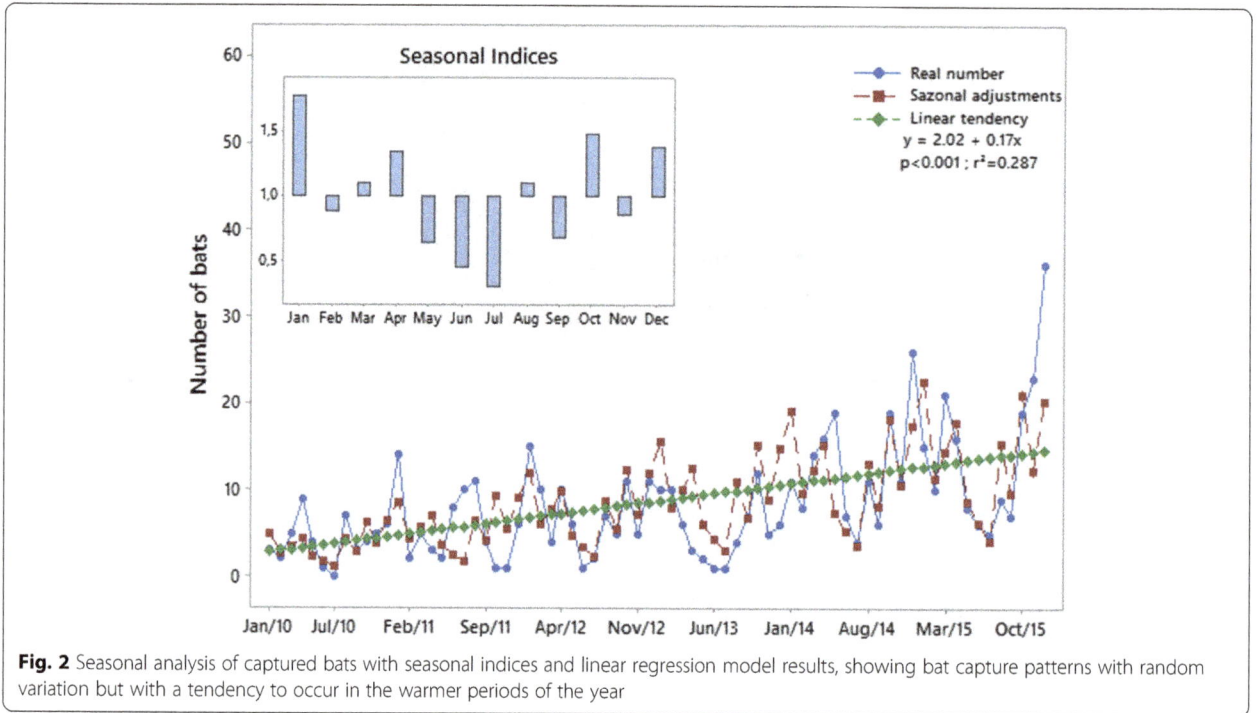

Fig. 2 Seasonal analysis of captured bats with seasonal indices and linear regression model results, showing bat capture patterns with random variation but with a tendency to occur in the warmer periods of the year

as the ceiling, attic, expansion joints, air conditioning, shutters boxes, and chimneys, among others (internal information not published).

The kernel density estimation has shown that the city's center-northern area may be characterized as a particular area for bats affected by rabies virus. Such a finding may note the importance of a monitoring service and local capturing of bats, since this service may prevent an accidental contact between an infected bat with a pet or human being. The geo-referencing may be an important tool used to identify the places where the bats were collected, providing a bat distribution overview throughout a region or city, which may be crossed with the geo-referencing of either rabies positive or negative pets, allowing health authorities to spatially combat the spread of the disease, particularly to other animal species and human beings.

Conclusion

This study showed that insectivorous bats (especially the *Molossidae* family) were important for rabies surveillance

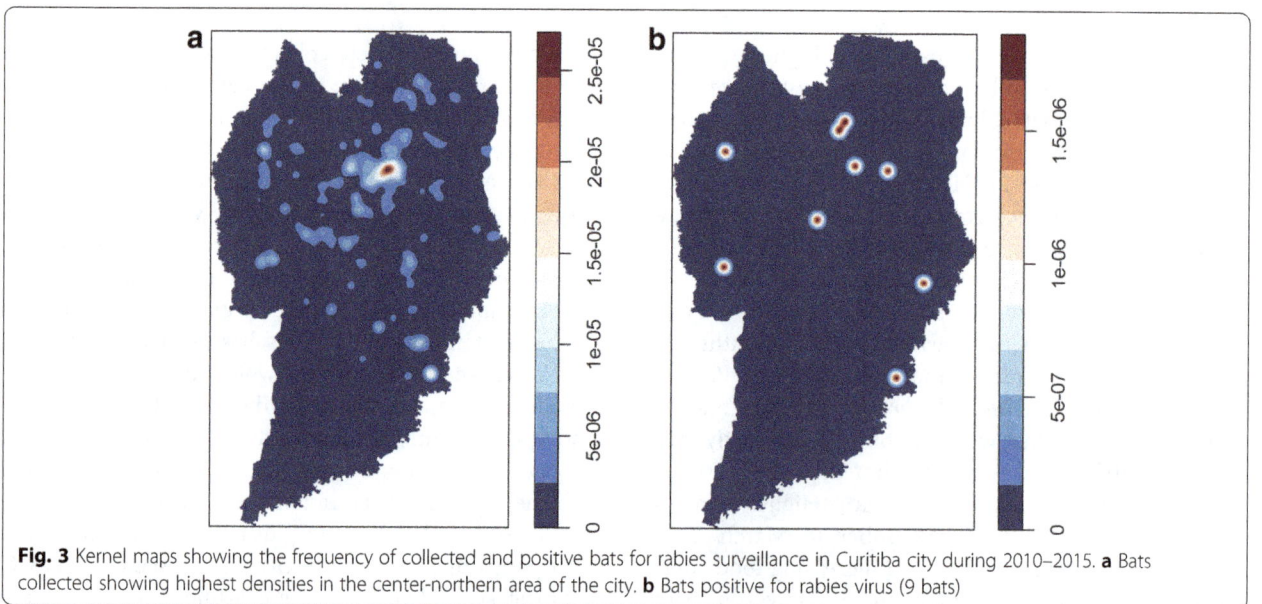

Fig. 3 Kernel maps showing the frequency of collected and positive bats for rabies surveillance in Curitiba city during 2010–2015. **a** Bats collected showing highest densities in the center-northern area of the city. **b** Bats positive for rabies virus (9 bats)

and transmission (positive bats and a cat spillover) during the period studied. There were zero positive dogs and only one positive cat, suggesting an increase in cat importance, and we recommend that public health authorities pay attention to mass vaccinations of cats in large urban centers. In addition, it is important that health services maintain and improve the monitoring of non-vampire bats in large urban centers, too.

Acknowledgements

The authors are grateful to the Municipal Secretary of Health of Curitiba, ZCC, for providing the data archive, to the University of State of Sao Paulo (UNESP - Botucatu) and University of Sao Paulo (USP) for financial and technical support, and for all work by ZCC that directly or indirectly contributed for this study.

Authors' contributions

JR participated in the study design, analysis and manuscript preparation. CS participated in bat collection, identification, discussion and analysis of the data about bats. CMM and FF participated with discussion and analysis of data and manuscript preparation. JP and LSU participated in the study design and manuscript preparation. AWB participated in the study design, coordination and supervision. All authors read and approved the final manuscript.

Competing interests

The authors declare that they have no competing interests.

Author details

[1]Graduate Program in Cellular and Molecular Biology, Federal University of Parana, Curitiba, Paraná 81531-990, Brazil. [2]Zoonoses Control Center, City Secretary of Health, Curitiba, Paraná 80060-130, Brazil. [3]Department of Preventive Veterinary Medicine and Animal Health, University of São Paulo, São Paulo 05508-270, Brazil. [4]UNESP – Univ. Estadual Paulista, Campus de Botucatu, Institute of Biotechnology, Botucatu, São Paulo, Botucatu, São Paulo 18607-440, Brazil. [5]Department of Veterinary Medicine, Federal University of Paraná, Rua dos Funcionários, 1540, Curitiba, Paraná 80035-050, Brazil.

References

1. Simmons NB. Order Chiroptera; pp. 312-529. In: Wilson DE, Reeder DM, editors. Mammals species of the world: a taxonomic and geographic reference. Baltimore: The John Hopkins University Press; 2005.
2. Wilson DE, Reeder DM. Mammal species of the world: a taxonomic and geographic reference. 3rd Ed. Baltimore, Maryland, The Johns Hopkins University Press; 2005.
3. Hutson A M, Mickleburgh SP, Racey PA. 2001. Microchiropteran bats: global status survey and conservation action plan. IUCN/SSC Chiroptera specialist group, IUCN, gland, x + 258 pp.
4. Bernard E, Aguiar LMS, Machado RB. Discovering the Brazilian bat fauna: a task for two centuries? Mammal Rev. 2010;41:23–39.
5. Nogueira MR, Lima IP, Moratelli R, Tavares VC, Gregorin R, Peracchi AL. Checklist of Brazilian bats, with comments on original records. Check List. 2014;10(4):808–21.
6. Dos Reis NR, Peracchi AL, Pedro WA, de Lima IP. [Bats of Brazil]. Londrina: Universidade Estadual de Londrina Press; 2007.
7. Johnson N, Aréchiga-Ceballos N, Aguilar-Setien A. Vampire bat rabies: ecology. Epidemiology and Control Viruses. 2014;6(5):1911–28.
8. Streicker DG, Lemey P, Velasco-Villa A, Rupprecht CE. Rates of viral evolution are linked to host geography in bat rabies. PLoSPathog. 2012;8(5):e1002720. https://doi.org/10.1371/journal. Pp at.1002720.
9. World Health Organization. WHO expert consultation on rabies. Second report. World Health Organ Tech Rep Ser. 2013;982(982):1–139.
10. Cordeiro RA, Duarte NFH, Rolim BN, Soares Júnior FA, Franco ICF, Ferrer LL, Almeida CP, Duarte BH, Araújo DB, Rocha MFG, Brilhante RSN, Favoretto SR, Sidrim JJC. The importance of wild canids in the epidemiology of rabies in Northeast Brazil: a retrospective study. Zoonoses Public Health. 2016;63:486–93. 10.1111 / zph.12253.
11. Sparkes JL, Fleming PJS, Ballard G, Scott-Orr H, Durr S, Ward MP. Canine rabies in Australia: a review of preparedness and research needs. Zoonoses Public Hlth. 2014;62:237–53.
12. Ellison JA, Gilbert AT, Recuenco S, Moran D, Alvarez DA, et al. Bat rabies in Guatemala. PLoSNegl Trop Dis. 2014;8(7):e3070. https://doi.org/10.1371/journal.pntd.0003070.
13. Da Rosa ES, Kotait I, Barbosa TF, Carrieri ML, Brandão PE, Pinheiro AS, et al. Bat-transmitted human rabies outbreaks, Brazilian Amazon. Emerg Infect Dis. 2006, 12(8): 1197 [PMC free article] [PubMed].
14. Mendes W. An outbreak of bat-transmitted human rabies in a village in the Brazilian Amazon. Rev Saude Publica. 2009;43:1075–7.
15. Kotait I, Carrieri ML, Carnieli Júnior P, Castilho JG, Oliveira RDN, Macedo CI, Achkar SM. Wildlife reservoirs of rabies virus: a new challenge to a public health. BEPA. Boletim Epidemiológico Paulista (Online). 2007;4(40):02–8.
16. Shi Z. Bat and virus. ProteinCell. 2010;1(2):109–14. https://doi.org/10.1007/s13238-010-0029-7.
17. Sodré MM, Gama ARD, Almeida MFD. Updated list of bat species positive for rabies in Brazil. Rev Inst Med Trop Sao Paulo. 2010;52(2):75–81.
18. Wada MY, Rocha SM, Maia-Elkhoury ANS. Rabies situation in Brazil, 2000 to 2009. Epidemiologia e Serviços de Saúde. 2011;20(4):509–18. https://doi.org/10.5123/S1679-49742011000400010
19. Frymus T, Addie D, Belák S, Boucraut-Baralon C, Egberink H, Gruffydd-Jones T, Hartmann K, Hosie MJ, Lloret A, Lutz H, Marsilio F, Pennisi MG, Radford AD, Thiry E, Truyen U, Horzinek MC. Feline rabies. ABCD guidelines on prevention and management. J Feline Med Surg. 2009;11:585–93. https://doi.org/10.1016/j.jfms.2009.05.007.
20. Morikawa VM, Ribeiro J, Biondo AW, Fellini A, Bier D, Molento MB. Cat infected by a variant of bat rabies virus in a 29-year disease-free urban area of southern Brazil. Rev Soc Bras Med Trop. 2012;45:255–6.
21. IBGE. The Brazilian Institute of Geography and Statistics. Statistical Yearbook. Brasília: Instituto Brasileiro de Geografia e Estatística. http://www.censo2010.ibge.gov.br/amostra/. Accessed 10 May 2018.
22. Curitiba. City Hall of Curitiba. Prefeitura Municipal de Curitiba http://www.curitiba.pr.gov.br/conteudo/meio-ambiente-de-curitiba/182. Accessed 10 May 2018.
23. Gardner AL. Mammals of South America. Vol. 1. Marsupials, xenarthrans, shrews, and bats. Chicago: University of Chicago Press; 2008.
24. Gregorin R, Taddei VA. Chave Artificial para a Identificação de Molossídeos Brasileiros (Mammalia, Chiroptera). MastoNeotrop. 2002;9:13–32.
25. Dean DJ, Abelseth MK, Atanasiu P. The fluorescent antibody test. Laboratory techniques in rabies. 1996;4:88–95.
26. Koprowski H. The mouse inoculation test. In: Meslin FX, Kaplan MM, Koprowski H, editors. Laboratory techniques in rabies. Geneva: World Health Organization; 1996. p. 80–7.
27. Morettin PA, Toloi CMC. Análise de séries temporais. 2a Ed. São Paulo: Edgard Blücher; 2006.
28. Ryan E, Andrew H. RDSTK: An R wrapper for the Data Science Toolkit API. R package version. 2013;1.1. https://cran.r-project.org/web/packages/RDSTK/index.html. Accessed 10 May 2018.
29. R Core Team. R: A language and environment for statistical computing. Vienna: R Foundation for Statistical Computing; 2015. https://www.R-project.org/. Accessed 10 May 2018.
30. ESRI. ArcGIS Desktop: Release 10. 2011; Redlands, CA: Environmental Systems Research Institute.
31. Anselin L. Exploratory spatial data analysis in a geocomputational environment. In: Longley, brooks, McDonnell, Geocomputation: a primer. London: Macmillan; 1998. p. 77–94.

32. Schneider, M. Estudo de avaliação sobre área de risco para a raiva no Brasil. Rio de Janeiro, 1990 (Doctoraldissertation, Dissertação de Mestrado-Escola Nacional de Saúde Pública-FIOCRUZ].[Links]).

33. Dacheux L, Larrous F, Mailles A, et al. European bat lyssavirus transmission among cats, Europe. Emerg Infect Dis. 2009;15(2):280–4. https://doi.org/10.3201/eid1502.080637.

34. Genaro G. Gato doméstico: futuro desafio para controle da raiva em áreas urbanas? Pesq Vet Bras. 2010;30(2):186–9. https://doi.org/10.1590/S0100-736X2010000200015.

35. Kotait I, Carrieri ML, Takaoka NY. Raiva: Aspectos gerais e clínica. In: Manual Técnico do Instituto Pasteur (no. 8). Instituto Pasteur; 2009.

36. Paez A, Polo L, Heredia D, Nuñez C, Rodriguez M, Agudelo C, Parra E, Paredes A, Moreno T, Rey G. Brote de rabia humana transmitida por gato enelmunicipio de Santander de Quilichao, Colombia, 2008. Revista de Salud Pública. 2009;11(6):931–43. https://doi.org/10.1590/S0124-00642009000600009

37. Bustos Claro MM, Ávila Álvarez AA, Carrascal B, José E, Aguiar Martínez LG, Meek Benigni E, et al. Encephalitis due to rabies secondary to the bite of a cat infected with a rabies virus of Silvester Origen. Infection. 2013;17(3):167–70.

38. Plowright RK, Eby P, Hudson PJ, Smith IL, Westcott D, Bryden WL, McCallum H. Ecological dynamics of emerging bat virus spillover. Proc R Soc B Biol Sci. 2014;282(1798):20142124. https://doi.org/10.1098/rspb.2014.2124.

39. Wood JLN, Leach M, Waldman L, MacGregor H, Fooks AR, Jones KE, et al. A framework for the study of zoonotic disease emergence and its drivers: spillover of bat pathogens as a case study. Philosophical Transactions of the Royal Society B: Biological Sciences. 2012;367(1604):2881–92. https://doi.org/10.1098/rstb.2012.0228.

40. Romano MC, Maidagan JI, Pire F, E. Behavior and demography in an urban colony of Tadarida brasiliensis (Chiroptera: Molossidae) in Rosario, Argentina. Rev Biol Trop. 1999;47(4):1121–7.

41. Esbérard C. Composição de colônia e reprodução de Molossusrufus (E. Geoffroy) (Chiroptera, Molossidae) em um refúgio no sudeste do Brasil. Revista Brasileira de Zoologia. 2002;19(4):1153–60. https://doi.org/10.1590/S0101-81752002000400021

42. Steece R, Altenbach JS. Prevalence of rabies specific antibodies in the Mexican free-tailed bat (Tadaridabrasiliensismexicana) at lava cave, New Mexico. J Wild I Dis. 1989;25:490–6.

43. De Lucca T, Rodrigues RCA, Castagna C, Presotto D, De Nadai V, Fagre A, Braga GB, Guilloux AGA, Alves AJS, Martins CM, Amaku M, Ferreira F, Dias RA. Assessing the rabies control and surveillance systems in Brazil:an experience of measures toward bats after the halt of massive vaccination ofdogs and cats in Campinas, Sao Paulo. Prev Vet Med. 2013;111(1–2):126–33.

44. Burles DW, Brigham RM, Ring RA, Reimchen TE. Influence of weather on two insectivorous bats in a temperature Pacific northwest rainforest. Can J Zool. 2009;87:132–8.

8

Temporal and spatial distribution of lumpy skin disease outbreaks in Ethiopia in the period 2000 to 2015

W. Molla[1,2]*, M. C. M. de Jong[1] and K. Frankena[1]

Abstract

Background: Lumpy skin disease (LSD) is an infectious viral disease of cattle caused by a virus of the genus *Capripoxvirus*. LSD was reported for the first time in Ethiopia in 1981 and subsequently became endemic. This time series study was undertaken with the aims of identifying the spatial and temporal distribution of LSD outbreaks and to forecast the future pattern of LSD outbreaks in Ethiopia.

Results: A total of 3811 LSD outbreaks were reported in Ethiopia between 2000 and 2015. In this period, LSD was reported at least once in 82% of the districts ($n = 683$), 88% of the administrative zones ($n = 77$), and all of the regional states or city administrations ($n = 9$ and $n = 2$) in the country. The average incidence of LSD outbreaks at district level was 5.58 per 16 years (0.35 year^{-1}). The incidence differed between areas, being the lowest in hot dry lowlands and highest in warm moist highland. The occurrence of LSD outbreaks was found to be seasonal. LSD outbreaks generally have a peak in October and a low in May. The trend of LSD outbreaks indicates a slight, but statistically significant increase over the study period. The monthly precipitation pattern is the reverse of LSD outbreak pattern and they are negatively but non-significantly correlated at lag 0 ($r = -0.05$, $p = 0.49$, Spearman rank correlation) but the correlation becomes positive and significant when the series are lagged by 1 to 6 months, being the highest at lag 3 ($r = 0.55$, $p < 0.001$). The forecast for the period 2016–2018 revealed that the highest number of LSD outbreaks will occur in October for all the 3 years and the lowest in April for the year 2016 and in May for 2017 and 2018.

Conclusion: LSD occurred in all major parts of the country. Outbreaks were high at the end of the long rainy season. Understanding temporal and spatial patterns of LSD and forecasting future occurrences are useful for indicating periods when particular attention should be paid to prevent and control the disease.

Keywords: Ethiopia, Lumpy skin disease, Time series, Spatial, Temporal, Forecast

Background

Lumpy skin disease (LSD) is an infectious viral disease of cattle caused by LSD virus of the genus *Capripoxvirus* and the disease often occurs as epidemics. It has spread from Zambia, where it was first observed in 1929 to most African countries (except Libya, Morocco, Algeria and Tunisia), Middle Eastern countries, and more recently also to European countries [1–5]. LSD can occur in diverse ecological zones from the very dry semi-desert, the wet and dry areas to the high altitude temperate areas [1].

LSD was introduced in Ethiopia, for the first time, through north-west (Gojjam and Gondar) in 1981 with subsequent introductions in the West (Wollega) in 1982 from Sudan and in the central part (Shewa) in 1983 [6]. After the introduction, the disease initially spread Eastwards, later to all directions and currently it has affected all regions and agro-climatic zones of the country [6–8]. The spread of LSD was enhanced by uncontrolled cattle movements, communal grazing and watering, and pastoralism [2, 7]. The poor animal health situation, inefficient prevention and control efforts in combination with late detection of the disease have further contributed to the spread of LSD in Ethiopia [2, 8].

In general, the temporal pattern of disease occurrence can be described with short-term, cyclical and seasonal,

* Correspondence: wassie.abebe@wur.nl; mollawassie@yahoo.com
[1]Quantitative Veterinary Epidemiology, Wageningen University & Research, Droevendaalsesteeg 1, 6708 PB Wageningen, The Netherlands
[2]Faculty of Veterinary Medicine, University of Gondar, P.O. Box 196, Gondar, Ethiopia

and long-term trends; time series analysis is a frequently used method to assess these temporal patterns [9]. The cyclical trends are associated with regular, periodic fluctuations in the level of disease occurrence. A seasonal trend is a special case of a cyclical trend, where the periodic fluctuations in disease incidence are related to particular seasons [9]. Seasonal variation in the occurrence of infectious diseases is a common phenomenon in both temperate and tropical climates. Seasonal changes in vector abundance are well-known causes of seasonality of vector-borne infections. A good knowledge on the seasonal variation of disease outbreaks has paramount importance for the understanding of the dynamics of the disease and in designing better control strategies [10].

Field observations and experimental studies indicate that blood feeding arthropods are involved as passive vectors in the transmission of LSD virus [2, 11]. The spread of LSD has been frequently associated with epidemics [12]. Epidemics of LSD occurred during the rainy season in which the arthropod vector populations are abundant while LSD incidence sharply drops during the dry and cold weather seasons [1, 13, 14]. Seasonal variation in the incidence of LSD outbreaks is common in Ethiopia in which it occurs most frequently between September and December [15]. Resurgence of the disease has been consistently associated with the high rainfall, emergence of large numbers of vectors and a low level of herd immunity [13, 16]. Epidemics of LSD were reported to recur at intervals of 5 or 6 years [13]. The reoccurrence of the disease in provisionally free area is possible when the infection is introduced into the population and the reproduction ratio (R), the average number of secondary cases caused by a single typical infectious individual, becomes greater than one [17].

Animal disease monitoring data is of fundamental importance to know the disease status of a country. In Ethiopia, the disease monitoring is mainly passive as most of the disease outbreaks reported to the federal veterinary services are based on clinical observations [18]. Monitoring of livestock diseases in the field is the responsibility of regional animal health services, regional veterinary laboratories and district animal health personnel. Disease investigations are generally conducted in response to reports of health problems from livestock owners. There is a regular follow up of disease outbreaks but the monthly livestock disease reporting rate is less than 47% which is below the required OIE (world organization for animal health) standards of at least 80% [19].

Assessing the spatial and temporal patterns is a prerequisite for guiding successful surveillance and control efforts in a country. Therefore, the objectives of this study were to evaluate the spatial and temporal distribution of LSD outbreaks and to forecast future patterns of outbreaks in Ethiopia based on data reported over the period 2000–2015.

Methods
Study area
Ethiopia is located in Eastern Africa bordering with Sudan, Eritrea, Djibouti, Somalia, Kenya, and South Sudan. It is a federation of nine member regional states (Tigray, Afar, Amhara, Oromia, Benshangul-Gumuz, Gambella, Southern Nations Nationalities and Peoples Region (SNNP), Harari, and Somali) and two city administrations (Addis Ababa and Dire Dawa). The regional states and city administrations are further divided into zones and the zones into woredas (districts), and the woredas into kebeles. As a whole there are about 15,000 kebeles (5000 urban dwellers associations in towns and 10,000 peasant associations in rural areas) in the country [20–22]. The country's territory presents a diverse topography, ranging from 116 m below sea level at the Dallol Depression, in the East, to 4620 m above sea level on the Ras Dashen in the North and covers an area of approximately 1.1 million km^2. Ethiopia is broadly divided into three climatic zones: "Kolla" (the hot lowland zone below 1500 m); "Weyna Dega" (mid highland zone between 1500 and 2400 m); and "Dega" (the cool highlands zone above 2400 m). Average daily temperature ranges from 20 °C to 30 °C. Rainfall ranges from 200 mm to 2000 mm per year. Ethiopia receives heavy rainfall in June, July and August and occasional showers in February and March. In general, the highlands of Ethiopia receive more rain than the lowlands [20, 21].

The total cattle population of the country is estimated to be about 56.71 million heads, mostly local breeds (98.7%); the remaining are hybrid (1.2%) and exotic breeds (0.1%) [23]. The livestock production system practiced in the country is usually extensive. In the highland and mid highland, it is highly integrated with crop production where cattle are primarily kept for traction purpose and to provide milk and meat as by-products. In the lowland, where no or little farming is practiced, pastoralists and agro-pastoralists keep cattle to provide mainly milk [23, 24].

Outbreak and weather data
LSD is a notifiable disease and it is required that all occurrences of this disease be reported. LSD outbreak data were obtained from the Federal Veterinary Services Directorate of Ethiopia for the period 2000–2015. The records contained information on place, time, number of cases, number of deaths and number of animals at risk for each month. The reporting format enables calculation of the temporal and spatial distribution of LSD. An outbreak is defined as one or more bovines showing LSD symptoms in a specified geographical area (usually Kebele). During the 16 years period, no significant changes in operation of the veterinary organization that could have affected the level of reporting from the field were noted.

The LSD outbreak incidence was established at district (woredas, n = 683) level using the 16 years outbreak data. The mean LSD outbreak incidence in a district was calculated by summing all reported LSD outbreaks in a district over the study period and divide it by 16. The geographical distribution of LSD outbreaks over the 16 years was mapped by administrative zone using GIS software QGIS 2.2 (QGIS developer team, Open Source Geospatial Foundation, 2014). The spread of the epidemic was also shown using SPMAP (South Platte Mapping and Analysis Program, Stata 14) by superimposing the yearly outbreak data onto Ethiopian Woreda 2008 shape files in Microsoft power point program.

The monthly mean precipitations for the period 1999–2013 were obtained from the Global weather data for SWAT (Soil, Water, and Air Team) website. From a meteorological point of view, three seasons can be distinguished in Ethiopia; 'Belg' (February to May), 'Kiremt' (June to September) and 'Bega' (October to January). 'Kiremt' is the main rainy season in which the magnitude of rainfall is highest as compared to the other seasons for many parts of the country [25].

Data analysis

Data on the number of LSD outbreaks reported each month during the 16-year study period were analysed to detect temporal trends and seasonal effects. A simple inspection of the graph of the original LSD outbreak time series was employed to appreciate the presence of a clear long-term trend or seasonal effect. The existence of a long-term trend in LSD outbreaks was modelled by linear regression (STATA version 14) using the number of LSD outbreaks (or trend component of the outbreak) as dependent variable and month of the outbreaks as explanatory variable. Spectral analysis with SAS (Statistical Analysis Software) 9.4 was performed to detect seasonality and cyclical patterns in the LSD outbreak time series.

Decomposition of LSD outbreak time series was performed using package 'TTR' (Technical Trading Rules) in R software, to identify and estimate the three components of the temporal additive model: seasonality, long-term trend, and irregularity [26].

The time series were also seasonally differenced (i.e. deducting the 12 months earlier observation value from each observation value) first followed by first order trend differencing (i.e. deducting the preceding observation value from each observation value) according to the procedure described by Allard [27] and Coghlan [26] to make the series stationary (diff function in R). Next autocorrelation analysis (Autocorrelation function in R) was used to assess the seasonality of the differenced time series. The autocorrelation function (ACF) enables to test the significance of seasonality in a time series by examining the ACF correlogram at lags of 12 month intervals [28]. The ACF estimates the correlation between the number of outbreaks reported in a given month and the number of outbreaks reported in each of the previous 1 to 192 months. The autocorrelations and partial autocorrelations values of various lags were used for the selection of terms to be included in the initial autoregressive integrated moving average (ARIMA) model (autocorrelations and partial autocorrelations functions in R).

The Holt-Winters exponential smoothing technique as described by Coghlan [26] was applied to make short term (36 months) forecasts using package 'forecast' in R software. The possibility of improving the predictive model was evaluated by making a correlogram and carrying out the Ljung-Box test on the in-sample forecast errors for evidence of non-zero autocorrelations at lags 1 to 20. In this method the estimates of the parameters alpha, beta, and gamma represents the level, the slope of the trend component, and the seasonal component, respectively at the current time point. All the three parameters have values between 0 and 1. Parameter values that are close to 0 indicate that relatively little weight is placed on the most recent observations while forecasting future values.

Exponential smoothing methods are useful for making forecasts, but it does not take into account the correlations between successive values of the time series. However, a better predictive model can be made by taking correlations in the data into account. ARIMA models include an explicit statistical model for the irregular component of a time series that allows for non-zero autocorrelations in the irregular component. An ARIMA $(1, 1, 1) \times (1, 1, 1)12$ model [26, 29] seemed a plausible model for the LSD outbreak stationary time series and this model was used to forecast the expected numbers of LSD outbreaks for a 36 month (January 2016 to December 2018) future time using the "forecast.Arima()" function in the "forecast" R package. Finally, it was investigated whether or not successive forecast errors of an ARIMA $(1, 1, 1) \times (1, 1, 1)12$ models were correlated by making a correlogram and carrying out the Ljung-Box test.

The association between monthly rainfall and monthly LSD outbreaks was tested with Spearman rank test (Stata version 14).

Results

Geographical distribution and incidence of LSD outbreaks

During the period 2000–2015, LSD has been reported from all regional states (n = 9) and city administrations (n = 2) of Ethiopia. About 82% of the districts (n = 683) and 88% of the administrative zones (n = 77) in the country reported at least one LSD outbreak in this time period. In total 3811 LSD outbreaks were reported in Ethiopia during the study period (Additional file 1: Table S1). Most of these outbreaks were from Oromia (54.5%), Amhara (27.9%), SNNP (10.1%) and Tigray regional states (3.6%) (Additional file 2: Figure S1).

The average incidence of LSD outbreaks at district level was 5.58 over all 16 years or 0.35 per district per year. The lowest incidences were observed in the eastern lowland (Afar and Somali), southern lowland (Liben), south-west (Benchi Maji) and North (North western zone of Tigray) areas whereas the highest number of outbreaks were documented in the north-west, central, West and south-western parts of the country (Fig. 1).

The data shows that LSD affects districts for one or two years and then spreads to other nearby districts/areas with a susceptible cattle population. In this fashion the disease moves from one geographical area to the other and circulates in the country (Additional file 3: Figure S2). The reoccurrence of the disease in the study districts varies from 1 year to 13 years, with average length of 4.54 years and median 4 years. The time between outbreaks was shorter in districts geographically located in the West, south-west and central part of the country.

LSD outbreak time series description and analysis

The monthly distribution of LSD outbreaks is presented in Fig. 2, Additional file 1: Table S1 and Additional file 4: Figure S3. It showed a slight increase in the number of monthly outbreaks which was statistically significant ($P < 0.05$) (Fig. 2). The seasonality in the numbers of outbreaks is apparent, which tend to be higher in the months following the long rainy season compared to other seasons (Fig. 2). The undecomposed and undifferenced original LSD outbreak time series was found seasonal by spectral analysis techniques (Fig. 3).

The trend, seasonal and irregular components of the LSD outbreak time series were estimated by decomposing the time series (Fig. 4). The estimated trend component

shows a decrease from about 20 outbreaks in 2002 to about 6 outbreaks in 2006, followed by a substantial increase to about 41 outbreaks in 2009, decrease to about 16 outbreaks at the end of 2013 and finally increase to about 26 outbreaks in 2014. Though the LSD outbreak pattern from the trend component appears to have a cycle with a periodicity of 5–7 years (peaks in 2002, 2009 and 2014) (Fig. 4) it was not established by spectral analysis. Linear regression on the trend component of the decomposed time series shows a statistically significant ($p < 0.001$) increase in monthly LSD outbreak numbers between 2000 and 2015.

The seasonal pattern of LSD outbreaks is clearly indicated in Figs. 3 and 4. Seasonal factors were estimated for each month over the 16 year period as the seasonal component of the decomposed LSD time series. The largest seasonal factor is recorded for October (about 16.8) and the lowest for May (about −12.1), indicating that number of LSD outbreaks peaks in October and has a low (trough) in May (Figs. 4 and 5). In general the number of LSD outbreaks was above average for the months September to January and below average for February to August (Fig. 5).

The rainfall season of Ethiopia is indicated in Figs. 2 and 5. The precipitation is above average from April to September and below average from October to March. The rainfall is high in July and August and low in December to February. The precipitation pattern is the reverse of LSD outbreak pattern (Fig. 5), resulting in a negative correlation coefficient ($r = -0.05$, $p = 0.49$, Spearman rank correlation) at lag 0 and positive correlation coefficients when the series were lagged by 1 to 6 months, the correlation at lag 3 being the highest ($r = 0.55$, $p < 0.001$).

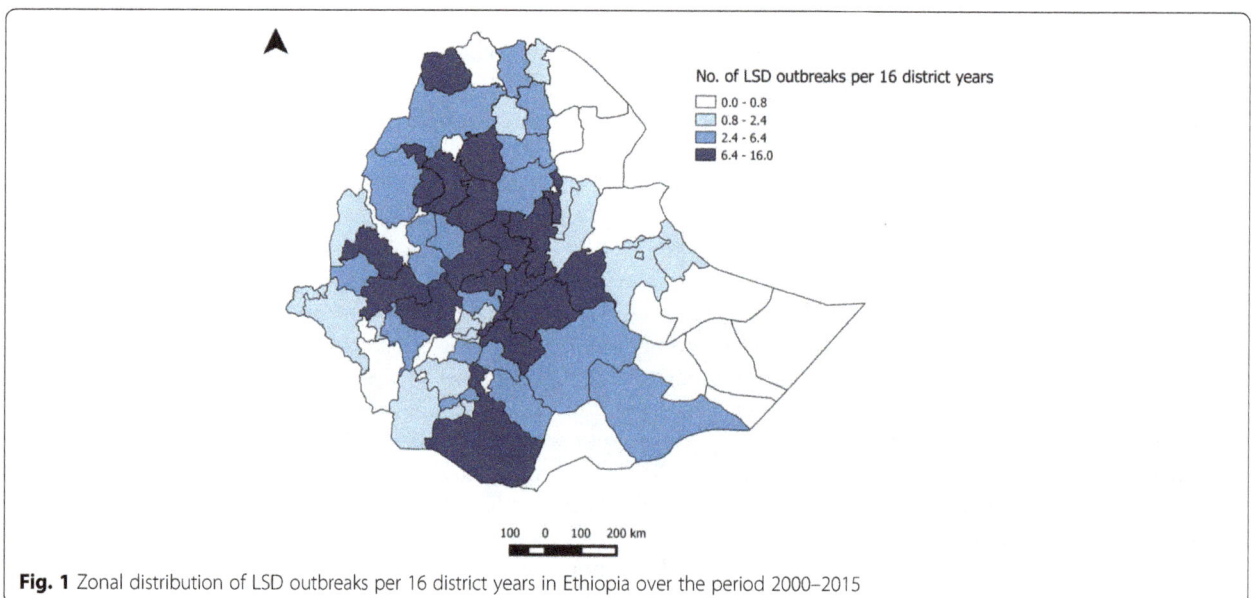

Fig. 1 Zonal distribution of LSD outbreaks per 16 district years in Ethiopia over the period 2000–2015

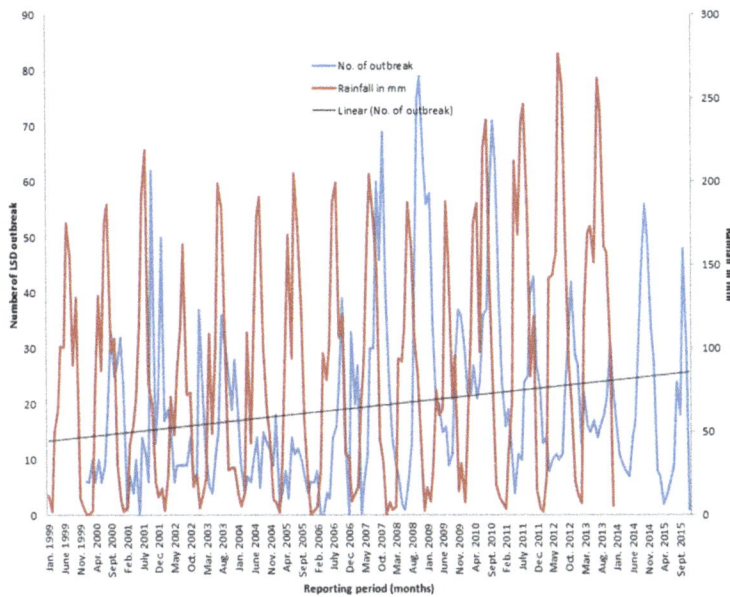

Fig. 2 Monthly outbreak and trend of LSD from 2000 to 2015 and monthly rainfall in millimetre from 1999 to 2013 in Ethiopia

LSD outbreak times series forecasting

For forecasting with Holt-Winters exponential smoothing, the three parameters: alpha, beta, and gamma which are important for forecasting future values were 0.56, 0.00, and 0.32, respectively. The original LSD outbreak times series and the forecasted values plotted using Holt's exponential smoothing is shown in Additional file 5: Figure S4. The future times, from January 2016 to December 2018 were also forecasted with Holt-Winters' exponential methods (Additional file 6: Figure S5). However, the correlogram and Ljung-Box test showed the presence of significant ($P = 0.002$) autocorrelations of the in-sample forecast errors at lags 1–20. This indicates that Holt-Winters exponential smoothing could not provide an adequate forecast.

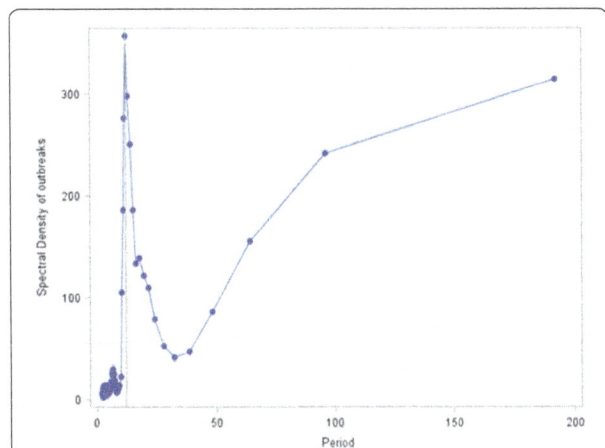

Fig. 3 Spectral density estimates of LSD outbreaks by month, the vertical reference line at the 12 month period shows the seasonality of the disease

The LSD outbreak time series was differenced for trend and seasonality, and the resulting series of first order differences appeared to be stationary in mean and variance. The ACF correlogram of first differenced LSD time series indicates significant autocorrelation at lag 1 (−0.349), 12 (−0.468), and 13 (0.273) (Additional file 7: Figure S6A). This demonstrates the seasonality of the series because the current monthly value is related to the value of 12 months earlier. The partial correlogram also shows that the partial autocorrelations at lags 1 (−0.349), 9 (−0.199), 12 (−0.487) and 15 (0.185) exceed the significance bounds (Additional file 7: Figure S6B). Hence, the ARIMA model (1, 1, 1) x (1, 1, 1)12 was used for making forecasts for the number of LSD outbreaks from January 2016 to December 2018 (Fig. 6). The correlogram for the forecasted value shows that none of the sample autocorrelations for lags 1–20 exceed the significance bounds, and the p-value for the Ljung-Box test is 0.107, so it can be concluded that there is little evidence for non-zero autocorrelations in the forecast errors at lags 1–20. Based on the forecast, the highest numbers of LSD outbreaks are expected in October for all predicted years and the lowest in April for 2016 and May for 2017 and 2018 (Additional file 8: Table S2).

Discussion

In the current study, LSD has been recorded from all regional states and city administrations in Ethiopia. A previous retrospective study that covered a period from January 2007 to December 2011 reported no outbreaks from Dire Dawa city administration and Harari regional state [15]. The present study, however, showed that they are affected by the disease.

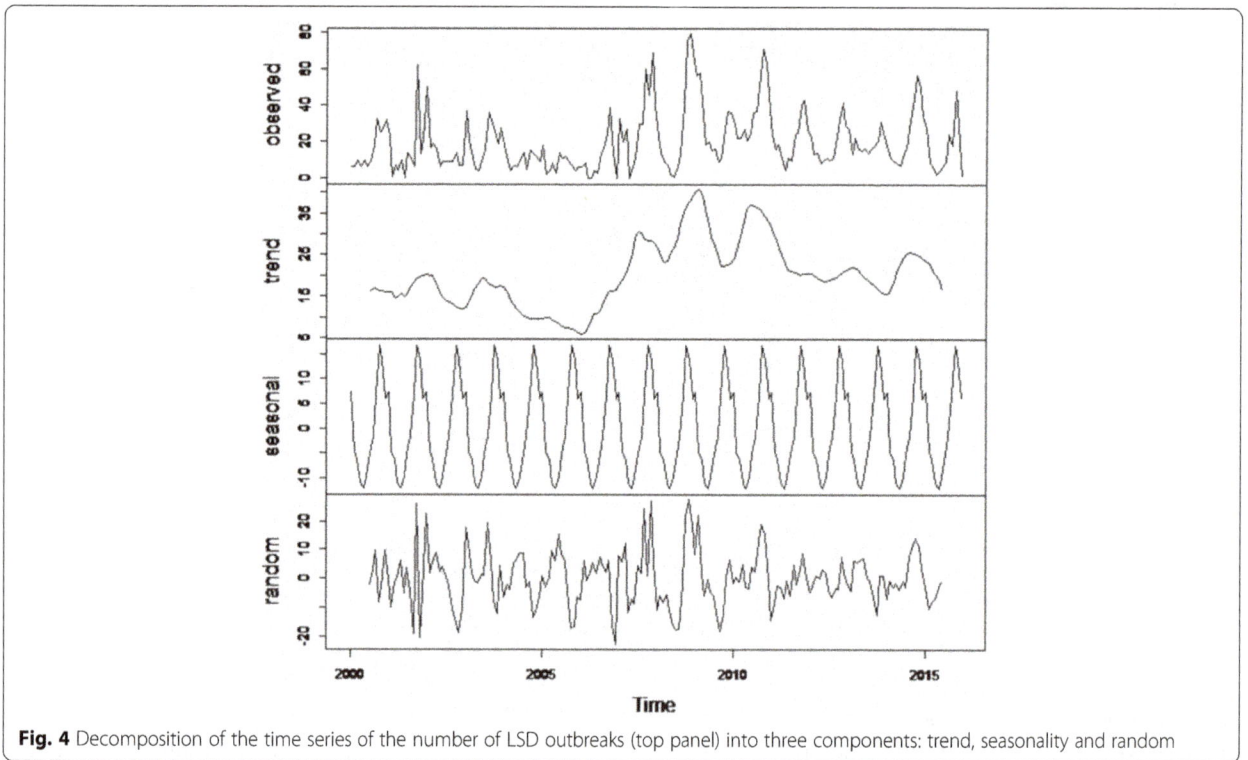

Fig. 4 Decomposition of the time series of the number of LSD outbreaks (top panel) into three components: trend, seasonality and random

Our spatial analysis have shown that distribution of incidence of LSD outbreaks vary among areas (Fig. 1). The highest LSD incidences were in warm moist highland and the lowest in hot dry lowland areas. This indicate that the parts of the country which receive relatively high rain fall for a reasonable period of time is conducive for the replication and survival of blood feeding arthropods and then for the spread of the disease in the geographical areas [7, 15, 30, 31]. The LSD outbreak incidence indicated for the different zones should be treated consciously because under reporting might result in an underestimated incidence.

In this study, it became clear that the occurrence of LSD in an area/districts is sporadic. However, endemicity of the disease is maintained in the country because the outbreaks in different districts/area do not occur at the same time (Additional file 3: Figure S2). The average time to reoccurrence was 4.54 years, in line with the 5 yearly reoccurrence of LSD epidemic in unvaccinated populations [31]. The reoccurrence was variable across the study districts. Some districts reported outbreaks after 1 year of quiescence, whereas others reported an outbreak after a much longer period (up to 13 years), which is in line with Gari et al. [7]. This indicates that the disease is not endemic in a district/an area but it occurs in an outbreak (epidemic) form after some years. The reoccurrence is only possible after the seroprevalence (herd immunity) dropped below the critical value and reproduction ratio

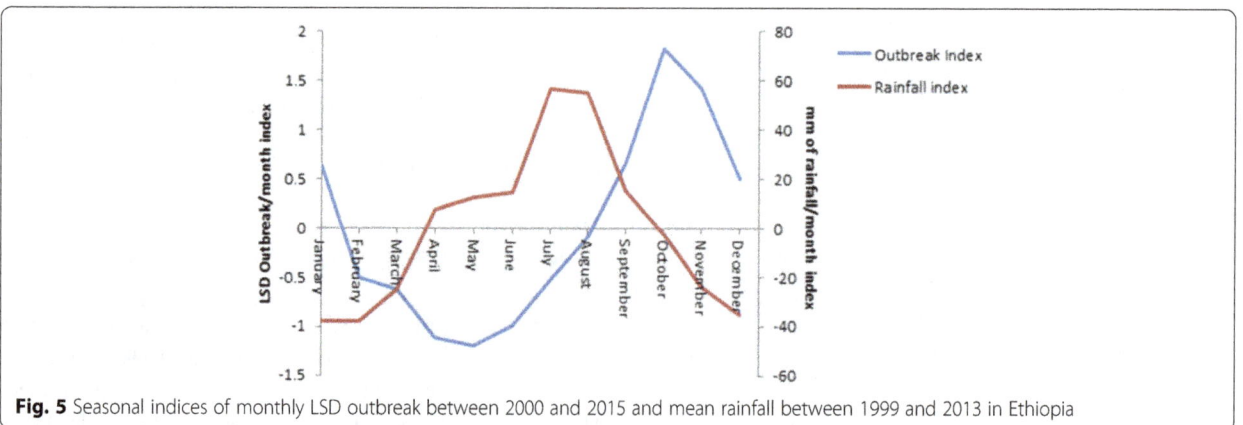

Fig. 5 Seasonal indices of monthly LSD outbreak between 2000 and 2015 and mean rainfall between 1999 and 2013 in Ethiopia

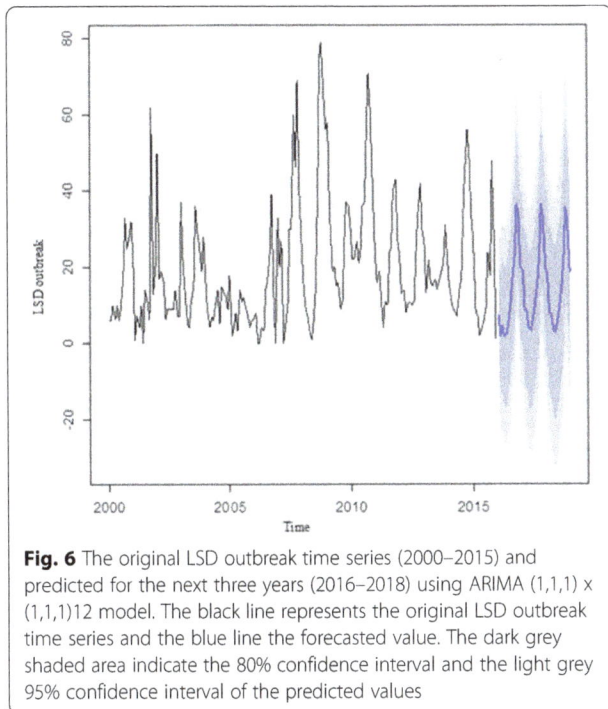

Fig. 6 The original LSD outbreak time series (2000–2015) and predicted for the next three years (2016–2018) using ARIMA (1,1,1) x (1,1,1)12 model. The black line represents the original LSD outbreak time series and the blue line the forecasted value. The dark grey shaded area indicate the 80% confidence interval and the light grey 95% confidence interval of the predicted values

(R) is above one. How long it will take depends on the rate at which LSD is introduced into the district/area (spark rate) and how far the R has increased above one [17]. The time between outbreaks was shorter in the West and south-west (where rainfall occurs for extended periods of time) and central (where live animal from different parts of the country cross through to the central market) parts of the country.

LSD outbreaks do also not occur at random in time and we demonstrated the seasonality by spectral analysis (Fig. 3) and estimated a significant autocorrelation between LSD outbreaks at lag 12, 24, 36, etc., indicating the seasonality of the disease. The seasonal pattern of the disease is also clearly indicated in Fig. 4. The seasonal LSD outbreak variation might be related to the variation in temperature and rainfall between seasons leading to varying arthropod densities in the environment. Seasonal variations in vector abundance, including mosquitoes, ticks, and flies, are well known causes of seasonality for vector borne diseases [10]. Identification of temporal patterns can indicate times when particular attention should be paid to control the disease [9].

The trend of LSD outbreaks from January 2000 to December 2015 indicates a slight, but statistically significant increase over the period (Figs. 2 and 4). This might be attributed to the absence of a specific national strategy for LSD control or eradication [18] and the increased tendency of using irrigation for crop cultivation that create favourable environmental conditions for vector borne diseases in the country. The implication of the

trend component is that the disease will continue to persist if the environmental circumstances and the poor disease control activities continue.

The positive and significant cross correlations between precipitation and increased LSD outbreaks at lag 3 ($r = 0.55$, $p < 0.001$) suggests that the rainfall in the previous months are an important factor for the occurrence of LSD outbreaks. The time delay for LSD outbreaks to occur might be justified by the time required for the build-up of arthropod population following the rains [32, 33], incubation period (2–4 weeks [34]) of the virus within the cattle host and delay in reporting.

Based on the 2000–2015 reports, the number of LSD outbreaks to occur in each month from 2016 to 2018 was forecasted. The forecast suggests that high number of LSD outbreak will occur from August to January and this is comparable with the available LSD outbreak time series data. The reappearance of the seasonality in the original time series again in the forecasts is an indication of the forecast is reasonable [27]. The wide confidence interval (Fig. 6) indicates the need of frequent updating of the model by incorporating the latest outbreak reports [27]. The confidence interval was even wider when Holt-Winters' exponential methods were used (Additional file 6: Figure S5). The wider confidence interval is related to a limitation of this method, i.e. it does not take the correlations between successive values of the time series into account. The ARIMA model, taking the correlations in the data into account, therefore, is the preferred model to get a reasonable forecast in this study [26]. The forecasting process can be continued to any point in the future, but will become less reliable for predictions further in time [29]. This means we can only gain advantage from the use of short term forecasts. The forecasted results of this study, therefore, will alert and help policy makers to focus on the unusual situations to decide whether any disease control intervention is required to halt the occurrence of the disease in the future.

Currently, Ethiopia has no a well-designed control strategy for LSD [18]. The animal health authority undertake reactive vaccination campaign using Kenyan sheep pox vaccine when an LSD outbreak is reported somewhere in the country. Vaccination is the only measure taken for LSD control. However, research findings indicate that the vaccine used in Ethiopia is not fully protective [35] which might be the reason for the increase in incidence of LSD outbreaks observed over the current study period. Because there is no regular vaccination program against LSD this might attribute to a drop of herd immunity below the critical point and for the reoccurrence of the disease. We now understood that LSD does not establish endemicity in an area, but it recurs as epidemic in, at average, every 5 years. Therefore, outbreaks might be prevented by bringing the herd immunity above the critical level through vaccination and by prohibiting the entrance of infected animals to the provisionally free area. Vaccination should be undertaken regularly ahead of the onset of the main rainy season with a high coverage. The vaccine currently in use shall be replaced by more competent homologous (Neethling virus) vaccine [36]. It is widely agreed that vaccination is the most manageable and realistic approach to control the disease in endemic and resource poor countries. However, to be more effective, the vaccination should be complemented by other additional measures such as movement control.

Conclusion

LSD is wide spread and well established in Ethiopia. It occurred in all regional states and city administrations in the time period between 2000 and 2015. LSD does not establish endemicity in a district, but it does in the country as a whole. It recurs in a district as epidemic, on average in 5 years period. The average incidence of LSD outbreaks at district level was 5.58 over all 16 years. The trend of LSD outbreaks increased over time. Outbreaks are seasonal and occurred more often in the months following the long rainy season. The results of the spatiotemporal analysis and the forecasted value may serve as a guide for the routine surveillance of LSD in the country.

Additional files

Additional file 1: Table S1. Number of LSD outbreaks reported monthly over the period 2000–2015 in Ethiopia. (DOCX 17 kb)

Additional file 2: Figure S1. Distribution of LSD outbreaks (n = 3811) over regional states and city administrations in the period 2000–2015. (DOCX 14 kb)

Additional file 3: Figure S2. Animation of the spread of lumpy skin disease epidemics in Ethiopia, 2000–2015. (DOCX 2749 kb)

Additional file 4: Figure S3. Annual course of LSD outbreaks in Ethiopia, 2000–2015. (DOCX 169 kb)

Additional file 5: Figure S4. The original LSD outbreak time series (black) and the predicted values (red) using Holt-Winters filtering. (DOCX 27 kb)

Additional file 6: Figure S5. LSD outbreak forecasts based on Holt-Winters analysis for January 2016 to December 2018. (DOCX 24 kb)

Additional file 7: Figure S6. ACF (A) and Partial ACF (B) correlogram after first order seasonal and trend differencing of the original LSD outbreak time series. (DOCX 21 kb)

Additional file 8: Table S2. 36 month forecast of the number of LSD outbreaks based on ARIMA (1, 1, 1) x (1, 1, 1)12. (DOCX 17 kb)

Abbreviations
ACF: Autocorrelation Function; ARIMA: Autoregressive Integrated Moving Average; LSD: Lumpy Skin Disease; OIE: Office International des Epizooties (world organization for animal health); R: Reproduction ratio; SAS: Statistical Analysis Software; SNNP: Southern Nations Nationalities and Peoples Region; SPMAP: South Platte Mapping and Analysis program; SWAT: Soil, Water, and Air Team; TTR: Technical Trading Rules

Acknowledgements
The authors are very much grateful to Nuffic for financing the study. We would like to acknowledge the Ethiopian Veterinary Services Directorate for allowing access to its LSD outbreak report records, and Dr. Bewket Sirawbezu, Dr. Getachew Gari and Dr. Yesmashewa Wogayehu for facilitating access to the database.

Funding
This work was financed by Nuffic (Netherlands organization for international cooperation in higher education). The funding organization has no role in the design of the study and collection, analysis, and interpretation of data and in writing the manuscript.

Authors' contributions
WM participated in the planning of the study, obtained the data, performed the analyses and drafted the manuscript. KF and MCMdJ participated in the planning of the study, analysed and interpreted the data, and revised the manuscript. All authors read and approved the final manuscript.

Consent for publication
Not applicable.

Competing interests
The authors declare that they have no competing interests.

References
1. Davies FG. Lumpy skin disease, an African capripox virus disease of cattle. Br Vet J. 1991;147:489–503.
2. Tuppurainen ES, Oura CA. Review: lumpy skin disease: an emerging threat to Europe, the Middle East and Asia. Transbound Emerg Dis. 2012;59:40–8.
3. Tuppurainen ES, Venter EH, Shisler JL, Gari G, Mekonnen GA, Juleff N, Lyons NA, De Clercq K, Upton C, Bowden TR et al. Review: Capripoxvirus diseases: current status and opportunities for control. Transbound Emerg Dis. 2015; doi:101111/tbed12444.
4. Tasioudi KE, Antoniou SE, Iliadou P, Sachpatzidis A, Plevraki E, Agianniotaki EI, Fouki C, Mangana-Vougiouka O, Chondrokouki E, Dile C. Emergence of lumpy skin disease in Greece, 2015. Transbound Emerg Dis. 2016;63:260–5.
5. WAHIS. Summary of immediate notifications and follow-ups. World animal health information database (WAHIS interface). 2016. http://www.oie.int/wahis_2/public/wahid.php/Diseaseinformation/Immsummary. Accessed 18 July 2016.
6. Mebratu GY, Kassa B, Fikre Y, Berhanu B. Observation on the outbreak of lumpy skin disease in Ethiopia. Rev Elev Méd Vét Pays Trop. 1984;37:395–9.
7. Gari G, Waret-Szkuta A, Grosbois V, Jacquiet P, Roger F. Risk factors associated with observed clinical lumpy skin disease in Ethiopia. Epidemiol Infect. 2010;138:1657–66.
8. APHRD. Ethiopia animal health yearbook 2011, Animal and plant health regulatory directorate (APHRD), Addis Ababa, Ethiopia. 2012.
9. Thrusfield M. Veterinary epidemiology. 3rd ed. Oxford, UK: Blackwell Science; 2007. p. 144–51.
10. Grassly NC, Fraser C. Seasonal infectious disease epidemiology. Proc R Soc B. 2006;273:2541–50.
11. Chihota CM, Rennie LF, Kitching RP, Mellor PS. Mechanical transmission of lumpy skin disease virus by Aedes aegypti (Diptera: Culicidae). Epidemiol Infect. 2001;126:317–21.
12. Carn VM. Control of capripoxvirus infections. Vaccine. 1993;11:1275–9.
13. Woods JA. Lumpy skin disease- A review. Trop Anim Hlth Prod. 1988;20:11–7.
14. Wainwright S, El Idrissi A, Mattioli R, Tibbo M, Njeumi F, Raizman E. Emergence of lumpy skin disease in the eastern mediterranean basin countries.FAO. 2013. http://www.fao.org/docrep/019/aq706e/aq706e.pdf. Accessed 20 August 2016.
15. Ayelet G, Haftu R, Jemberie S, Belay A, Gelaye E, Sibhat B, Skjerve E, Asmare K. Lumpy skin disease in cattle in central Ethiopia: outbreak investigation and isolation and molecular detection of lumpy skin disease virus. Rev Sci Tech Off Int Epiz. 2014;33:877–87.

16. Hunter P, Wallace D. Lumpy skin disease in southern Africa: a review of the disease and aspects of control. J S Afr Vet Ass. 2001;72:68–71.

17. Dibble CJ, O'Dea EB, Park AW, Drake JM. Waiting time to infectious disease emergence. J R Soc Interface. 2016;13:20160540.

18. APHRD. Biannual epidemiology newsletter, animal and plant health regulatory directorate (APHRD), Ministry of Agriculture, Addis Ababa, Ethiopia. 2012.

19. APHRD. Ethiopia animal health yearbook (2009/10), animal and plant health regulatory directorate (APHRD), Ministry of Agriculture, Addis Ababa, Ethiopia. 2010.

20. Tadesse D, Desta A, Geyid A, Girma W, Fisseha S, Schmoll O. Rapid assessment of drinking water quality in the Federal Democratic Republic of Ethiopia: country report of the pilot project implementation in 2004–2005. World Health Organization (WHO) and UNICEF. 2010; http://www.wssinfo.org/fileadmin/user_upload/resources/RADWQ_Ethiopia.pdf. Accessed 9 July 2016

21. GoE. Ethiopian government portal: official web gateway to the government of Ethiopia (GoE). 2016. http://www.ethiopia.gov.et. Accessed 8 November, 2016.

22. Mbogo CM. Current status of entomological monitoring and surveillance for an effective delivery of vector control interventions in Ethiopia. Produced for review by the United States Agency for International Development 2012. http://pdf.usaid.gov/pdf_docs/PA00J33M.pdf. Accessed 9 July 2016.

23. CSA: Agricultural sample survey, 2014/15 (2007 E.C.), volume II: report on livestock and livestock characteristics (private peasant holdings). Statistical bulletin 578. Central statistical agency (CSA), Federal Democratic Republic of Ethiopia, Addis Ababa. 2015.

24. Gari G, Grosbois V, Waret-Szkuta A, Babiuk S, Jacquiet P, Roger F. Lumpy skin disease in Ethiopia: seroprevalence study across different agro-climate zones. Acta Trop. 2012;123:101–6.

25. NMA. Annual climate bulletin, national meteorological agency (NMA), Addis Ababa Ethiopia. 2013. http://www.ethiomet.gov.et/bulletins/view_pdf/348/2013__annual__bulletin.pdf. Accessed 30 June, 2016.

26. Coghlan A. A little book of R for time series, Release 0.2. 2015. https://media.readthedocs.org/pdf/a-little-book-of-r-for-timeseries/latest/a-little-book-of-r-for-time-series.pdf. Accessed 18 May 2016.

27. Allard R. Use of time-series analysis in infectious disease surveillance. Bull WHO. 1998;76:327–33.

28. Courtin F, Carpenter TE, Paskin RD, Chomel BB. Temporal patterns of domestic and wildlife rabies in central Namibia stock-ranching area, 1986-1996. Prev Vet Med. 2000;43:13–28.

29. Emel K. Time-series analysis. https://datajobs.com/data-science-repo/Time-Series-Analysis-Guide.pdf. Accessed 10 October 2016.

30. Ecotravelworldwide. Ethiopia weather and climate zones. http://www.nationalparks-worldwide.info/eaf/ethiopia/ethiopia-weather.html. Accessed 18 May 2016.

31. Woods JA. Lumpy skin disease virus. In: Dinter Z, Morein B, editors. Virus infections of ruminants. Amsterdam: Elsevier Science publishers BV; 1990. p. 53–67.

32. Linthicum KJ, Davies FG, Bailly CL, Kairo A. Mosquito species succession in a dambo in an east African forest. Mosq News. 1983;43:464–70.

33. Stewart Ibarra AM, Ryan SJ, Beltrán E, Mejía R, Silva M, Munõz A. Dengue vector dynamics (*Aedes aegypti*) influenced by climate and social factors in Ecuador: implications for targeted control. PLoS One. 2013;8:11.

34. Radostits OM, Gay CC, Hinchcliff KW, Constable PD. Veterinary medicine: a textbook of the diseases of cattle, sheep, pigs, goats and horses. 10th ed. Spain: Sounders Elsevier; 2007. p. 1424–6.

35. Ayelet G, Abate Y, Sisay T, Nigussie H, Gelaye E, Jemberie S, Asmare K. Lumpy skin disease: preliminary vaccine efficacy assessment and overview on outbreak impact in dairy cattle at Debre Zeit, central Ethiopia. Antivir Res. 2013;98:261–5.

36. Ben-Gera J, Klement E, Khinich E, Stram Y, Shpigel NY. Comparison of the efficacy of Neethling lumpy skin disease virus and x10RM65 sheep-pox live attenuated vaccines for the prevention of lumpy skin disease - the results of a randomized controlled field study. Vaccine. 2015;33:4837–42.

9

Genetic heterogeneity of dolphin morbilliviruses detected in the Spanish Mediterranean in inter-epizootic period

Consuelo Rubio-Guerri[1,2*] (iD), M. Ángeles Jiménez[3], Mar Melero[1], Josué Díaz-Delgado[4], Eva Sierra[4], Manuel Arbelo[4], Edwige N. Bellière[5], Jose L. Crespo-Picazo[2], Daniel García-Párraga[2,6], Fernando Esperón[5] and Jose M. Sánchez-Vizcaíno[1]

Abstract

Background: In the last 20 years, Cetacean Morbillivirus (CeMV) has been responsible for many die-offs in marine mammals worldwide, as clearly exemplified by the three dolphin morbillivirus (DMV) epizootics of 1990–1992, 2006–2008 and 2011 that affected Mediterranean striped dolphins (*Stenella coeruleoalba*). Systemic infection caused by DMV in the Mediterranean has been reported only during these outbreaks.

Results: We report the infection of five striped dolphins (*Stenella coeruleoalba*) stranded on the Spanish Mediterranean coast of Valencia after the last DMV outbreak that ended in 2011. Animal 1 stranded in late 2011 and Animal 2 in 2012. Systemic infection affecting all tissues was found based on histopathology and positive immunohistochemical and polymerase chain reaction positive results. Animal 3 stranded in 2014; molecular and immunohistochemical detection was positive only in the central nervous system. Animals 4 and 5 stranded in 2015, and DMV antigen was found in several tissues. Partial sequences of the DMV phosphoprotein (P), nucleoprotein (N), and hemagglutinin (H) genes were identical for Animals 2, 3, 4, and 5, and were remarkably different from those in Animal 1. The P sequence from Animal 1 was identical to that of the DMV strain that caused the epizootic of 2011 in the Spanish Mediterranean. The corresponding sequence from Animals 2–5 was identical to that from a striped dolphin stranded in 2011 on the Canary Islands and to six dolphins stranded in northeastern Atlantic of the Iberian Peninsula.

Conclusions: These results suggest the existence of an endemic infection cycle among striped dolphins in the Mediterranean that may lead to occasional systemic disease presentations outside epizootic periods. This cycle involves multiple pathogenic viral strains, one of which may have originated in the Atlantic Ocean.

Keywords: Paramyxoviridae, Endemic ocurrence, Dolphin morbillivirus, *Stenella coeruleoalba*, Mediterranean striped dolphin

Background

Cetacean morbillivirus (CeMV) is an enveloped, negative-strand RNA virus within the genus *Morbillivirus* and family *Paramyxoviridae* [1] that may cause serious respiratory, lymphoid and central nervous system (CNS) disease in susceptible cetacean species, leading to strandings and death. Three main lineages of CeMV have been described, dolphin morbillivirus (DMV) [2], porpoise morbillivirus (PMV) [1], and pilot whale morbillivirus (PWMV). Dolphin and porpoise morbilliviruses caused mass mortalities in several cetacean species, and pilot whale morbillivirus has been reported sporadically in pilot whales [3–5], although this species has been reported also to be infected by DMV [6]. In addition to these three lineages of CeMVs, three more recent strains have been identified in stranded cetaceans [7–9].

* Correspondence: crubio@oceanografic.org
[1]VISAVET Center and Animal Health Department, Veterinary School, Complutense University of Madrid, Avda. Puerta del Hierro s/n, 28040 Madrid, Spain
[2]Fundación Oceanografic de la Comunitat Valenciana, C/. Eduardo Primo Yúfera (Científic) 1B, 46013 Valencia, Spain
Full list of author information is available at the end of the article

The first recognized morbilliviral epizootic of cetaceans (actually the first reported for any marine mammal) occurred in 1987–88 on the Atlantic coast of the USA, killing an estimated 50% of the regional bottlenose dolphins (*Tursiops truncatus*) [10]. In 1990 approximately 1000 Spanish Mediterranean striped dolphins (*Stenella coeruleoalba*) were killed by a DMV epizootic [11]. At least another 100 CeMV-caused deaths of bottlenose dolphins occurred in 1993 in the Gulf of Mexico [12, 13]. Epizootic episodes occurred again in the Spanish Mediterranean in 2007 and 2011, with over 200 and 50 deaths of striped dolphins, respectively [14–17], caused by two variants of the same DMV strain (judged from viral phosphoprotein sequences).

Since 2011, CNS-restricted morbilliviral infection (MI) was reported in several animals stranded on the Italian Mediterranean coast in 2012 [18–20] and mass mortality of striped dolphins and of sperm whales due to sytemic infection were recorded in Italy in 2013 and 2016, respectively [21, 22]. Although isolated cases of systemic infection (several tissues affected) and CNS-restricted infection have been reported in dolphins in the Atlantic Ocean and in Italy in a fin whale [2, 23–25], no systemic infection has been reported in dolphins in the Spanish Mediterranean outside mass outbreaks. Here we report some cases of systemic DMV infection in striped dolphins in the Spanish Mediterranean in the 2011–2016 period, exploiting sequence comparisons of DMV phosphoprotein (P) sequences to suggest potential Atlantic origins for the cases detected from 2012 till now.

Methods

Animals

Of 92 dolphins stranded on the coast of the Valencian Community of the Spanish Mediterranean that were necropsied between 2011 and 2016, only five animals, all of them striped dolphins, were DMV-positive. Of these five, animals 2, 4 and 5 stranded dead, whereas the other two animals died shortly after being found. Table 1 and Fig. 1 give the dates and locations of the strandings and relevant features of the stranded animals.

Necropsy and tissue sampling

Postmortem examinations were performed within 24 h after stranding. Samples of lung, kidney, pulmonary, prescapular and mesenteric lymph nodes, pharyngeal tonsils, urinary bladder, spleen, skin, cerebellum and cerebrum were collected. All these organs, except the mesenteric lymph node and the urinary bladder, were investigated also by RT-PCR for CeMV.

Histopathology and immunohistochemistry

Samples from virtually all organs and tissues were collected and fixed in 10% neutral buffered formalin, processed for histopathology using routine methods, and stained with hematoxylin and eosin (HE) according to standard laboratory procedures. Immunohistochemistry to detect morbilliviral antigen was done on cerebrum, lung, kidney, urinary bladder, stomach and intestine of animal 1; lung and cerebrum of animals 2 and 3; lung and pharyngeal tonsil of animal 4; and lung of animal 5. It was carried out on formalin-fixed, paraffin-embedded tissue (FFPE) sections, using IgG2B-isotype monoclonal antibody against Canine Distemper Virus nucleoprotein (CDV-NP; VMRD® Inc., USA), known to cross-react with CeMV [4, 16, 26]. Sections of FFPE canine brain with known CDV infection and immunopositivity were used as positive controls. For negative controls, the primary antibody was replaced with normal mouse serum [27].

Molecular diagnosis and phylogenetic analysis

For DMV molecular diagnoses we used RNA isolated from homogenized tissues (Bullet Blender™homogeneizer, Next Advance, Averill Park, NY, USA) with the NucleoSpin RNA

Table 1 Animals and stranding information

Animal #	Species	Date and place of stranding		Age Sex	Comments
1	*Stenella coeruleoalba*	July 2011	Alicante 38° 11′ 32.26″ N 0° 33′ 18.75″ W	Adult Female	Stranded alive 3 months after outbreak end
2	*Stenella coeruleoalba*	October 2012	Alcossebre 40° 14′ 44.75″ N 0° 16′ 34.14″ E	Juvenile Male	Dead when found
3	*Stenella coeruleoalba*	July 2014	Nules 39° 50′ 41.69″ N 0° 5′ 56.50″ W	Adult Male	Alive with tremors
4	*Stenella coeruleoalba*	November 2015	Sueca 39° 12′ 12″ N 0° 18′ 40.78″ W	Adult Male	Dead when found
5	*Stenella coeruleoalba*	December 2015	Torrevieja 37° 59′ 5.00′′ N 0° 40′ 51.00′′ W	Calf Female	Dead when found

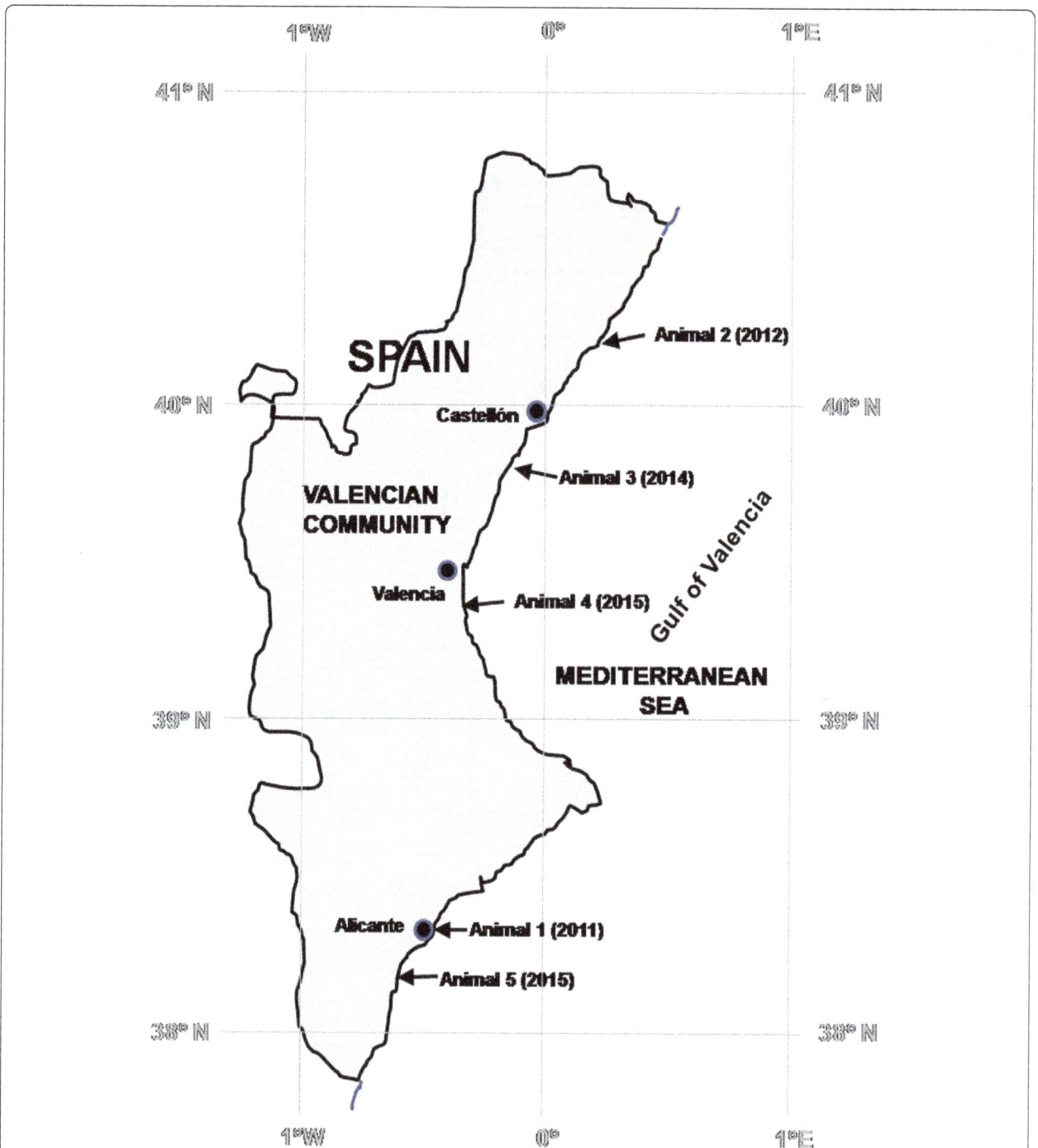

Fig. 1 Schematic cartographic representation of the coastal area with the sites of stranding of the five DMV-positive dolphins studied here. Arrows mark the locations of the strandings, giving the identification numbers of the animals and the year of stranding (between parentheses)

II kit (from Macherey-Nagel, Duren, Germany), utilizing our recently described Universal Probe Library (UPL) reverse transcription PCR (RT-PCR) assay, which targets the gene for the viral fusion protein [28]. This real-time PCR assay was performed on the tissue samples routinely used to detect CeMV (described above in the section "tissue sampling"). Afterwards, portions of the viral genes for phosphoprotein (P) [29, 30], nucleoprotein (N) [31] and hemagglutinin (H) [4] were amplified in conventional RT-PCR assays using as templates pure RNA from DMV-positive lung samples from animals 1, 2, 4 and 5, and from a DMV-positive cerebrum sample from animal 3. RNA extracted from the brain of a DMV-positive striped dolphin stranded in 2007 on the Spanish Mediterranean coast served as a positive control. As a negative control, nuclease-free water was used instead of tissue-derived template.

The PCR-amplified regions of the P, N and H genes were purified with the QIAquick PCR purification kit (QIAGEN, Hilden, Germany) and Sanger-sequenced in an ABI Prism 3730 apparatus (Applied Biosystems, Foster City, CA, USA) by a commercial sequencing service (Secugen, CIB-CSIC, Madrid, Spain). These sequences were compared using BLASTN (http://blast.ncbi.nlm.nih.gov) with all the sequences of CeMV and other morbillivirus species deposited in the GenBank. Phylogenetic analysis was carried out using MEGA 4.0 software [32]. P-distance matrices were calculated and tree topology was inferred by the neighbor-joining method based on p-distances. Topology reliability was tested by bootstrapping 2000 replicates generated with a random seed. The partial P gene sequences obtained from the five striped dolphins were aligned using ClustalX to check for differences.

Diagnostic tests for *Brucella* spp. and *Toxoplasma gondii* were carried out on DNA extracted from brain homogenates from animals 1 to 4, all of which had histopathological traits of non-suppurative encephalitis because there have been some cases of combination of these infectious agents with CeMV in Italy [33, 34]. All these animals tested negative for *Brucella* spp. and *Toxoplasma gondii*. The *Brucella spp* assay was based on TaqMan Real time PCR amplification of the brucellar insertion sequence IS711 [35]. For *T. gondii*, nested PCR was used to target a sequence of the repetitive gene *B1* of this microorganism [36].

Results

Necropsy examination

The most relevant gross findings in animal 1 were marked prescapular and pulmonary lymphadenomegaly, pharyngeal tonsil enlargement, and multifocal pulmonary atelectasis. In Animal 2, the lungs were bilaterally expanded with noticeable rib impressions on the dorsolateral surfaces and were diffusely mottled pale gray to dark red. The meninges were diffusely congested and prescapular lymph nodes were enlarged. Animal 3 had focal ventral pulmonary atelectasis and mild caudolateral emphysema. Both prescapular lymph nodes were diffusely congested. The pulmonary-associated lymph nodes were enlarged, edematous and congested. Animal 4 had external cutaneous lesions consistent with interspecific aggressive interaction, throughout the whole body. Multiple organs and cavities were hemorrhagic because of evidence of internal lesions possibly derived from this aggressive interaction. Animal 5 was poorly preserved, and the majority of organs were severely autolyzed. However, there was clear evidence that the lungs were atelectatic and that the meninges were diffusely congested, being difficult to dissect.

Histologic and immunohistochemical examination

Animal 1 had diffuse interstitial pneumonia with type II pneumocyte hyperplasia and syncytial cells. The spleen, prescapular, mesenteric and pulmonary-associated lymph nodes were diffusely and moderately depleted of lymphocytes and showed scattered foci of lymphocytolysis. Animal 2 also had severe multifocal bronchointerstitial pneumonia with type II pneumocyte hyperplasia and alveolar septa thickened by infiltrates of lymphocytes, plasma cells, and reactive fibroblasts. Alveoli contained macrophages and syncytial cells mixed with edema fluid. Multiple round, acidophilic intranuclear and cytoplasmic inclusion bodies were observed in syncytial cells, macrophages and type II pneumocytes. There was also evidence of multifocal lymphoplasmacytic encephalitis, with perivascular cuffing, multifocal acute neuronal necrosis, astrocytosis, and gitter cell infiltration. There was marked lymphoid depletion in the spleen and in pulmonary lymph nodes, which also presented sinus histiocytosis and syncytia. In animal 3 there was multifocal lymphoplasmacytic and histiocytic meningoencephalitis with perivascular cuffing. It also presented lymphoplasmacytic and histiocytic bronchointerstitial pneumonia with mild fibrosis, edema and a focal granuloma, and, in the examined lymph nodes, marked diffuse lymphoid depletion with sinus edema and histiocytosis. In animal 4 mixed interstitial bronchopneumonia with intralesional syncitia and both intranuclear and cytoplasmic eosinophilic inclusion bodies were observed. There was also neutrophilic and histiocytic lymphadenitis and pharyngeal tonsillitis, with syncitia and inclusion bodies. Although in animal 5 autolysis masked cellular detail preventing histological evaluation in most tissues, moderate multifocal interstitial bronchopneumonia was evidenced.

The bronchointerstitial pneumonia, syncitia, inclusion bodies and encephalitis in these animals were indicative of morbilliviral infection, which was confirmed by immunohistochemistry for morbilliviral antigen. In Animal 1, the lung was moderately immunopositive (syncytial cells, type II pneumocytes, alveolar and interstitial macrophages, Fig. 2a), and there was intranuclear and cytoplasmic positivity in mononuclear and multinucleated giant and syncytial cells of the prescapular lymph node (Fig. 2b) and in brain oligodendroglia and astrocytes (Fig. 2c). Mild cytoplasmic immunopositivity was observed in the transitional epithelium of the urinary bladder and in kidney's cortical and medullary tubular epithelium. In animal 2 immunopositivity was observed in pulmonary syncytial cells, type II pneumocytes and macrophages (Fig. 2d and e), in cerebral neurons (Fig. 2f) and gitter cells, whereas in animal 3 neuronal cell bodies and dendritic processes were intensely immunoreactive (Fig. 2g and h), with no immunoreactivity detected in lung tissue. In animal 4 pulmonary pneumocytes and syncitia

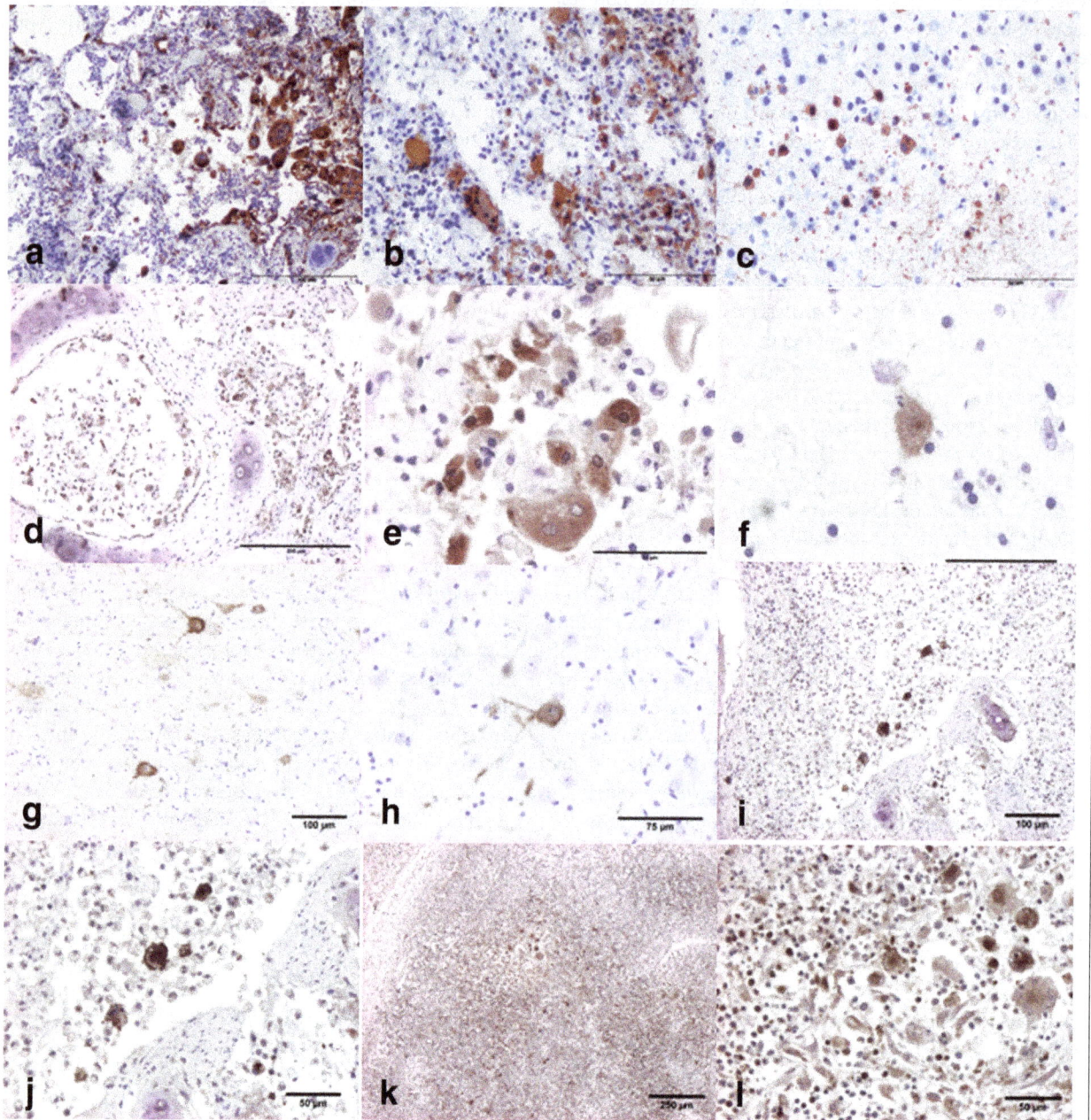

Fig. 2 Immunohistochemistry of animals 1 (**a-c**), 2 (**d-f**), 3 (**g-h**) and 4 (**i-l**) using avidin-biotin-peroxidase and Harris hematoxylin counterstain. **a** Lung. Intense positive immunoperoxidase staining of morbilliviral antigen (red) within hyperplastic type II pneumocytes and multinucleated syncytial cells. **b** Prescapular lymph node. Positive intranuclear and intracytoplasmic immunoperoxidase staining of morbilliviral antigen in mononuclear and multinucleated giant and syncytial cells. **c** *Cerebrum*. Strong immunolabeling of morbilliviral antigen in neurons, glial cells, and neuronal processes. **d** Lung. Intense cytoplasmic immunolabeling of alveolar syncytia, pneumocytes and sloughed bronchiolar epithelium. **e** Lung. Detail of positive cytoplasm immunolabeling of alveolar syncytia and type II pneumocytes. **f** Cerebrum. Detail of intense immunolabeling of a neuronal body. **g** Cerebrum. Strong immunolabeling of neurons (× 10). **h** Cerebrum. Detail of intense immunolabeling of a neuronal body (× 20). **i**. Lung. (× 10) and **j** Lung. (× 20). Alveolar spaces containing large numbers of histiocytes, syncytia and sloughed type II pneumocytes with positive cytoplasmic staining against morbillivirus antigen. **k** Pharyngeal tonsils (× 4) and **l** Pharyngeal tonsils (× 20). Multifocal histiocytic infiltrates and syncitya with intensely stained cytoplasm against morbillivirus antigen

and histiocytes of lung (Fig. 2i and j) and pharyngeal tonsils (Fig. 2k and l) were strongly labelled. The lung from animal 5 was clearly immunopositive for morbilliviral antigen.

Molecular and phylogenetic analyses

As expected for systemic morbilllivirus infection, all tissues tested from animals 1, 2, 4 and 5 were strongly positive to the fusion protein gene by RT-PCR (assayed by UPL

RT-PCR) (Table 2) and for the P, N and H genes (assayed in conventional PCR assays). In contrast, in animal 3 only CNS samples were strongly positive for mobillivirus by these same criteria, whereas all other tissues proved negative for these four genes (Table 2), indicating CNS-restricted infection in this animal.

Subsequent Sanger sequencing confirmed that all amplicons were mobilliviral sequences. The nucleotide sequences from the P gene, N gene and H gene of animal 1 corresponded to those reported in animals of the 2011 Spanish Mediterranean outbreak (P gene, GenBank JN210891; N gene MG773794; H gene, MG773792), indicating that the DMV responsible for that outbreak caused the systemic infection of animal 1. However, the P, N and H sequences derived from animals 2 to 5 differed from those of animal 1, being identical in these four animals (P gene, GenBank KC572861; N gene, GenBank MG773795; H gene, MG773793) (Figs. 3 and 4).

Discussion

The results of this study show that five striped dolphins stranded in the Spanish Mediterranean between 2011 and 2015 had an infectious disease presenting with variable degrees of bronchointerstitial pneumonia, non-suppurative encephalitis and multicentric lymphoid depletion. In animals 1, 2 and 4, which had good tissue preservation, the pathological observations in several organs (Fig. 2) were consistent with acute systemic morbilliviral infection, with immunohistochemical confirmation. In addition, molecular tests proved the presence of DMV in multiple tissues of these animals and of the poorly preserved animal 5 (Table 2). In this last animal, lung preservation was sufficient to also reveal typical mobilliviral pathology and immunohistochemistry (Fig. 2). These animals are the first reported cases in the Spanish Mediterranean Sea of acute systemic DMV infection between epizootic outbreaks. In

contrast to the other four dolphins, in animal 3, which presented meningoencephalitis, the immunohistochemistry for the viral antigen and the molecular tests for DMV were only positive in cerebral tissue (Table 2). Although this animal also had bronchointerstitial pneumonia, the signs of chronicity for this derrangement, together with the negativity of the lung for the laboratory tests for DMV did not support active pulmonary disease due to this virus. At stranding, the DMV infection appears limited in this animal to the CNS, corresponding to the chronic encephalitic form of DMV infection [37].

Comparisons of viral sequences suggest that the same DMV strain infected animals 2–5, and that this strain differs from the one in animal 1, as well as from the ones detected in the Spanish Mediterranean outbreaks of 1990, 2007 and 2011 (Fig. 3) [2, 15, 16]. Instead, the strain infecting animals 2–5 appears to be identical (p-distance, 0.000) to a strain identified in the Atlantic Ocean in dolphins stranded in 2011–2013, one in the Canary Islands (GenBank KF695110) and six in the Northwestern coast of the Iberian Peninsula (GenBank, KP836003, KT878656, KT878657, KT878658, KT878660, KT878661). Thus, it appears that the DMV virus infecting animals 2–5 had an Atlantic origin [23, 25]. This conclusion is supported also by the observation that the P gene sequence of animals 2–5 is phylogenetically very close to the P gene sequence of the DMV responsible for the 1990 Spanish Mediterranean outbreak (GenBank AJ608288; p-distance, 0.012), known to have Atlantic origin [11]. This close relation was also found (Fig. 3) for the N gene sequence (p-distance 0.015) (GenBank AJ608288). P-distances for the H gene (1990; GenBank AY586536) also support the Atlantic origin of the strain found in animals 2–5. Unfortunately, the lack of reporting of these sequences for Atlantic DMV isolates prevented similar comparisons with the

Table 2 Samples analyzed and results obtained in the molecular and immunohistochemical studies of our animals[a]

Animal	Conventional RT-PCR for CeMV	UPL RT-PCR for CeMV	Immunohistochemistry for CDV
1	Skin, lung, prescapular lymph node, pulmonary lymph node, kidney, pharyngeal tonsils, cerebrum, cerebellum	Skin, lung, prescapular lymph node, pulmonary lymph node, kidney, pharyngeal tonsils, cerebrum, cerebellum	Lung, cerebrum, kidney, *intestine, stomach,* urinary bladder
2	Skin, lung, prescapular lymph node, pulmonary lymph node, kidney, pharyngeal tonsils, cerebrum, cerebellum	Skin, lung, prescapular lymph node, pulmonary lymph node, kidney, pharyngeal tonsils, cerebrum, cerebellum	Cerebrum, lung
3	*Skin, lung, prescapular lymph node, pulmonary lymph node, kidney, pharyngeal tonsils,* cerebrum, cerebellum	*Skin, lung, prescapular lymph node, pulmonary lymph node, kidney, pharyngeal tonsils,* cerebrum, cerebellum	*Lung,* cerebrum
4	Skin, lung, prescapular lymph node, pulmonary lymph node, kidney, pharyngeal tonsils, cerebrum, cerebellum	Skin, lung, prescapular lymph node, pulmonary lymph node, kidney, pharyngeal tonsils, cerebrum, cerebellum	Lung, pharyngeal tonsils
5	Skin, lung, prescapular lymph node, pulmonary lymph node, kidney, pharyngeal tonsils, cerebrum, cerebellum	Skin, lung, prescapular lymph node, pulmonary lymph node, kidney, pharyngeal tonsils, cerebrum, cerebellum	*Cerebrum,* lung

[a]The indicated tissues or organs were those analyzed. Italic type highlights tissues that tested negative

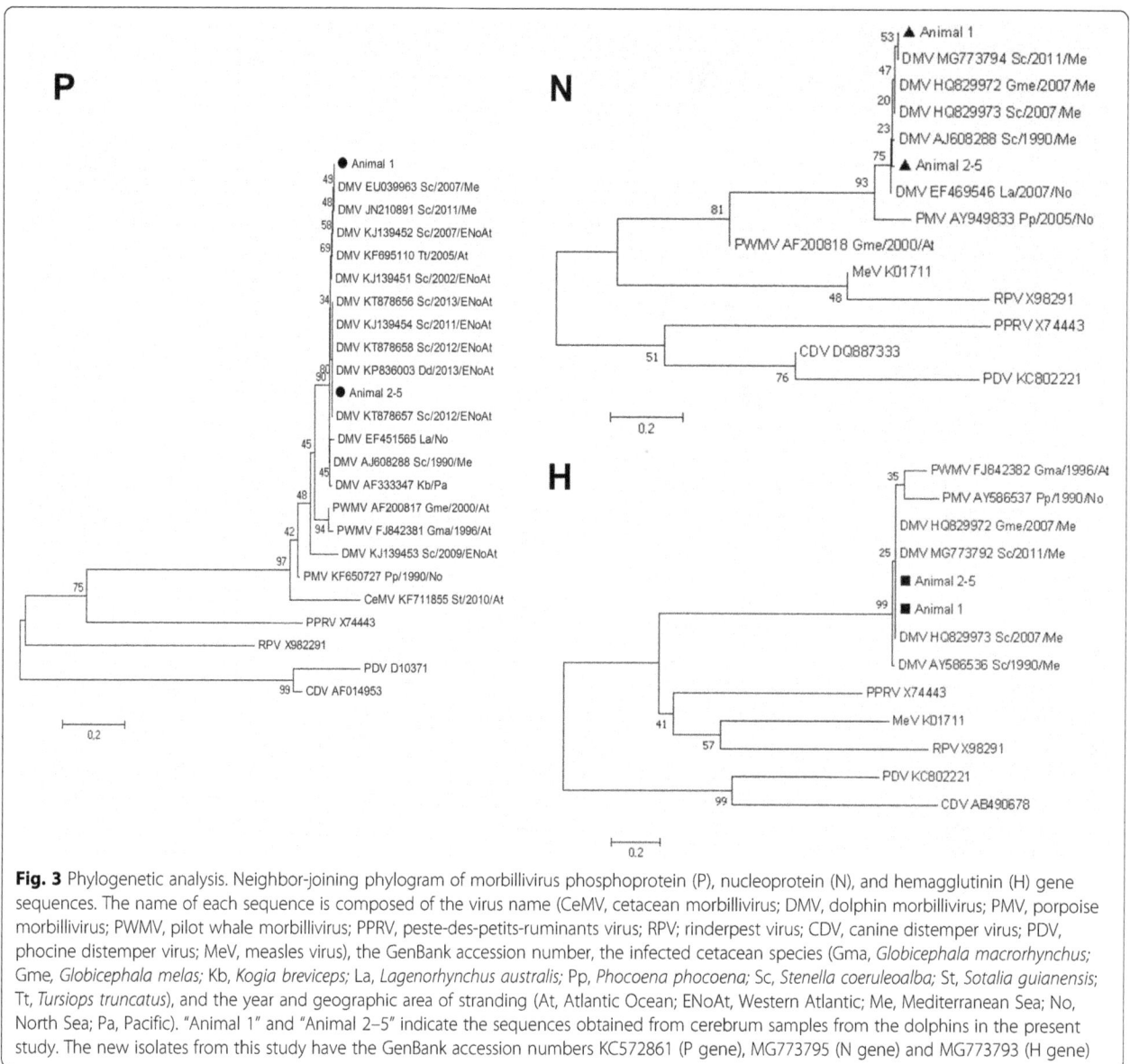

Fig. 3 Phylogenetic analysis. Neighbor-joining phylogram of morbillivirus phosphoprotein (P), nucleoprotein (N), and hemagglutinin (H) gene sequences. The name of each sequence is composed of the virus name (CeMV, cetacean morbillivirus; DMV, dolphin morbillivirus; PMV, porpoise morbillivirus; PWMV, pilot whale morbillivirus; PPRV, peste-des-petits-ruminants virus; RPV; rinderpest virus; CDV, canine distemper virus; PDV, phocine distemper virus; MeV, measles virus), the GenBank accession number, the infected cetacean species (Gma, *Globicephala macrorhynchus*; Gme, *Globicephala melas*; Kb, *Kogia breviceps*; La, *Lagenorhynchus australis*; Pp, *Phocoena phocoena*; Sc, *Stenella coeruleoalba*; St, *Sotalia guianensis*; Tt, *Tursiops truncatus*), and the year and geographic area of stranding (At, Atlantic Ocean; ENoAt, Western Atlantic; Me, Mediterranean Sea; No, North Sea; Pa, Pacific). "Animal 1" and "Animal 2–5" indicate the sequences obtained from cerebrum samples from the dolphins in the present study. The new isolates from this study have the GenBank accession numbers KC572861 (P gene), MG773795 (N gene) and MG773793 (H gene)

viruses of the 2011–2013 strandings in the Canary Islands and the Nothwestern Iberian coast.

Concerning animal 1, since it was found to contain the same DMV strain that caused the 2011 outbreak and it stranded only three months after that outbreak was declared terminated [16], it cannot be excluded that it belongs to the trailing edge of the 2011 outbreak. However, systemic cases were found in Italy in 2011 to 2013 in one fin whale and in numerous sperm whales due to DMV strains closely related to the viruses responsible for the 2007 and 2011 Spanish outbreaks [22, 24], and DMV-positive striped dolphin mass mortality also occurred in Italy in 2013 that could possibly be related to these previous Mediterranean strains [21]. Thus, as already proposed [3, 16] there appears to be endemic viral circulation within the Mediterranean, possibly within the abundant striped dolphin population,

although the potential roles of whales in the transmission and maintenance of DMV strains, possible in principle given the reports of pilot whales and a Cuvier's beaked whale (*Ziphius cavirrostris*) infected with DMV [6, 20, 38], appear unlikely given the low density of these species in the Mediterranean. The fact that the Valencian Community was the epicenter of the previous three DMV epizootics in the Mediterranean makes conceivable that this region could harbor local DMV reservoirs, although this possibility requires further investigation, since other factors like abundance of striped dolphins in the area, and efficient and active surveillance networks may be responsible for the early detection of the epizootics in this region.

Even though most cases of DMV infection outside outbreak periods appear to be of the chronic encephalitic form [30, 39], our findings highlight the need to be

```
Animal 1    CAAGAGGTTGAAGGAGTCAAGGATGCTGACCTGCTCGTGGTTCCAACAGGTAGTGATGAT 60
Animal 2-5  CAAGAGGTTGAAGGAGTCAAGGATGCTGACCTGCTCGTGGTTCCAACAGGCAGTGATGAT 60
            ************************************************ ** *********

Animal 1    GATGCAGAATTCAGAGACGGAGATGAGAGCTCTCTCGAGAGCGATGGTGAATCTGGCACT 120
Animal 2-5  GATGCAGAATTCAGAGACGGAGATGAGAGCTCTCTTGAGAGCGATGGTGAATCTGGCACT 120
            *********************************** ************************

Animal 1    GTTGATACCAGAGGAACTTCTTCCTCTAACAGGGGATCTGCTCCCAGGATTAAGGTCGAG 180
Animal 2-5  GTTGATACCAGAGGAAATTCTTCCTCTAACAGGGGATCTGCTCCCAGGATTAAGGTCGAG 180
            ****************  *****************************************

Animal 1    AGATCGTCTGACGTTGAGACTATAAGCAGTGAAGAGCTACAAGGACTGATTAGATCTCAG 240
Animal 2-5  AGATCGTCTGACGTTGAGACTATAAGCAGTGAAGAGCTACAAGGACTGATTAGATCTCAG 240
            ************************************************************

Animal 1    AGTCAAAAACATAATGGATTTGGAGTAGACAGATTCTTAAAAGTCCCACCAATTCCAACT 300
Animal 2-5  AGTCAAAAACATAATGGATTTGGAGTCGACAGATTCTTAAAGGTCCCACCAATTCCAACC 300
            ************************** ************** ***** **********

Animal 1    TCAGTGCCGCTGGACCCCGCTCCCAAATCCATTAAAAAGGGCACAGGAGAGAG 353
Animal 2-5  TCAGTGCCGCTGGACCCCGCTCCCAAATCCATTAAAAAGGGCACAGGAGAGAG 353
            ****************************************************
```

Fig. 4 Alignment of partial P gene sequences from animal 1 and from animals 2–5. Identical regions are shadowed gray and marked with asterisks at the bottom. Differences are highlighted using black shadowing over white lettering

prepared to deal with possible cases of systemic infection within these inter-epizootic periods. In addition, our finding of sequences of apparent Atlantic origin in four of the animals dealt with here stress the importance of sequencing standard fragments of viral genes to make inferences that could suggest paths of virus circulation and provide epidemiological insight, even giving clues for early detection of impending outbreaks.

Comparisons of viral sequences from the three previous Spanish Mediterranean outbreaks suggest that the outbreaks were caused by closely related DMV strains [6, 16]. For example, our results indicate a close relationship between sequences linked to the Spanish outbreaks of 2007 and 2011. Sequence comparisons extending over 9.050 kb [6] suggest that the Spanish 2007 DMV strain is derived from the Spanish 1990 DMV strain. Our finding that the sequences from animal 1 are more closely related to the 2011 sequence (P gene, GenBank JN210891; N gene, GenBank MG773794; H gene, GenBank MG773792) than to the sequences from 1990 or 2007 could be consistent with a sequential lineage (1990 → 2007 → 2011 → DMV in animal 1) or with re-introduction from the Atlantic in 2007. In contrast to the case of animal 1, partial P, N and H gene sequences from animals 2, 3, 4 and 5 were identical to each other and appear to cluster separately from the sequences from the 1990, 2007 and 2011 isolates. Our finding that animal 2, which stranded in 2012, contained an identical P sequence to that of DMV identified in the Atlantic in a dolphin that stranded in 2011 [25], suggests an Atlantic origin for this DMV strain, which was subsequently found also in animals 3, 4 and 5. Consistent with this is the fact that animal 2 had acute disseminated disease, with the virus detected in the lung and the nervous

system, which would be expected if the Mediterranean striped dolphin population in 2012 was naïve to this Atlantic strain. The fact that we found the same strain in animal 3 that stranded in 2014 with CNS-restricted infection indicates that this strain can induce both systemic and chronic infection.

Conclusions

In summary, our results suggest that multiple DMV strains are circulating in striped dolphins and causing systemic infection in the western Mediterranean during inter-epizootic periods. At least one of these strains can produce both systemic and chronic infection and appears to have originated in the Atlantic Ocean. These findings warrant further studies aiming at clarifying how the virus circulates and causes epidemics in the Mediterranean Sea.

Abbreviations
CDV-NP: Canine Distemper Virus nucleoprotein; CeMV: Cetacean morbillivirus; CNS: Central nervous system; DMV: Dolphin morbillivirus; FFPE: Formalin-fixed, paraffin-embedded; H: Haemagluttinin; HE: Hematoxylin and eosin; MI: Morbillivirus infection; N: Nucleoprotein; P: Phosphoprotein; PMV: Porpoise morbillivirus; PWMV: Pilot whale morbillivirus; RT-PCR: Reverse transcription PCR; UPL: Universal probe library

Acknowledgments
The authors thank Belen Rivera and Rocío Sánchez for technical assistance and Prof. Vicente Rubio (IBV-CSIC, Valencia) for critical reading of the manuscript and for help with writing of the revised versions of this paper. This work was supported by a collaborative agreement involving the Conselleria de Agricultura, Medio ambiente, Cambio Climático y Desarrollo Rural de la Generalitat Valenciana, the Oceanogràfic Aquarium of the Ciudad de las Artes y las Ciencias of Valencia and the VISAVET Center of the Complutense University of Madrid. This work was cofinanced by Project CGL2009-08125 of the Spanish National Research Plan. Mar Melero is the recipient of a PhD student grant from the University Complutense of Madrid. Consuelo Rubio-Guerri and Josué Diaz-Delgado are recipients of FPU grants from the Spanish Ministry of Education. This research did not receive any other specific grant from funding agencies in the public, commercial, or not-for-profit sectors.

Authors' contributions

Necropsy and sample taking were performed by CRG, JLC, MM; histopathology and immunohistochemistry were performed by JDD, ES, MA and MAJ; viral study and phylogenetic study were analyzed by CRG, ENB, MM and FE; the manuscript was prepared and critically discussed by CRG, DGP, and JMSV with the contributions of all the remaining authors. All authors have read and approved the final manuscript.

Competing interests

The authors declare that they have no competing interests.

Author details

[1]VISAVET Center and Animal Health Department, Veterinary School, Complutense University of Madrid, Avda. Puerta del Hierro s/n, 28040 Madrid, Spain. [2]Fundación Oceanografic de la Comunitat Valenciana, C/. Eduardo Primo Yúfera (Científic) 1B, 46013 Valencia, Spain. [3]Medicine and Surgery Department (Anatomic Pathology), Veterinary School, Complutense University of Madrid, 28040 Madrid, Spain. [4]Unit of Histology and Veterinary Pathology, Institute for Animal Health, Veterinary School, University of Las Palmas de Gran Canaria, Trasmontaña, s, /n 35416 Arucas (Las Palmas), Canary Islands, Spain. [5]National Institute for Agricultural and Food Research and Technology, Ctra. de Algete a El Casar s/n, 28130 Madrid, Spain. [6]Veterinary Services, Avanqua Oceanogràfic S.L., C/ Eduardo Primo Yúfera (Científic) 1B, 46013 Valencia, Spain.

References

1. Kennedy S. Morbillivirus infections in aquatic mammals. J Comp Pathol. 1998;119:201–25.
2. Domingo M, Ferrer L, Pumarola M, Marco A, Plana J, Kennedy S, McAliskey M, Rima BK. Morbillivirus in dolphins. Nature. 1990;348:21.
3. Bellière EN, Esperón F, Fernandez A, Arbelo M, Muñoz MJ, Sanchez-Vizcaino JM. Phylogenetic analysis of a new Cetacean morbillivirus from a short-finned pilot whale stranded in the Canary Islands. Res Vet Sci. 2011;90:324–8.
4. Sierra E, Fernández A, Suárez-Santana C, Xuriach A, Zucca D, Bernaldo de Quirós Y, García-Álvarez N, de la Fuente J, Sacchini S, Andrada M, Díaz-Delgado J, Arbelo M. Morbillivirus and pilot whale deaths, Canary Islands, Spain. Emerg Infect Dis. 2016;22:740–2.
5. Taubenberger JK, Tsai MM, Atkin TJ, Fanning TG, Krafft AE, Moeller RB, Kodsi SE, Mense MG, Lipscomb TP. Molecular genetic evidence of a novel morbillivirus in a long-finned pilot whale (Globicephalus melas). Emerg Infect Dis. 2000;6:42–5.
6. Bellière EN, Esperon F, Sanchez-Vizcaino JM. Genetic comparison among dolphin morbillivirus in the 1990-1992 and 2006-2008 Mediterranean outbreaks. Infect Genet Evol. 2011;11:1913–20.
7. Groch KR, Colosio AC, Marcondes MC, Zucca D, Diaz-Delgado J, Niemeyer C, Marigo J, Brandao PE, Fernandez A, Luiz C-DJ. Novel cetacean morbillivirus in Guiana dolphin. Brazil Emerg Infect Dis. 2014;20:511–3.
8. Stephens N, Duignan PJ, Wang J, Bingham J, Finn H, Bejder LS, Patterson AP, Holyoake C. Cetacean morbillivirus in coastal indo-Pacific bottlenose dolphins. Western Australia Emerg Infect Dis. 2014;20:666–70.
9. West KL, Levine G, Jacob J, Jensen B, Sanchez S, Colegrove K, Rotstein D. Coinfection and vertical transmission of Brucella and morbillivirus in a neonatal sperm whale (Physeter macrocephalus) in Hawaii, USA. J Wildl Dis. 2015;51:227–32.
10. Lipscomb TP, Schulman FY, Moffett D. Kennedy SMorbilliviral disease in Atlantic bottlenose dolphins (Tursiops truncatus) from the 1987-1988 epizootic. J Wild Dis. 1994;30:567–71.
11. Aguilar A, Raga JA. The striped dolphin epizootic in the Mediterranean Sea. Ambio. 1993:524–8.
12. Krafft AE, Lichy JH, Lipscomb TP, Klaunberg BA, Kennedy S, Taubenberger JK. Postmortem diagnosis of morbillivirus infection in bottlenose dolphins (Tursiops truncatus) in the Atlantic and Gulf of Mexico epizootics by a polymerase chain reaction-based assay. J Wild Dis. 1995;1:410–5.
13. Lipscomb TP, Kennedy S, Moffett D, Krafft A, Klaunberg BA, Lichy JH, Regan GT, Worthy GA, Taubenberger JK. Morbilliviral epizootic in bottlenose dolphins of the Gulf of Mexico. J Virol Methods. 1996;8:283–90.
14. Keck N, Kwiatek O, Dhermain F, Dupraz F, Boulet H, Danes C, Laprie C, Perrin A, Godenir J, Micout L, Libeau G. Resurgence of morbillivirus infection in Mediterranean dolphins off the French coast. Vet Rec. 2010;166:654–5.
15. Raga JA, Banyard A, Domingo M, Corteyn M, Van Bressem MF, Fernandez M, Aznar FJ, Barrett T. Dolphin morbillivirus epizootic resurgence, Mediterranean Sea. Emerg Infect Dis. 2008;14:471–3.
16. Rubio-Guerri C, Melero M, Esperon F, Belliere EN, Arbelo M, Crespo JL, Sierra E, Garcia-Parraga D, Sanchez-Vizcaino JM. Unusual striped dolphin mass mortality episode related to cetacean morbillivirus in the Spanish Mediterranean Sea. BMC Vet Res. 2013;9:106.
17. Soto S, González R, Alegre F, González B, Medina P, Raga JA, Marco A, Domingo M. Epizootic of dolphin morbillivirus on the Catalonian Mediterranean coast in 2007. Vet Rec. 2011;169:101.
18. Di Guardo G, Di Francesco CE, Eleni C, Cocumelli C, Scholl F, Casalone C, Peletto S, Mignone W, Tittarelli C, Di Nocera F, Leonardi L, Fernandez A, Marcer F, Mazzariol S. Morbillivirus infection in cetaceans stranded along the Italian coastline: pathological, immunohistochemical and biomolecular findings. Res Vet Sci. 2013;94:132–7.
19. Mazzariol S, Marcer F, Mignone W, Serracca L, Goria M, Marsili L, Di Guardo G, Casalone C. Dolphin morbillivirus and Toxoplasma gondii coinfection in a Mediterranean fin whale (Balaenoptera physalus). BMC Vet Res. 2012;8:20.
20. Mazzariol S, Peletto S, Mondin A, Centelleghe C, Di Guardo G, Di Francesco CE, Casalone C, Acutis PL. Dolphin morbillivirus infection in a captive harbor seal (Phoca vitulina). J Clin Microbiol. 2013;51:708–11.
21. Casalone C, Mazzariol S, Pautasso A, Di Guardo G, Di Nocera F, Lucifora G, Ligios C, Franco A, Fichi G, Cocumelli C, Cersini A, Guercio A, Puleio R, Goria M, Podestà M, Marsili L, Pavan G, Pintore A, De Carlo E, Eleni C, Caracappa S. Cetacean strandings in Italy: an unusual mortality event along the Tyrrhenian Sea coast in 2013. Dis Aquat Org. 2014;109:81–6.
22. Mazzariol S, Centelleghe C, Di Provvido A, Di Renzo L, Cardeti G, Cersini A, Fichi G, Petrella A, Di Francesco CE, Mignone W, Casalone C, Di Guardo G. Dolphin morbillivirus associated with a mass stranding of sperm whales, Italy. Emerg Infect Dis. 2017;23:144–6.
23. Bento MC, Eira CI, Vingada JV, Marçalo AL, Ferreira MC, Fernandez AL, Tavares LM, Duarte AI. New insight into dolphin morbillivirus phylogeny and epidemiology in the Northeast Atlantic: opportunistic study in cetaceans stranded along the Portuguese and Galician coasts. BMC Vet Res. 2016;12:176.
24. Mazzariol S, Centelleghe C, Beffagna G, Povinelli M, Terracciano G, Cocumelli C, Pintore A, Denurra D, Casalone C, Pautasso A, Di Francesco CE, Di Guardo G. Mediterranean fin whales (Balaenoptera physalus) threatened by dolphin morbillivirus. Emerg Infect Dis. 2016;22:302–5.
25. Sierra E, Zucca D, Arbelo M, García-Álvarez N, Andrada M, Déniz S, Fernández A. Fatal systemic morbillivirus infection in bottlenose dolphin, Canary Islands, Spain. Emerg Infect Dis. 2014;20:269–71.
26. Stone BM, Blyde DJ, Saliki JT, Blas-Machado U, Bingham J, Hyatt A, Wang J, Payne J, Crameri S. Fatal cetacean morbillivirus infection in an Australian offshore bottlenose dolphin (Tursiops truncatus). Aust Vet J. 2011;89:452–7.
27. Ramos-Vara JA. Principles and methods of immunohistochemistry. Methods of Mol Biol. 2011;691:83–96.
28. Rubio-Guerri C, Melero M, Rivera-Arroyo B, Bellière EN, Crespo JL, García-Párraga D, Esperón F, Sánchez-Vizcaíno JM. Simultaneous diagnosis of cetacean morbillivirus infection in dolphins stranded in the Spanish Mediterranean Sea in 2011 using a novel universal probe library (UPL) RT-PCR assay. Vet Microbiol. 2013;165:109–14.
29. Barrett T, Visser IK, Mamaev L, Goatley L, van Bressem MF, Osterhaust AD. Dolphin and porpoise morbilliviruses are genetically distinct from phocine distemper virus. Virology. 1993;193:1010–2.
30. Yang WC, Pang VF, Jeng CR, Chou LS, Chueh LL. Morbilliviral infection in a pygmy sperm whale (Kogia breviceps) from Taiwanese waters. Vet Microbiol. 2006;116:69–76.
31. Van de Bildt MW, Kuiken T, Osterhaus AD. Cetacean morbilliviruses are phylogenetically divergent. Arch Virol. 2005;150:577–83.
32. Tamura K, Peterson D, Peterson N, Stecher G, Nei M, Kumar S. MEGA5: molecular evolutionary genetics analysis using maximum likelihood, evolutionary distance, and maximum parsimony methods. Mol Biol Evol. 2011;28:2731–9.

33. Profeta F, Di Francesco CE, Marsilio F, Mignone W, Di Nocera F, De Carlo E, Lucifora G, Pietroluongo G, Baffoni M, Cocumelli C, Eleni C, Terracciano G, Ferri N, Di Francesco G, Casalone C, Pautasso A, Mazzariol S, Centelleghe C, Di Guardo G. Retrospective seroepidemiological investigations against morbillivirus, Toxoplasma gondii and Brucella spp. in cetaceans stranded along the Italian coastline (1998-2014). Res Vet Sci. 2015;101:89–92.

34. Sierra E, Sánchez S, Saliki JT, Blas-Machado U, Arbelo M, Zucca D, Fernández A. Retrospective study of etiologic agents associated with nonsuppurative meningoenceph-alitis in stranded cetaceans in the Canary Islands. J Clin Microbiol. 2014;52:2390–7.

35. Hinic V, Brodard I, Thomann A, Cvetnic Z, Makaya PV, Frey J, Abril C. Novel identification and differentiation of Brucella melitensis, B. abortus, B. suis, B. ovis, B. canis, and B. neotomae suitable for both conventional and real-time PCR systems. J Microbiol Methods. 2008;75:375–8.

36. Montoya A, Miró G, Blanco MA, Fuentes I. Comparison of nested PCR and real-time PCR for the detection of Toxoplasma gondii in biological samples from naturally infected cats. Res Vet Sci. 2010;89:212–3.

37. Van Bressem MF, Duignan PJ, Banyard A, Barbieri M, Colegrove KM, De Guise S, Di Guardo G, Dobson A, Domingo M, Fauquier D, Fernández A, Goldstein T, Grenfell B, Groch KR, Gulland F, Jensen BA, Jepson PD, Hall A, Kuiken T, Mazzariol S, Morris SE, Nielsen O, Raga JA, Rowles TK, Saliki J, Sierra E, Stephens N, Stone B, Tomo I, Wang J, Waltzek T, Wellehan,JF. Cetacean morbillivirus: current knowledge and future directions. Viruses 2014; 6: 5145–5181.

38. Centelleghe C, Beffagna G, Palmisano G, Franzo G, Casalone C, Pautasso A, Giorda F, Di Nocera F, Iaccarino D, Santoro M, Di Guardo G, Mazzariol S. Dolphin morbillivirus in a Cuvier's beaked whale (Ziphius cavirostris), Italy. Front Microbiol. 2017;8:111.

39. Domingo M, Vilafranca M, Visa J, Prats N, Trudgett A, Visser I. Evidence for chronic morbillivirus infection in the Mediterranean striped dolphin (Stenella coeruleoalba). Vet Microbiol. 1995;44:229–39.

10

Pathobiological investigation of naturally infected canine rabies cases from Sri Lanka

S. Beck[1*], P. Gunawardena[2], D. L. Horton[4], D. J. Hicks[1], D. A. Marston[4], A. Ortiz-Pelaez[3], A. R. Fooks[4] and A. Núñez[1]

Abstract

Background: The recommended screening of rabies in 'suspect' animal cases involves testing fresh brain tissue. The preservation of fresh tissue however can be difficult under field conditions and formalin fixation provides a simple alternative that may allow a confirmatory diagnosis. The occurrence and location of histopathological changes and immunohistochemical (IHC) labelling for rabies in formalin fixed paraffin embedded (FFPE) canine brain is described in samples from 57 rabies suspect cases from Sri-Lanka. The presence of Negri bodies and immunohistochemical detection of rabies virus antigen were evaluated in the cortex, hippocampus, cerebellum and brainstem. The effect of autolysis and artefactual degeneration of the tissue was also assessed.

Results: Rabies was confirmed in 53 of 57 (93%) cases by IHC. IHC labelling was statistically more abundant in the brainstem. Negri bodies were observed in 32 of 53 (60.4%) of the positive cases. Although tissue degradation had no effect on IHC diagnosis, it was associated with an inability to detect Negri bodies. In 13 cases, a confirmatory Polymerase chain reaction (PCR) testing for rabies virus RNA was undertaken by extracting RNA from fresh frozen tissue, and also attempted using FFPE samples. PCR detection using fresh frozen samples was in agreement with the IHC results. The PCR method from FFPE tissues was suitable for control material but unsuccessful in our field cases.

Conclusions: Histopathological examination of the brain is essential to define the differential diagnoses of behaviour modifying conditions in rabies virus negative cases, but it is unreliable as the sole method for rabies diagnosis, particularly where artefactual change has occurred. Formalin fixation and paraffin embedding does not prevent detection of rabies virus via IHC labelling even where artefactual degeneration has occurred. This could represent a pragmatic secondary assay for rabies diagnosis in the field because formalin fixation can prevent sample degeneration. The brain stem was shown to be the site with most viral immunoreactivity; supporting recommended sampling protocols in favour of improved necropsy safety in the field. PCR testing of formalin fixed tissue may be successful in certain circumstances as an alternative test.

Keywords: Rabies canine histopathology immunohistochemistry hemi-nested reverse transcription polymerase chain reaction

Background

Rabies is a highly neurotropic virus which is thought to have the potential to infect most mammalian species [10]. There is an almost universal case fatality rate (approaching 100%) in humans, for whom the predominant source of exposure leading to death is dog bites [14, 27]. Rabies continues to be a significant threat to human health in developing countries, with most human cases occurring in Africa and Asia [11]. There is also a need for effective rabies surveillance and diagnosis in Europe where there is a low potential risk of transmission to humans from the 7064 domestic animal and wildlife cases reported in 2013 (WHO Rabies bulletin) [20]. Even countries that are officially free of rabies such as the UK report rare, imported (quarantined) cases [21]. The ability to diagnose rabies virus in domestic animals and wildlife is the basis of most national reporting programs.

Canine rabies cannot be diagnosed by clinical presentation alone, which is often varied and may include non-specific signs such as pyrexia, altered neurological function, ataxia and paralysis prior to death [29].

---
* Correspondence: sbeck@rvc.ac.uk
[1]Pathology Department, Animal and Plant Health Agency, Weybridge, UK
Full list of author information is available at the end of the article

Seroconversion is not a useful antemortem test because it occurs late in the disease process and there are no specific gross lesions at post-mortem [36]. Confirmation of rabies infection requires laboratory testing of brain tissue. The standard methods for the diagnosis of rabies virus in animals are recommended by the OIE (World Health Organisation for Animal Health) [31, 36]. Histopathologic examination of infected brain tissue can be used to directly identify Negri bodies in neuronal cytoplasm, which is considered pathognomonic for rabies virus infection [23]. Negri bodies are observed concurrently with perivascular, non-suppurative inflammation, glial proliferation, neuronal degeneration and necrosis. The reduced sensitivity in comparison to immunological and molecular tests has resulted in this technique no longer being recommended for primary diagnosis by the WHO or OIE [12, 36].

Many new diagnostic techniques are available for rabies diagnosis [9, 12] but the recommended ('gold-standard') OIE-prescribed diagnostic assay for statutory diagnosis of rabies virus is the fluorescent antibody test (FAT). This test is a direct technique undertaken on impression smears of fresh brain tissue labelling viral nucleocapsid protein [3, 15, 26]. Obtaining suitable fresh samples without a cold chain and performing testing that requires specialised local equipment remains a challenge [25, 36]. Tissue fixation provides a cheap and simple solution, by preserving samples taken in remote areas prior to laboratory analysis and also allows safe inactivation of the test material. The FAT can also be undertaken on glycerol preserved and formalin fixed tissue, with additional washing or proteolytic enzyme steps respectively, however this is less reliable then using fresh samples [36]. Immunohistochemical (IHC) methods on fixed tissues offer an alternative or additional confirmatory method to the FAT.

Heminested reverse transcription polymerase chain reaction (HnRT-PCR) and real time, Taqman reverse transcription polymerase chain reaction (Taqman RT-PCR) assays have been described [16, 17, 34]. These methods, have been reported to give 100% specificity and sensitivity [2] and Taqman RT-PCR is considered capable of detecting divergent, novel Lyssaviruses [16, 24]. PCR techniques are important confirmatory tests, often used in tandem with FAT [21], which may also be used when brain tissue is severely autolysed [8, 25]. At present PCR is not recommended for routine diagnosis by the WHO/OIE because of the requirement for exacting standards of quality control, but it has a crucial role in epidemiological analysis of virus type [13, 34, 36]. PCR is usually undertaken using fresh tissue, although HnRT-PCR has previously been unsuccessfully attempted using formalin fixed (non-paraffin embedded) tissue in a single case study [7]. PCR using viral RNA extracted from FFPE

tissue would be a practical tool for rabies diagnosis where this is the only tissue available or further diagnostic confirmation is required.

A final consideration in rabies diagnosis is that viral antigen is not equally distributed throughout the brain tissue of rabies positive animals. A large case series involving multiple naturally infected species from South Africa, was tested by FAT. The brainstem was most often antigen positive in canine cases, while the hippocampus contained the largest deposits of antigen when positive [5]. Two small (10 naturally infected canids) case series suggested that viral antigen distribution is greater in the brainstem, compared to other supratentorial structures [30, 32]. Another small case series (3 naturally infected canids) suggested that the hippocampus contains the most intense signal [31]. The general anatomic location of viral antigen has also been assessed by IHC labelling in 21 rabies positive canine cases, in which the cerebellum and hippocampus most often contained positive neurones [1]. It is unknown if this variation is due to study design, stage of infection, strain differences or another factor. In murine models, infected with rabies virus or European bat lyssavirus type 1 and 2, IHC labelling revealed variation in viral antigen distribution, with the brainstem most often being positive [18, 19]. Street and laboratory ("fixed") rabies virus have been compared using a murine model, also with variation in antigen distribution and inflammation [33]. Sample site selection is therefore crucial. The current OIE recommendation is to collect the whole brain in a necropsy room and sample hippocampus, thalamus, cerebral cortex and medulla oblongata: a pool of tissues including the brain stem is advocated for FAT sampling [5, 36]. Opening the cranium carries obvious risk for the operator and for this reason 'straw' techniques have been described for sampling in the field [36].

Methods

The aim of this study is to assess in a large series of naturally infected canine rabies cases the general anatomical location of classic histopathological changes and rabies viral antigen labelling using FFPE tissue, with an analysis of the effect of artefact on detecting these features. A subset of cases were tested for rabies viral RNA by Taqman RT-PCR and HnRT-PCR using both fresh and FFPE tissue.

Case collection

FFPE brain tissue was available from fifty seven canids submitted to the University of Peradeniya, Sri Lanka between the years of 2007-2011 (Table 1). The time for which this material was fixed prior to processing is unknown. The cases were suspected to be infected with rabies virus because of observed neurologic changes but no further clinical information was provided. The areas of the brain available for analysis varied between cases; fresh frozen tissue was available from thirteen cases.

Histopathology and IHC
Serial sections were cut at 4 μm and either stained with haematoxylin and eosin (HE) for histopathological examination or labelled immunohistochemically for viral antigen as described [19].

Histopathological (HE) analysis
The general anatomic locations available for each case were recorded and evaluated separately as cerebrum, cerebellum, hippocampus and brain stem (including pons and medulla). The following histological features, observed in each anatomical location, were semi-quantitatively graded 0-3 by one observer (SB); (i) Perivascular cuffing (lymphocytes within the Virchow-Robins space); (ii) Gliosis (glial nodule formation); (iii) Satellitosis and neuronophagia and (iv) Artefactual degradation (freezing and autolysis) [Fig. 1]. In addition the presence or absence of Negri bodies (intracytoplasmic, 1-27 μm round, eosinophilic inclusions) [23] was recorded for each location.

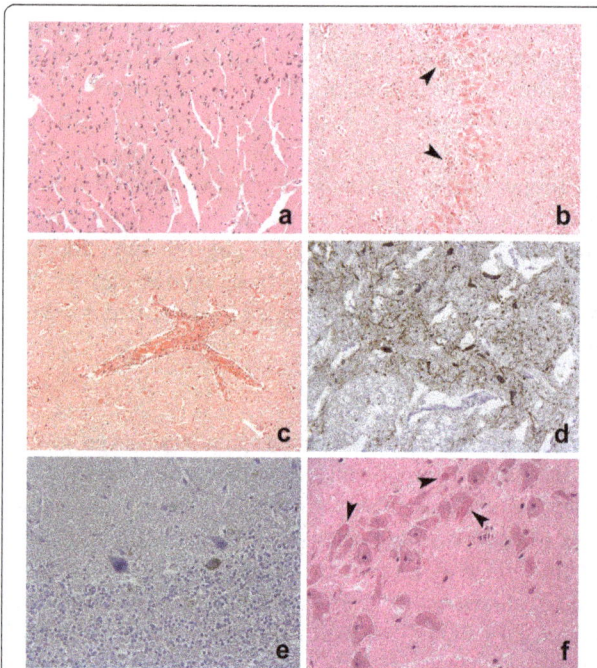

Fig. 1 Selected histopathology and immunohistochemistry photomicrographs. **a**; Cerebrum, case 39, tissue artefact score grade 3 (HE ×10) **b**; Hippocampus, case 27, multifocal glial nodules grade 3 (HE × 10) **c**; Brainstem, case 14, perivascular cuffing grade 2 (HE ×10) **d**; Brainstem, case 29, positive grade 3 immunolabelling in grade 2 artefact tissue (IHC × 10) **e**; Cerebellum, case 3, positive grade 1 immunolabelling in grade 2 artefact tissue (IHC × 10) **f**; Hippocampus, case 12, neuronal intracytoplasmic eosinophilic viral inclusion (Negri body) in grade 2 artefact tissue (IHC × 40)

IHC analysis
Each anatomical location was semi-quantitatively graded 0-3 for the presence of specific neuronal immunolabelling by one observer (SB) [Fig. 1]. In the absence of suitable material on which to perform FAT, an IHC grade above 0 was considered a positive rabies diagnosis.

PCR analysis (Taqman and HnRT-PCR)
One assay of each kind was performed per fresh/frozen tissue pool (13 cases) or FFPE samples (Table 1). RNA was extracted from fresh brain tissue using the TriZol™ (Invitrogen, Life technologies, Paisley, UK) method following the manufacturer's instructions. RNA from FFPE samples was extracted from 80 μm of each individual paraffin block available from those same 13 cases using the RecoverAll™ kit (Ambion, Life technologies, Paisley, UK), and each block was tested separately one to three times. FFPE controls included a negative canine brain and two samples from a canine case that died in UK quarantine [21]. The pan-*lyssavirus* primer Jw12 was used to generate cDNA by reverse transcription. This was used to perform HnRT-PCR assays with primers Jw12 and Jw6dpl and Taqman RT-PCR assays with primers JW12 and N165-146, as previously described [16, 17, 34].

Statistical analysis
The sensitivity of Negri body detection for the diagnosis of rabies and the presence of perivascular cuffing, gliosis and satellitosis/neuronophagia were calculated per case and by general anatomic area. The median IHC labelling and HE scores for the general anatomic areas available from positive cases were compared using a Kruskal-Wallis H test. Statistical significance was set at $p < 0.05$. Post hoc pairwise comparisons were performed using Dunn's (1964) procedure with a Bonferroni correction for multiple comparisons; statistical significance was set at $p < 0.0083$ and only corrected p-values are reported. The association between artefact score and IHC labelling score was analysed using a Spearman's rank-order correlation. A binary logistic regression was performed to ascertain the effect of artefact on the likelihood of detecting Negri bodies. PCR and IHC case results were compared but no statistical tests were required.

Results
The general anatomic areas available for examination from each case are contained in Table 1. There was a mean of 2.3 anatomic areas per case (range 1-4) with 22 cases having a single anatomic area studied (20 positive, 2 negative), 9 with 2 anatomic areas (9 positive, 0 negative), 15 with 3 (14 positive, 1 negative) and 11 with 4 (10 positive, 1 negative). Fifty three of fifty seven cases

Table 1 Summary of results for individual animals

Case	Cerebrum			Hippocampus			Cerebellum			Brainstem			Pooled Fresh PCR
	NB[b]	IHC[c]	Path[d]	NB	IHC	Path	NB	IHC	Path	NB	IHC	Path	
1	+	++	1/2/3/2	N/a[a]	N/a	N/a	N/a	N/a	N/a	N/a	N/a	N/a	N/a
2	–	+	1/3/2/2	N/a	N/a	N/a	N/a	N/a	N/a	N/a	N/a	N/a	N/a
3	N/a	N/a	N/a	N/a	N/a	N/a	–	+	2/0/1/2	N/a	N/a	N/a	N/a
4	+	++	0/3/2/2	+	++	0/3/2/2	+	+++	1/2/1/0	–	++	0/3/2/2	N/a
5	–	+	1/2/2/1	–	++	1/2/2/1	–	+	2/1/1/2	N/a	N/a	N/a	N/a
6	–	–	0/2/2/2	–	–	0/2/2/1	–	–	0/0/0/2	–	–	0/1/1/1	–
7	N/a	N/a	N/a	–	+	1/2/1/2	N/a	N/a	N/a	–	+++	1/2/1/2	N/a
8	–	–	0/1/1/2	N/a	N/a	N/a	–	–	0/2/2/0	–	–	0/1/1/2	N/a
9	–	–	0/1/1/2	N/a	N/a	N/a	N/a	N/a	N/a	–	+	2/2/2/1	+
10	N/a	N/a	N/a	–	++	0/2/1/1	–	++	0/1/0/1	+	+++	1/2/1/2	N/a
11	N/a	N/a	N/a	N/a	N/a	N/a	N/a	N/a	N/a	–	+++	1/2/2/1	+
12	N/a	N/a	N/a	+	+++	1/2/2/2	N/a	N/a	N/a	N/a	N/a	N/a	N/a
13	–	++	2/1/1/2	N/a	N/a	N/a	N/a	N/a	N/a	N/a	N/a	N/a	+
14	N/a	N/a	N/a	–	–	1/2/2/0	–	++	1/1/1/1	–	++	2/2/2/1	N/a
15	–	+++	1/1/1/2	N/a	N/a	N/a	+	+++	1/2/1/2	+	+++	1/1/1/1	N/a
16	–	+	0/2/1/2	–	++	2/3/2/2	–	+	2/2/1/1	–	++	2/3/3/0	N/a
17	–	++	0/2/1/1	–	++	0/2/1/1	–	++	0/1/1/1	–	+++	2/3/2/1	N/a
18	–	+++	1/2/2/0	N/a	N/a	N/a	+	++	1/1/1/1	–	+++	2/2/2/0	+
19	–	+++	0/2/1/1	–	++	0/1/2/0	N/a	N/a	N/a	N/a	N/a	N/a	+
20	–	+	0/1/1/1	N/a	N/a	N/a	N/a	N/a	N/a	N/a	N/a	N/a	+
21	+	++	1/2/3/1	+	+++	0/2/1/0	N/a	N/a	N/a	+	+++	0/2/2/0	+
22	N/a	N/a	N/a	N/a	N/a	N/a	N/a	N/a	N/a	+	+++	2/2/2/1	N/a
23	–	–	2/1/2/2	+	++	2/2/2/2	–	–	0/1/1/2	+	++	2/2/2/2	N/a
24	+	++	2/2/3/1	N/a	N/a	N/a	N/a	N/a	N/a	N/a	N/a	N/a	N/a
25	–	+++	2/2/1/1	+	+++	2/2/2/1	–	+++	3/1/1/0	–	+++	2/2/2/1	+
26	+	++	3/2/1/3	N/a	N/a	N/a	N/a	N/a	N/a	N/a	N/a	N/a	+
27	N/a	N/a	N/a	+	+++	0/3/3/0	+	+++	0/3/2/0	–	++	1/3/3/0	+
28	–	++	3/2/2/2	N/a	N/a	N/a	N/a	N/a	N/a	N/a	N/a	N/a	N/a
29	N/a	N/a	N/a	+	+++	1/3/2/2	N/a	N/a	N/a	+	+++	1/3/2/2	N/a
30	+	+++	3/3/2/0	N/a	N/a	N/a	N/a	N/a	N/a	N/a	N/a	NA	N/a
31	–	++	2/3/2/2	+	++	3/2/2/2	–	++	1/2/1/2	N/a	N/a	N/a	N/a
32	–	++	1/2/2/3	+	++	2/2/2/2	N/a	N/a	N/a	–	++	2/2/2/2	N/a
33	–	+	0/2/2/3	–	+	1/0/0/3	–	+	0/0/0/3	–	++	1/0/0/3	N/a
34	+	++	2/3/3/0	N/a	N/a	N/a	+	++	2/2/2/0	+	++	2/2/2/0	+
35	–	++	1/3/2/2	+	++	2/1/1/2	–	+++	2/2/2/3	–	++	2/1/1/2	N/a
36	+	++	0/1/1/0	+	+++	0/1/1/0	+	++	0/2/1/2	NA	NA	NA	N/a
37	–	–	0/1/1/0	–	++	0/2/2/0	–	+	0/0/0/1	–	++	3/3/2/0	N/a
38	–	+++	0/0/0/3	–	++	0/0/0/3	N/a	N/a	N/a	N/a	N/a	N/a	N/a
39	–	+	0/0/0/3	N/a	N/a	N/a	N/a	N/a	N/a	N/a	N/a	N/a	N/a
40	–	+	0/2/2/3	–	++	2/3/2/3	N/a	N/a	N/a	N/a	N/a	N/a	+
41	–	++	0/2/2/3	+	++	0/1/1/3	N/a	N/a	N/a	–	+++	1/3/3/2	N/a
42	–	++	2/2/0/3	+	++	2/2/1/2	N/a	N/a	N/a	–	++	3/3/1/2	N/a
43	+	+++	1/3/3/2	N/a	N/a	N/a	N/a	N/a	N/a	+	+++	1/3/2/2	N/a

Table 1 Summary of results for individual animals *(Continued)*

44	–	+	1/2/2/1	N/a	N/a	N/a	N/a	N/a	N/a	N/a	N/a	N/a	N/a
45	–	+	0/2/1/2	N/a	N/a	N/a	N/a	N/a	N/a	N/a	N/a	N/a	N/a
46	–	+	2/3/3/1	–	+	1/2/2/1	–	+	2/1/1/0	–	+++	3/3/3/1	N/a
47	+	++	2/2/2/2	N/a	N/a	N/a	N/a	N/a	N/a	N/a	N/a	N/a	N/a
48	+	++	3/3/3/0	N/a	N/a	N/a	N/a	N/a	N/a	N/a	N/a	N/a	N/a
49	+	+++	3/3/3/1	N/a	N/a	N/a	+	++	3/2/2/0	N/a	N/a	N/a	N/a
50	+	++	3/3/3/1	+	+++	3/3/3/0	+	+++	1/2/2/2	N/a	N/a	N/a	N/a
51	+	+++	3/3/3/3	N/a	N/a	N/a	N/a	N/a	N/a	N/a	N/a	N/a	N/a
52	–	–	0/3/3/1	N/a	N/a	N/a	N/a	N/a	N/a	N/a	N/a	N/a	N/a
53	+	++	2/2/1/2	N/a	N/a	N/a	N/a	N/a	N/a	N/a	N/a	N/a	N/a
54	+	+++	2/3/3/2	N/a	N/a	N/a	N/a	N/a	N/a	–	+++	1/2/2/2	N/a
55	–	+++	0/3/3/0	+	++	0/3/3/0	+	+++	1/2/2/0	+	+++	0/3/3/0	N/a
56	+	+++	2/3/3/0	N/a	N/a	N/a	N/a	N/a	N/a	N/a	N/a	N/a	N/a
57	–	–	0/0/0/3	N/a	N/a	N/a	N/a	N/a	N/a	N/a	N/a	N/a	N/a

Where two or more of the same anatomical area were available for analysis the highest scores are presented

[a]: Not available

[b]: Negri body observed (– or +)

[c]: Viral antigen immunohistochemical labelling score (– to +++)

[d]: Perivascular cuffing score/ Gliosis score/ Satellitosis and neuronophagia score/ Artefact score

(93%) were positive by IHC for rabies viral antigen in at least one anatomical area.

Histopathology

Perivascular cuffing (47/53; 88.7%), glial proliferation (50/53; 94.3%) and satellitosis/neuronophagia (51/53; 96.2%) were frequently found in rabies positive cases. The mean rank of perivascular cuffing scores were not statistically significantly different, χ^2 [3] = 6.054, p = .109. Glial ($\chi^2(3)$ = 18.376, p = < .001) and satellitosis/neuronophagia ($\chi^2(3)$ = 17.686, p = .001) scores were statistically significantly different between general anatomical areas. Post hoc analysis revealed statistically significant differences in glial scores between the cerebellum (mean rank =45.90) and hippocampus (75.65) (p = .050), cerebellum and cerebrum (82.04) (p = .001) and cerebellum and brainstem (mean rank =90.51) (p = < .001). Overall the cerebellum had the lowest glial score in rabies positive cases. There were no statistically significant differences between other anatomical areas. Post hoc analysis revealed statistically significant differences in satellitosis/neuronophagia scores between the cerebellum (mean rank =46.50) and cerebrum (mean rank =84.12) (p = .001) and cerebellum and brainstem (mean rank =88.59) (p = .001).

The overall sensitivity of Negri body detection in rabies positive cases was 60.4% (32/53). Negri body detection was most sensitive in the hippocampus (15/30; 50.0%) and least sensitive in the brain stem (9/34; 26.5%). Inversely, perivascular cuffing in rabies positive cases was most and least commonly observed in the brainstem (30/34; 88.2%) and hippocampus (17/30; 56.7%) respectively. Negri bodies were not found in the negative IHC cases, however in

one negative case there was histopathological evidence of granulomatous meningoencephalitis. In the remaining three negative cases a cause for clinical suspicion of rabies was not determined.

Association between artefact grade and histopathology scores

A binary logistic regression was performed to ascertain the effect of the presence of artefact on the likelihood of detecting Negri bodies. The artefact score was dichotomised as either no artefact (score 0) or artefact present (scores 1, 2 and 3). The logistic regression model was statistically significant, $\chi2$ [1] = 8.672, (p = .003). The model explained 7.1% (Nagelkerke R2) of the variance in Negri body detection and correctly classified 67.7% of cases. Negri bodies were 0.34 times less likely to be detected in tissue with artefactual degeneration. Negri body detection sensitivity was almost twice as great (56.4% versus 30.8%) when there was no artefact (grade 0) compared to artefact (grade 1,2 or 3). The sensitivities of perivascular cuffing, glial proliferation and satellitosis/neuronophagia varied by less than 10% between grades when there was artefact compared to when there was none. Therefore artefact did not significantly affect HE diagnosis of encephalitis, but reduced Negri body detection.

IHC results analysis

Specific immunolabelling was found within the perikaryon of infected neurones and within the neuropil in association with axonal processes. There was inter and intracase variability in specific patterns of labelling: including morphology, location and the number of neurones labelled, but detailed analysis was not appropriate in this series. A Kruskal-Wallis H test was conducted to determine if there were differences in rabies antigen IHC labelling scores between general anatomical areas: the brainstem (n = 34), cerebrum (n = 65), cerebellum (n = 24) and hippocampus (n = 30). IHC scores were statistically significantly different between the different general anatomic areas, χ^2 [3] = 15.859, p = .001. Post hoc analysis revealed statistically significant differences in rabies antigen IHC labelling scores between the brainstem (mean rank =101.37) and hippocampus (73.88) (p = .045), brainstem and cerebellum (66.94) (p = .010) and the brainstem and cerebrum (69.41) (p = .001). Overall, there were significantly greater IHC scores within the brainstem as compared to other areas. The differences between cerebrum, cerebellum and hippocampus were not statistically significant.

Association between artefact grade and IHC

A Spearman's rank-order correlation was performed to assess the relationship between artefact score and rabies virus IHC score. There was a moderate negative correlation between increasing artefact score and IHC score, r_s (162) = −.260 (p = .001). Therefore higher artefact scores are associated with lower IHC scores. When a Spearman's rank-order correlation was run to assess the association between artefact score and rabies virus immunoreactivity (positive/negative) there was no statistically significant correlation between increasing artefact score and overall IHC score, r_s (162) = −.089 (p = .255). Therefore case diagnosis was not affected by increasing artefact score.

PCR assays

Of the 13 cases tested using fresh tissue, 12 were positive and one was negative for rabies. There was total agreement between IHC diagnosis (positive/negative) and the fresh tissue PCR test results. Despite multiple attempts, HnRT-PCR and Taqman RT-PCR assays were negative in the matching sample sets using FFPE tissue. Taqman RT-PCR, but not HnRT-PCR, produced positive results in the two FFPE samples used as positive technique control.

Discussion

The purpose of this study was to assess the distribution of rabies viral antigen and inflammatory changes within the general anatomical areas of the brain alongside an evaluation of the effect of artefact and suitability of FFPE tissue for molecular diagnosis of rabies virus. In this large canine case series Negri bodies were present in 60% (32 of 53) of the positive cases, which is similar to a relatively recent report [1]. Similarly, as previously reported in canine brain, Negri bodies were most often detected in the hippocampus [3, 23]. There have been

reports of rare Negri body-like inclusions occurring in rabies negative canine brain, however in this series inclusions were not observed in IHC negative cases [26, 28].

The observation of Negri bodies was associated with lower scores of autolysis and freeze thaw artefact. Therefore, whilst of some value if no other method is available; the sensitivity of Negri body detection is such that other complementary tests must be employed to confirm or refute a diagnosis particularly in autolytic specimens.

The anatomical distribution of inflammation observed with HE staining is relevant to sampling site identification, pathogenesis and immunological studies. The pons and medulla have previously been identified as the site of greatest inflammation in a murine model and the brainstem has been described as the site of most intense inflammation in naturally infected paralytic forms of canine rabies [18, 19, 30]. In this series, rabies positive cases also exhibited perivascular cuffing, most frequently in the brain stem and least often within the hippocampus, where Negri bodies were more frequently detected. Negri bodies were least often identified in the brainstem; potentially this is because the inflammatory response is destructive and may obscure Negri body observation. The relationship between inflammation and paralytic or furious clinical presentation has previously been described [30] but was not possible in this study because the clinical information was unavailable.

The cerebellar glial score was significantly lower than other areas and satellitosis/neuronophagia score was significantly greater in the cerebrum and brainstem. The relative increase in inflammation within the brainstem, as compared to other structures, may be related to the natural time course of infection, whereby it is infected via retrograde transport of virus antigen from a distal site of inoculation (eg bite wound) before the cerebellum/hippocampus. Therefore, earlier local infection results in increased upregulation of inflammatory cytokines and lymphocyte perivascular cuffing, resulting in more severe leukocyte recruitment and inflammation. This is supported by previous work where mice were inoculated in the foot pad with *Lyssavirus*: increased immunolabelling of CCL2, CCL5 and CXCL10 was identified in caudal brain regions, particularly the medulla and pons associated with more severe inflammatory changes [18].

The IHC labelling score for rabies viral antigen was significantly greater within the brainstem. This is in agreement with previous investigations using fresh tissue with FAT and IHC labelling [5, 30] although contrary to two much smaller canine series [1, 31]. Viral entry into the brain is expected via the brainstem because natural canine infections occur from bites, either directly from a facial injury or via the spinal cord when limbs are affected. Other authors have suggested that the viral location in cholinergic rich areas of the brain is an intrinsic property of the rabies virus [32]. Subsequent further local spread to higher structures such as the cerebrum, hippocampus and cerebellum would then logically follow.

Viral antigen labelling score is greater within the brain stem, while Negri bodies are most frequently observed within the hippocampus. The reasons for this are unclear but this change may reflect the local destructive effects of inflammation within the brainstem (as previously discussed) that obscures Negri body visualisation, or simply the convenience of observing viral inclusions within large Purkinje neurones that are present in high numbers within the dentate gyrus. An undefined predisposition of Purkinje neurones to accumulate viral antigen, without an accompanying florid inflammatory response, cannot be excluded; but it would seem more likely that inflammation would follow viral antigen accumulation. It is therefore possible that the clinical effects of brain stem inflammation have supervened before hippocampal lesions have had time to progress and therefore the observed differences in the hippocampus and brain stem areas reflects the natural course of infection.

Less invasive necropsy techniques such as sampling the brain using a straw via the occipital route [5], has been validated for use in the field where full brain exposure and removal of the encephalon is impractical and carries an associated operator risk [36].

Artefactual degeneration was associated with reduced immunoreactivity, but it did not affect the overall positive or negative diagnosis. Formalin fixed tissue can be stored without undergoing further bacterial and autolytic degeneration prior to accessing laboratory facilities. Excessive cross-linking from prolonged formalin fixation may affect antigen retrieval, but this was not significant in the studied series [29]. Therefore IHC labelling of FFPE tissue offers an ideal method for rabies diagnosis when artefactual degeneration of fresh brain tissue is inevitable or where fresh tissue is not available from suspect cases.

Although FAT is the OIE prescribed tool for rabies diagnosis in animals there are reported weaknesses of the FAT: autolysis and bacterial contamination are known to substantially reduce sensitivity [1, 2, 12] and formalin fixation may make a sample unsuitable for routine testing [29, 36]. IHC labelling of brain tissue has previously been reported to be as sensitive as FAT, with possibly increased sensitivity in autolysed tissue [1, 36]. IHC labelling has been used as an additional confirmatory test in clinical cases [21, 35] but is not a routine diagnostic technique endorsed by the OIE [36]. The time taken to acquire a result, as compared to FAT, is relatively slow [29] alongside the potential for false positive

or negative results as a consequence of relying on poorly trained staff [36].

Rabies viral RNA extraction from available fresh and FFPE tissue was undertaken with subsequent successful HnRT-PCR and Taqman RT-PCR testing of fresh tissue. The results from this were in full agreement with the IHC diagnoses, providing evidence of concordance between the two methods in the absence of FAT testing. The artefact score of the tested samples encompassed the full range of observed preservation (score 0 to 3). PCR has previously been reported as a very reliable method for rabies diagnosis despite sample deterioration, which appears to have been the case in this study [8, 25].

The PCR tests undertaken using FFPE tissues from Sri Lanka were negative. The primers used (JW12-JW6dpl) for HnRT-PCR have an analytical sensitivity of 0.01 FFD_{50} [2] but the amplified fragment is relatively large at 606 base pairs [21]. Samples that have been fixed in formalin, and particularly prolonged fixation at warm temperatures, are subject to RNA fragmentation: this may be from formalin induced cross linking, RNAse activation or both [4, 6, 22]. This fragmentation means that retrieval of intact large sequences above approximately 200-300 bp from FFPE tissue has been shown to be unlikely to succeed [22]. This may explain the negative results using HnRT-PCR and implies that this technique is unlikely to ever succeed on formalin fixed tissue unless alternative extraction methods, which reverse formalin induced cross linking, are developed.

The Taqman RT-PCR (JW12 and N165-146) is reported to be 200 fold more sensitive than the HnRT-PCR, with a much shorter amplicon of approximately 100 base pairs [16, 34]. This assay is therefore more likely to provide positive results from infected FFPE tissue. The Sri Lankan FFPE test material was also negative in this series however APHA processed control material was positive. The reasons for this may potentially lie in the age of the tissue, method of storage and time from processing to analysis. The Sri Lankan test material was up to 2 years older than the control tissue, subject to an unknown length of fixation and type of storage, whereas APHA material had been fixed for 7 days, processed to wax and stored in optimal humidity and temperature controlled conditions. The effect of wax embedding on RNA degradation is also unknown. Previously, real time PCR has been used to amplify a 126 bp sequence successfully for typing of a specific African *Lyssavirus* (isolate 864/09) from formalin fixed (but not paraffin embedded) tissue [7]. This might mean that results using formalin fixed, but unprocessed material may be more likely to be successful. This would reflect commonly encountered diagnostic field samples.

In this series, detailed anatomical and cellular analysis was not appropriate because of sample degeneration;

however this reflects the nature of 'field' cases. Additionally every anatomical region was not available for each case. The cerebrum was over represented, possibly because the most dorsal structure is large and therefore relatively easy to sample when the entire brain is exposed at necropsy or if the prosector is unfamiliar with rabies diagnosis; for similar reasons the brain stem is under represented.

Conclusions

HE examination is essential to define the differential diagnoses of behaviour modifying conditions in rabies virus negative cases and for future analysis of viral pathogenesis, but it is unreliable as the sole method for rabies diagnosis, particularly where artefactual change has occurred. Formalin fixation and paraffin embedding does not prevent detection of rabies virus via IHC labelling even where artefactual degeneration has occurred. This could represent a pragmatic secondary assay for rabies diagnosis in the field because formalin fixation is a convenient and economical way to prevent sample degeneration. The brain stem was shown to be the site with most viral immunoreactivity, supporting recommended sampling protocols in favour of improved necropsy safety in the field. The use of HnRT-PCR for testing of FFPE tissue is unlikely to be successful but Taqman RT-PCR may offer a better chance for success.

Abbreviations
FAT: Fluorescent antibody test; FFPE: Formalin fixed paraffin embedded; HE: Haematoxylin and eosin; HnRT-PCR: Hemi nested reverse transcription polymerase chain reaction; IHC: Immunohistochemical; PCR: Polymerase chain reaction; RT-PCR: Reverse transcription polymerase chain reaction

Acknowledgements
The authors wish to thank Katja Voller for excellent technical assistance.

Funding
The authors acknowledge the support of Commonwealth Scholarship Commission through the Academic Fellowship LKCF-2011-210 for this study (Sri Lanka) and UK Defra, Scottish Government and Welsh Government under project SE0426 (APHA). The funding bodies had no role in the design of the study, collection, analysis or interpretation of data nor in writing the manuscript.

Authors' contributions
SB carried out histopathological, immunohistochemical and statistical analysis and drafted the manuscript. PG carried out histopathological and immunohistochemical analysis. DLH participated in the acquisition of funding, conception and coordination of the study including molecular testing. DJH participated in the conception and coordination of the study including immunohistochemical testing. DM carried out molecular testing. AOP participated in the conception of the study; specifically the statistical analysis and carried out statistical tests. ARF participated in the acquisition of funding, conception, design and coordination of the study. AN conceived of the study, participated in its design and coordination and helped to draft the manuscript. All authors read and approved the final manuscript.

Competing interests
The authors declare that they have no competing interests.

Consent to publication
Not applicable.

Ethics approval and consent to participate
All canine tissue was made available from the archives of the Division of Veterinary Pathology of Faculty of Veterinary Medicine and Animal Science, University of Peradeniya. The dogs were submitted for examination as stray rabies suspects (non-experimental) as part of local human health protection measures and therefore consultation with an ethics committee was not required. Diagnostic material was retained, however applicable consent was not required because owners do not exist.

Author details
[1]Pathology Department, Animal and Plant Health Agency, Weybridge, UK. [2]Department of Veterinary Pathobiology, University of Peradeniya, Peradeniya, Sri Lanka. [3]Animal and Plant Health Agency, Weybridge, UK. [4]Wildlife Zoonoses and Vector Borne Diseases Research Group, Animal and Plant Health Agency, Weybridge, UK.

References
1. Arslan A, Saglam Y, Temur A. Detection of rabies viral antigens in non-autolysed and autolysed tissues by using an immunoperoxidase technique. Vet Rec. 2004;155:550–2.
2. Babu A, Manoharan S, Ramadass P, Chandran N. Evaluation of RT-PCR assay for routine laboratory diagnosis of rabies in post mortem brain samples from different species of animals. Indian J Virol. 2012;23:392–3963.
3. Beauregard M, Boulanger P, Webster W. The use of fluorescent antibody staining in the diagnosis of rabies. Can J Comp Med Vet Sci. 1965;29:141–7.
4. Ben-Ezra J, Johnson D, Rossi J, et al. Effect of fixation on the amplification of nucleic acids from paraffin embedded material by the polymerase chain reaction. J Histochem Cytochem. 1991;39:351–4.
5. Bingham J, van der Merwe M. Distribution of rabies antigen in infected brain material: determining the reliability of different regions of the brain for the rabies fluorescent antibody test. J Virol Methods. 2002;101:85–94.
6. Bussolati G, Annaratone L, Medico E, et al. Formalin fixation at low temperature better preserves nucleic acid integrity. PLoS One. 2011; doi:10.1371/journal.pone.0021043.
7. Coertse J, Nel L, Sabeta C, et al. A case study of rabies diagnosis from formalin-fixed brain material. J S Afr Vet Assoc. 2011;82:250–3.
8. David D, Yakobson B, Rotenberg D, et al. Rabies virus detection by RT-PCR in decomposed naturally infected brains. Vet Microbiol. 2002;87:111–8.
9. Dyer J, Niezgoda M, Orciari L, et al. Evaluation of an indirect rapid immunohistochemistry test for the differentiation of rabies virus variants. J Virol Methods. 2013;190:29–33.
10. Finnegan C, Brookes S, Johnson N, et al. Rabies in North America and Europe. J R Soc Med. 2002;95:9–13.
11. Fooks A, Banyard A, Horton D, et al. Current status of rabies and prospects for elimination. Lancet. 2014; doi:10.1016/S0140-6736(13)62707-5.
12. Fooks A, Johnson N, Freuling C, et al. Emerging technologies for the detection of rabies virus: challenges and hopes in the 21st century. PLoS Negl Trop Dis. 2009; doi:10.1371/journal.pntd.0000530.
13. Fooks A, McElhinney L, Horton D, et al. Molecular tools for rabies diagnosis in animals. In: Fooks A, Müller T, editors. OIE, compendium of the OIE global conference on rabies control. Paris: World Organisation for Animal Health; 2012. p. 75–87.
14. Franka R, Smith T, Dyer J, et al. Current and future tools for global canine rabies elimination. Antivir Res. 2013;100:220–5.
15. Goldwasser R, Kissling R. Fluorescent antibody staining of street and fixed rabies virus antigens. Proc Soc Exp Biol Med. 1958;98:219–23.
16. Hayman D, Banyard A, Wakeley P, et al. A universal real-time assay for the detection of Lyssaviruses. J Virol Methods. 2011;177:87–93.
17. Heaton P, Johnstone P, McElhinney L, et al. Heminested PCR assay for detection of six genotypes of rabies and rabies related viruses. J Clin Microbiol. 1997;35:2762–6.

18. Hicks D, Nunez A, Banyard A, et al. Differential chemokine responses in the murine brain following lyssavirus infection. J Comp Pathol. 2013;149:446–62.
19. Hicks D, Nunez A, Healy D, et al. Comparative pathological study of the murine brain after experimental infection with classical rabies virus and European bat lyssaviruses. J Comp Pathol. 2009;140:113–26.
20. Johnson N, Freuling C, Horton D, et al. Imported rabies. European Union and Switzerland, 2001 - 2010. Emerg Infect Dis. 2011;17:753–4.
21. Johnson N, Nunez A, Marston D, et al. Investigation of an imported case of rabies in a juvenile dog with atypical presentation. Animals. 2011;1:402–13.
22. Kashofer K, Viertler C, Pichler M, Zatloukal K. Quality control of RNA preservation and extraction from paraffin embedded tissue: implications for RT-PCR and microarray analysis. PLoS One. 2013; doi:10.1371/journal.pone.0070714.
23. Leach C. Comparative methods of diagnosis of rabies in animals. Am J Public Health. 1938;28:162–6.
24. Marston D, Horton D, Ngeleja C, et al. Ikoma lyssavirus, highly divergent novel lyssavirus in an African civet. Emerg Infect Dis. 2012;18:664–7.
25. McElhinney L, Marston D, Brookes S, Fooks A. Effects of carcase decomposition on rabies virus infectivity and detection. J Virol Methods. 2014;207:110–3.
26. McQueen J, Lewis A, Schneider N. Rabies diagnosis by fluorescent antibody- its evaluation in a public health laboratory. Am J Public Health. 1960;50:1743–52.
27. Moore D, Sischo W, Hunter A, Miles T. Animal bite epidemiology and surveillance for rabies post exposure prophylaxis. J Am Vet Med Assoc. 2000;217:190–4.
28. Nietfield J, Rakich P, Tyler D, Bauer R. Rabies-like inclusions in dogs. J Vet Diagn Investig. 1989;1:333–8.
29. Rupprecht C, Hanlon C, Hemachudha T. Rabies re-examined. Lancet Infect Dis. 2002;2:327–43.
30. Shuangshoti S, Thepa N, Phukpattaranont P, et al. Reduced viral burden in paralytic compared to furious canine rabies is associated with prominent inflammation at the brain stem level. BMC Vet Res. 2013;9:31–41.
31. Stein L, Rech R, Harrison L, Brown C. Immunohistochemical study of rabies virus within the central nervous system of domestic and wildlife species. Vet Pathol. 2010;47:630–6.
32. Suja M, Mahadevan A, Madhusudhana S, et al. Neuroanatomical mapping of rabies nucleocapsid viral antigen distribution and apoptosis in pathogenesis in street dog rabies – an immunohistochemical study. Clin Neuropathol. 2009;28:113–24.
33. Suja S, Mahadevan A, Madhusudana S, Shankar S. Role of apoptosis in rabies viral encephalitis: a comparative study in mice, canine and human brain with a review of literature. Pathol Res Int. 2011; doi:10.4061/2011/374286.
34. Wakeley P, Johnson N, McElhinney L, et al. Development of a real-time, TaqMan reverse transcription-PCR assay for detection and differentiation of lyssavirus genotypes 1, 5, and 6. J Clin Microbiol. 2005;43:2786–92.
35. White J, Taylor S, Wolfram K, O'Conner B. Case report: rabies in a 10 week old puppy. Can Vet J. 2007;48:931–4.
36. World Organisation for Animal Health (OIE). Rabies. In: Manual of Diagnostic Tests and Vaccines for Terrestrial Animals. OIE. 2015. http://www.oie.int/international-standard-setting/terrestrial-manual/access-online/ Accessed 2 Jan 2016.

11

Development of monoclonal antibodies and serological assays including indirect ELISA and fluorescent microsphere immunoassays for diagnosis of porcine deltacoronavirus

Faten Okda[1,2], Steven Lawson[1*], Xiaodong Liu[1], Aaron Singrey[1], Travis Clement[1], Kyle Hain[1], Julie Nelson[1], Jane Christopher-Hennings[1] and Eric A. Nelson[1]

Abstract

Background: A novel porcine deltacoronavirus (PDCoV), also known as porcine coronavirus HKU15, was reported in China in 2012 and identified in the U.S. in early 2014. Since then, PDCoV has been identified in a number of U.S. states and linked with clinical disease including acute diarrhea and vomiting in the absence of other identifiable pathogens. Since PDCoV was just recently linked with clinical disease, few specific antibody-based reagents were available to assist in diagnosis of PDCoV and limited serological capabilities were available to detect an antibody response to this virus. Therefore, the overall objective of this project was to develop and validate selected diagnostic reagents and assays for PDCoV antigen and antibody detection.

Results: The nucleoprotein of PDCoV was expressed as a recombinant protein and purified for use as an antigen to immunize mice for polyclonal, hyperimmune sera and monoclonal antibody (mAb) production. The resulting mAbs were evaluated for use in fluorescent antibody staining methods to detect PDCoV infected cells following virus isolation attempts and for immunohistochemistry staining of intestinal tissues of infected pigs. The same antigen was used to develop serological tests to detect the antibody response to PDCoV in pigs following infection. Serum samples from swine herds with recent documentation of PDCoV infection and samples from expected naïve herds were used for initial assay optimization. The tests were optimized in a checkerboard fashion to reduce signal to noise ratios using samples of known status. Statistical analysis was performed to establish assay cutoff values and assess diagnostic sensitivities and specificities. At least 629 known negative serum samples and 311 known positive samples were evaluated for each assay. The enzyme linked immunosorbent assay (ELISA) showed diagnostic sensitivity (DSe) of 96.1 % and diagnostic specificity (DSp) of 96.2 %. The fluorescent microsphere immunoassay (FMIA) showed a DSe of 95.8 % and DSp of 98.1 %. Both ELISA and FMIA detected seroconversion of challenged pigs between 8–14 days post-infection (DPI). An indirect fluorescent antibody (IFA) test was also developed using cell culture adapted PDCoV for comparative purposes.

(Continued on next page)

* Correspondence: steven.lawson@sdstate.edu
[1]Veterinary & Biomedical Sciences Department, South Dakota State University, Brookings, SD, USA
Full list of author information is available at the end of the article

(Continued from previous page)

Conclusion: These new, specific reagents and serological assays will allow for improved diagnosis of PDCoV. Since many aspects of PDCoV infection and transmission are still not fully understood, the reagents and assays developed in this project should provide valuable tools to help understand this disease and to aid in the control and surveillance of porcine deltacoronavirus outbreaks.

Keywords: Porcine deltacoronavirus (PDCoV), Monoclonal antibodies, Serology, ELISA, Fluorescent microsphere immunoassay (FMIA)

Background

Coronaviruses are enveloped, positive sense RNA viruses divided among several genera, including *Alphacoronavirus*, *Betacoronavirus*, *Gammacoronavirus* and the recently described genus *Deltacoronavirus*. A novel porcine deltacoronavirus (PDCoV) was reported in China in 2012 and designated HKU15 [1]. Other important porcine coronaviruses include porcine epidemic diarrhea virus (PEDV), transmissible gastroenteritis virus (TGEV) and porcine respiratory coronavirus (PRCV); and are members of the genus *Alphacoronavirus* [2]. In February 2014, the Ohio Department of Agriculture announced the identification of PDCoV in swine feces at five farms in Ohio and associated with enteric disease similar to PEDV in the U.S. [3]. Since then, PDCoV has been identified in numerous U.S. states and Canada, linked with apparent clinical disease including acute diarrhea and vomiting in the absence of other identifiable pathogens. According to field observations in the U.S., PDCoV infections cause less severe clinical disease than PEDV, but analysis of the field data is complicated since co-infections with PEDV or other pathogens are common. PDCoV is currently diagnosed by real time PCR and clinical symptoms [1, 4].

The severity of disease in both gnotobiotic and conventional piglets has further defined the pathogenicity and pathogenesis of the virus [5–7]. PDCoV causes diarrhea and vomiting in all age groups and mortality in nursing pigs but the mortality rates are less than that shown in cases of PEDV. Previously, there was little information about deltacoronavirus infections in pigs and only one surveillance study from Hong Kong reported its detection in pigs prior to its emergence in the U.S. The virus had not been reported to be associated with clinical disease in China. The newly emergent strain found on the Ohio farms, PorCoV HKU15 OH 1987, is closely related to the 2 strains from China, but it is unknown how this virus was introduced into the US [3].

Recently, Jung et al. [7] developed in-situ hybridization and immunofluorescence staining techniques to demonstrate the areas of PDCoV replication in tissues of infected pigs. The OH-FD22 and OH-FD100 PDCoV strains were confirmed as causing an acute infection through the entire intestine, but primarily the jejunum and ileum, and clinically lead to severe diarrhea and vomiting. Clinical signs

and pathological features of PDCoV-infected pigs resemble those of PEDV and TGEV infections. Effective differential diagnosis between PDCoV, PEDV, and TGEV is important to control the diseases.

Polymerase chain reaction (PCR) assays were quickly developed for the detection of PDCoV infections following the initial U.S. identification in 2014 but available serological assays are limited. Thachil et al. [8] developed an indirect anti-PDCoV IgG enzyme-linked immunosorbent assay (ELISA) based on the S1 portion of the spike protein. Although this assay was shown to be a highly sensitive (91 %) and specific test (95 %), there is need for other ELISAs utilizing alternative antigen targets, such as the nucleoprotein of PDCoV, to serve as primary serological surveillance or confirmatory assays. As noted in Thachil's research, several serum samples collected in 2010 were found positive for PDCoV antibody by their ELISA, but all those collected in 2011 and 2012 tested negative by that assay. This finding is interesting since PDCoV was not thought to be circulating in North America prior to late 2013 [7, 9]. Therefore, availability of several serological assay formats targeting different viral antigens can be valuable as confirmatory assays in the investigation of unexpected laboratory findings.

Since no specific antibody-based reagents were available to assist in diagnosis of PDCoV, one purpose of the current study was to develop readily available reagents for detection of PDCoV antigen in diagnostic tests, such as virus isolation, immunohistochemistry and fluorescent antibody techniques. Serological tests for the detection of antibody responses to PDCoV were also very limited. Therefore, another objective of this study was to develop and optimize several serological assays including an indirect ELISA, a fluorescent microsphere immunoassay (FMIA), and an indirect fluorescent antibody (IFA) test.

Both specific antibody-based reagents and serological tests are essential for the further study and control of PDCoV and the differentiation of PDCoV infection from other related diseases such as PEDV or TGEV. The tools developed during the course of this study can be applied to many ongoing and future studies to better understand and control PDCoV.

Methods

Serum samples

Samples used for optimization and validation of the PDCoV ELISA and FMIA assays included samples from a large PDCoV challenge study associated with National Pork Board (NPB) research project 14–182. These samples and samples from another group of 30 pigs which were collected near the time of initial field exposure to PDCoV and 28 days later were used in a time course study.

For further validation and assessment of diagnostic sensitivity (DSe) and diagnostic specificity (DSp), samples of known PDCoV serostatus were used ($n = 940$). The expected positive samples were submitted field serum samples ($n = 311$) from herds previously testing PDCoV positive by PCR at least 3 weeks prior to sample collection. Expected negative samples included archived experimental serum collected prior to 2009 ($n = 108$) and field samples from high-health herds with no known history of PDCoV exposure ($n = 521$). The total number of expected negative samples was 629.

Viruses and cells

Swine testicle (ST) cells were cultured in Eagle's minimum essential medium (MEM; Gibco BRL Life Technologies) supplemented with 10 % fetal bovine serum (FBS), and antibiotics (100 units/ml penicillin and 20 g/ml streptomycin). Cells were maintained at 37 °C in a humidified 5 % CO_2 incubator. Cell culture adapted PDCoV was provided by the National Veterinary Services Laboratories, designated porcine coronavirus HKU15 strain Michigan/8977/2014 (GenBank accession KM012168). PDCoV was propagated on ST cells utilizing 0.8 µg/ml trypsin (TPCK-treated, bovine derived (Sigma, St. Louis, MO)) in the inoculation and maintenance media. Virus infected cells were harvested 24–48 h after inoculation, when significant cytopathic effect (CPE) was noted.

Antigen production

The antigen used for the FMIA and indirect ELISA validation was a recombinantly expressed, full length, PDCoV nucleoprotein (NP). RNA isolated from semi-purified cultured virus corresponding to the PDCoV-NP was amplified by reverse transcriptase polymerase chain reaction (RT-PCR) and resulting DNA cloned into the pET-28a protein expression vector (Novagen, Madison, WI). Primers used for amplification of the nucleocapsid region are described:

PDCoV-NP fwd (5′-CGCGGATCCATGGCCGCACC AGTAGTC - 3′);

PDCoV-NP rev (5′-CACACTCGAGCGCGCTGCTG ATTCCTGCTT- 3′).

The NP gene was prokaryotically expressed as an insoluble 41 kDa, 6x polyhistidine-tagged, fusion protein then purified according to previously described methods [10]. Purified protein was analyzed by sodium dodecyl sulfate- polyacrylamide gel electrophoresis (SDS-PAGE) to determine purity and linear integrity. The expressed PDCoV NP was recognized in Western blotting by convalescent serum and two separate monoclonal antibodies developed in our laboratory were used to confirm the specificity of the proteins.

Refolding of the PDCoV NP purified protein

The purified protein was refolded by first solubilizing the recombinant protein expressed as insoluble inclusion bodies in *E. coli*. Briefly, the protein was solubilized in 50 mM 3-(Cyclohexylamino)-1-propanesulfonic acid (CAPS buffer, pH 11.0) containing 1.0 % N-lauroylsarcosine and 1.0 mM DTT. The protein was then dialyzed overnight at 4 °C in 20 mM Tris–HCl pH 8.5 containing 0.1 mM DTT to encourage correct disulfide bond formation and subsequent refolding of the protein. A second dialysis step was done in phosphate buffered saline (PBS) to remove excess reducing agent. Testing of the re-folded NP-based ELISA and FMIA began with checkerboard titrations of both the antigen and PDCoV convalescent sera to determine optimum concentrations of each. Depending on the calculation of signal to noise ratios, optimum concentration of NP antigen was identified for coating of the ELISA plates and coupling FMIA beads. We also identified optimum test serum dilutions and blocking agents.

Development and diagnostic application of rabbit antisera and monoclonal antibodies

Rabbits and mice were immunized with selected recombinant PDCoV proteins for production of hyperimmune antisera and monoclonal antibodies (mAbs) as previously described [11–13]. Immunoglobulin isotyping of the resulting mAbs was performed using a commercial lateral flow assay (Serotec, Raleigh, NC).

Indirect fluorescent antibody assay

An IFA assay was developed for reference purposes using pig serum of known serostatus. ST cells were grown in cultures for 2 to 3 days to 80 % confluence on 96-well plates. Odd numbered lanes were infected with PDCoV (approximately 1000 50 % tissue culture infective doses ($TCID_{50}$)/ml) in MEM supplemented with 0.8 ug/ml TPCK-treated trypsin. The plates were incubated for 18 to 24 h. then fixed with 50 % (vol/vol) acetone/methanol for 20 min at –20 °C, air dried, and frozen with a desiccant at –20 °C until they were used. Serum dilutions of 1:20 and 1:40 were applied to infected and control wells of the IFA plates and incubated 1 h. After washing three times with 300 µl of phosphate buffered saline (PBS), 40 µl of fluorescein isothiocyanate (FITC)-

labeled goat anti-swine immunoglobulin G (41.7 g/ml; KPL, West Chester, PA) was added to each well. The plates were incubated at 37 °C for 1 h and washed with PBS three times. The cells were examined for specific fluorescence with an inverted microscope and a UV light source (Nikon Eclipse TS100). Serum samples were considered positive if PDCoV specific fluorescence was observed at the 1:20 serum dilution.

Antibody detection indirect ELISA

The refolded PDCoV NP antigen-based indirect ELISA was performed using methods previously described in Okda et al. [13]. Briefly, alternate wells of Immulon 1B, 96-well, microtiter plates (Thermo Labsystems, Franklin, MA) were coated for 1 h at 37 °C with 200 ng/well of purified, refolded PDCoV-NP antigen diluted in 15 mM sodium carbonate-35 mM sodium bicarbonate, antigen coating buffer (ACB) pH 9.6. Next, non-bound antigen was poured off and the plates washed 3X with PBS plus 0.05 % tween-20 (PBST), then the remaining free-binding sites were blocked with 200 μl of sample milk diluent ((SMD)-PBST plus 5 % nonfat dry milk) and incubated overnight at 4 °C. Test and control sera were diluted 1:50 in SMD, and 100 μl of the solution was added to each well of a washed plate. The plates were incubated for 1 h at 22 °C. Next, 100 μl of biotinylated, FC-specific, goat anti-swine detection antibody (Bethyl Laboratories, Montgomery, TX) was diluted 1:4000 in PBST and allowed to incubate at 22 °C for 1 h. Plates were washed 3X, then 100 μl of streptavidin-horseradish peroxidase conjugate (Pierce, Rockford, IL, diluted 1:4000) was added and incubated for 1 h at 22 °C, then washed and developed using TMB (Surmodics, Eden Prairie, MN). Colorimetric development was stopped using 2 N H_2SO_4, then OD's were quantified spectrophotometrically at 450 nm with a ELx800 microplate reader (BioTek Instruments Inc., Winooski, VT). The raw OD's were normalized and the corrected S/P values were calculated as follows: S/P = (OD of sample - OD of buffer)/(OD of positive control - OD of buffer).

The optimal dilution of the recombinant protein and secondary detection antibody was determined by a checkerboard titration that gave the highest signal to noise ratio. In addition, a single lot of pooled convalescent serum from PDCoV infected pigs was used to generate quality control standards that gave high, medium and low Sample to Positive (S/P) values. The negative, low and medium samples served as internal quality standards while the high standard served as a serum constant to mathematically calculate S/P values of individual unknowns. For the ELISA, a high positive S/P = 0.8–1.0; medium S/P = 0.6–0.8; low S/P = 0.4–0.6; and negative S/P < 0.2.

Microsphere coupling and FMIA procedure

The coupling of purified, recombinant, refolded PDCoV-NP antigen to fluorescent microspheres was performed using a two-step, carbodiimide coupling reaction as previously described [14]. Prior to performing large scale coupling reactions for test validation, the optimization of the amount of antigen used was obtained by performing a checkerboard titration of antigen-coupled microspheres against a two-fold titration of swine serum. It was found that initiating a coupling reaction having 12.5ug of purified protein per 3.125×10^6 microspheres was optimal in obtaining the highest signal-to-noise fluorescence ratio. The performance of the FMIA test was described in detail previously by Okda et al. [11]. In the initial optimization of the FMIA test, we performed two-fold serial dilutions of swine serum and concluded that a dilution of 1:50 provided the highest signal-to-noise ratio. For the generation of sample fluorescence, antigen-coupled microsphere/antibody complexes were analyzed through a dual-laser Bio-Rad Bio-Plex 200 instrument. The median fluorescent intensity (MFI) for 100 microspheres corresponding to each individual bead analyte was recorded for each well. All reported MFI measurements were normalized by calculating individual S/P values using the following formula: S/P = MFI of sample - MFI of buffer control/MFI of high positive control - MFI of buffer control. The buffer control equated to the background signal determined from the fluorescence measurement of antigen-coated beads. Lastly, a single lot of pooled convalescent serum from PDCoV infected pigs was used to generate quality control standards that gave high, medium and low S/P values. The negative, low and medium samples served as internal quality standards while the high standard served as a serum constant to mathematically calculate S/P values of individual unknowns. For the FMIA, a high positive S/P = 0.8–1.0; medium S/P = 0.6–0.8; low S/P = 0.4–0.6; and negative S/P < 0.2, is consistent with the data of the ELISA standards.

Antibody capture efficacy comparison between refolded vs linear PDCoV-NP antigen

An antibody capture titration assay was employed to compare the efficacy of refolded vs. linear antigen to capture anti-PDCoV-NP specific antibody in swine serum. Wells of a 96-well microtiter plate were coated with 10-fold decreasing concentrations of either linearly expressed or refolded PDCoV-NP antigen in ACB, pH 9.6., then allowed to incubate for 1 h at 37 °C. Each well was then blocked with 200 μl of SMD and allowed to incubate overnight at 4 °C. The following day, the plates were washed 3X with 300 μl of PBST. Well characterized positive control sera having a "high" positive OD was diluted 1/50 in SMD, mixed, and 100 μl of the solution was added to each well. The ELISA was continued pursuant to the stated protocol and the OD was recorded at each titration point, then a logarithmic regression curve was generated. Relative capture efficiencies for each antigen-coated well was determined by analyzing the OD at each dilution point and position under the linear portion of the curve.

Assay validation

(i) Cutoff determination, DSe and DSp. To accurately assess the DSe and DSp of the assays, the assays were validated using known seronegative and seropositive samples from distinct animal populations. The expected positive samples used were field serum samples submitted to the ADRDL from herds previously testing PDCoV positive by PCR. Expected negative samples included archived experimental serum collected prior to 2009 and field samples from high-health herds with no known history of PDCoV exposure. Receiver operating characteristic (ROC) analysis was performed using MedCalc version 11.1.1.0 (MedCalc software, Mariakerke, Belgium).

(ii) Measurement of repeatability. The repeatability of the FMIA and ELISA was assessed by running the same lot of internal quality control serum standards multiple times on the same plates and on different plates over time. The intra-assay repeatability was calculated using 36 replicates on a single plate and then repeated over a 3-day period for inter-assay repeatability assessment. Each assay was run in a single-plex format, and median fluorescence intensity values were expressed as percent coefficient of variation (CV) for repeated measurements. Percent CV was calculated using a method described earlier [15].

Statistical analyses and measurement of testing agreement

Multiple comparison, inter-rater agreement (kappa measure of association) was calculated among all three tests (ELISA, FMIA and IFA) using IBM, SPSS version 20 software (SPSS Inc., Chicago, IL). The sample cohort used included a set of archived serum samples collected from PDCoV "positive testing" experimentally infected pigs over time and from archived experimental control uninfected PDCoV "negative testing" animals. The interpretation of kappa can be rated as follows: Kappa less than 0.0, "poor" agreement; between 0.0 and 0.20, "slight" agreement; between 0.21 and 0.40, "fair" agreement; between 0.41 and 0.60, "moderate" agreement; between 0.61 and 0.80, "substantial" agreement; and between 0.81 and 1.0, "almost perfect" agreement [16].

Validation of the tests was performed by ROC analysis using Medcalc statistical software. Correlations between the tests and scatterplots for seroconversion were performed using SPSS 20.

Results
Expression of recombinant full-length nucleoprotein

The full-length NP of PDCoV was cloned and expressed in *E. coli* as a polyhistidine fusion protein. Antigen purity was then evaluated using SDS-PAGE in which the His-tagged recombinant NP migrated through the gel according to its predicted molecular mass of 41 kDa upon staining with Coomassie brilliant blue R250 (Fig. 1a). The recombinant protein was expressed in the form of insoluble inclusion bodies. It was purified by Nickel-NTA affinity column chromatography and yielded a calculated concentration of approximately 10 mg PDCoV-NP/liter of 2XYT medium and having a purity greater than 95 % as measured by the Lowry protein assay. The protein was subsequently refolded back to its soluble, conformational structure and its specificity was tested via Western blotting (Fig. 1b). The figure illustrates the migration pattern and antigen specificity of the refolded PDCoV-NP/polyhistidine fusion protein as compared to the adjacent lane loaded with semi-purified, concentrated PDCoV. Both the rPDCoV-NP and native virus are recognized with equal intensity by a PDCoV-NP-specific monoclonal antibody (SD55-197) developed in our laboratory. Also, the recombinant nucleocapsid protein is shown to have a higher molecular mass than the native virus nucleocapsid due to its dual amino and carboxy terminus polyhistidine tags.

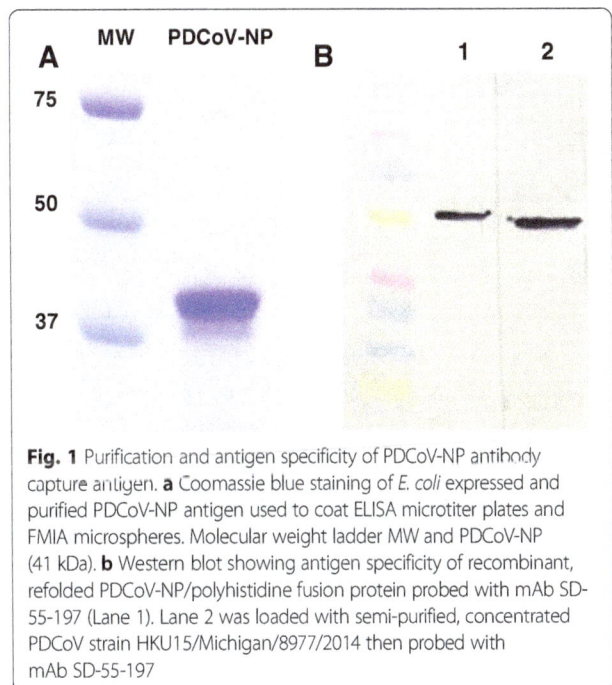

Fig. 1 Purification and antigen specificity of PDCoV-NP antibody capture antigen. **a** Coomassie blue staining of *E. coli* expressed and purified PDCoV-NP antigen used to coat ELISA microtiter plates and FMIA microspheres. Molecular weight ladder MW and PDCoV-NP (41 kDa). **b** Western blot showing antigen specificity of recombinant, refolded PDCoV-NP/polyhistidine fusion protein probed with mAb SD-55-197 (Lane 1). Lane 2 was loaded with semi-purified, concentrated PDCoV strain HKU15/Michigan/8977/2014 then probed with mAb SD-55-197

Experiments were conducted to assess the immunore-activity of refolded vs non-refolded PDCoV-NP antigen used for both tests. Specifically, an antibody capture ti-tration ELISA was employed to compare the ability of a refolded and non-refolded version of the antigen to cap-ture antibodies within swine serum. The immunoreactiv-ity of antigen was determined by end-point titration and relative absorbance values as we observed differences in immunoreactivity based upon the conformational state of antigen tested. Figure 2 demonstrates in a dose-dependent fashion, that as the concentration of each antigen coated well decreases, the refolded antigen im-parts a greater degree of antibody capture efficacy of swine antibodies than the non-refolded version. Spe-cifically, a 27 fold difference was calculated at the end of the linear portion of the curve indicating that the refolded protein maintained a marked enhance-ment of immunoreactivity resulting in a greater dy-namic range of the assay.

Fluorescent microsphere immunoassay and indirect ELISA development

(i) Establishment of control standards: ELISA and FMIA test reference standards for PDCoV were developed for each of the respective prototype assays (Fig. 3a and b). A control high positive standard was established by pooling several lots of serum collected from convalescent, seropositive pigs, and used to mathematically calculate S/P values of each assay. The medium and low positive samples served as internal quality control standards. The negative standard was pooled from a set of known

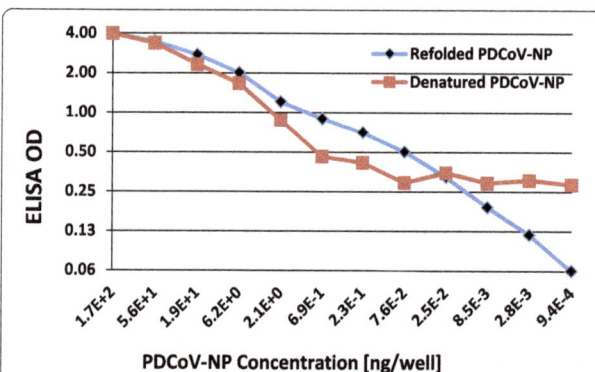

Fig. 2 Antibody capture efficacy comparison between refolded vs linear PDCoV-NP. An ELISA antibody capture titration assay was employed to compare the ability of refolded vs. linear antigen to capture anti-PDCoV-NP specific antibody in swine serum. Wells of a 96-well microtiter plate were coated with decreasing concentrations of either linearly expressed or refolded PDCoV-NP antigen. Refolded antigen demonstrated greater dynamic range of the assay and capture efficacy of swine antibody

seronegative pigs from a herd with no known prior PDCoV infection that also tested negative by PDCoV IFA and virus neutralization.

(ii) Test optimization: A series of coupling processes were performed using a two-fold titration of antigen to determine the optimum coupling concentration. A total of 3.125×10^6 beads, were incubated with various concentrations (100 µg, 50 µg, 25 µg, and 12.5 µg) of purified NP. Based upon the highest signal-to-noise ratio reflecting the detection of PDCoV-specific antibodies in standard serum, 12.5 µg per reaction was the optimal concentration for microsphere coupling. ELISA microtiter plates were coated with an optimized concentration of 200 ng antigen per well.

Testing the re-folded, NP antigen-based ELISA and FMIA began with checkerboard titration of both antigen and PDCoV convalescent and naïve swine serum to de-termine the optimum signal-to-noise ratio of each test. We also identified optimum test serum dilutions and blocking agents. The optimal serum dilution of each assay was determined by diluting serum samples two-fold in their respective blocking/detergent buffer diluent. For both the FMIA and ELISA, a serum dilution of 1:50 was shown to produce an optimal signal-to-noise ratio (Fig. 3c and d). Testing field samples of known sero-logical status was performed to gauge initial sensitivity of the assay. Our results showed a positive to negative sample ratio (P/N) of greater than 16-fold with ELISA and a P/N of greater than 40-fold with FMIA. The P/N is a relative measure of PDCoV antibody concentration between seropositive and seronegative samples. Having a diagnostic P/N of greater than 10 is highly desirable with any serological assay.

Assessment of test repeatability

The intra-assay repeatability was calculated for 36 repli-cates on a single plate and then repeated over a 3-day period for inter-assay repeatability assessment. Internal control serum standards were used to determine the precision of each FMIA and ELISA. The inter-assay and intra-assay repeatability of each test demonstrated a co-efficient of variation of less than 8.6 %. These results confirmed that the serological tests are highly repeatable in diagnostic applications.

Validation methods and cutoff determination

ROC analysis of both FMIA and ELISA was per-formed using MedCalc software to calculate an opti-mized cutoff value that maximizes the DSe and DSp of each assay. Using known seronegative and sero-positive serum samples ($n = 940$), the expected posi-tive samples used were submitted as field serum

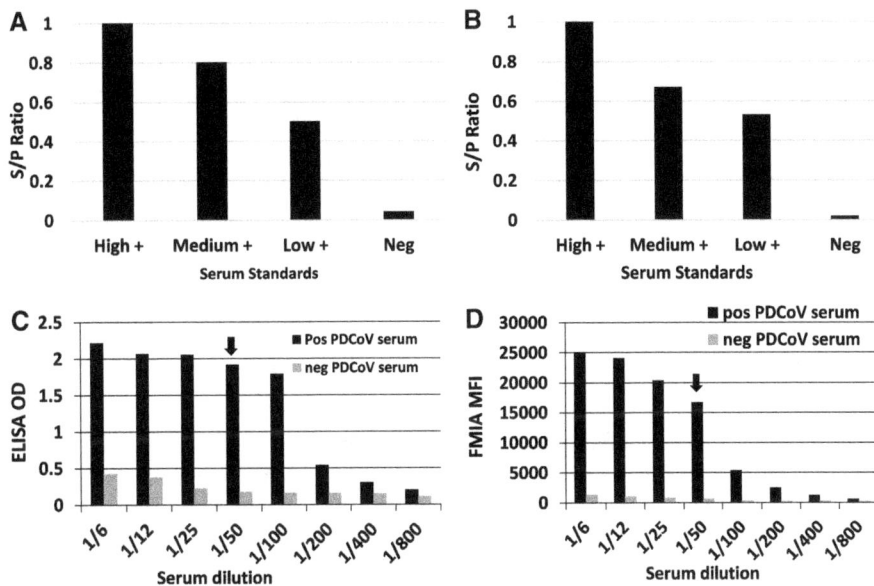

Fig. 3 Production and reactivity of serological reference standards and internal quality control standards for both ELISA (**a**) & FMIA (**b**). Reference serum standards were titrated 2-fold in antigen coated wells at a fixed concentration [250 ng/well] in order to gauge a maximum signal-to-noise ratios for each assay **c** ELISA, and **d** FMIA. Arrows indicate the optimal serum dilution that resulted in the greatest positive to negative (P/N) ratio for the test

samples ($n = 311$) from herds previously testing PDCoV positive by PCR. Expected negative samples included archived experimental serum collected prior to 2009 ($n = 108$) and field samples from high-health herds with no known history of PDCoV exposure ($n = 521$). ROC analysis was performed and DSe and DSp were shown to be 96.1 and 96.2 % respectively for the ELISA; and 95.8 and 98.1 % respectively for the FMIA (Fig. 4). The similar cutoff values of both assays confirm the utility of these new diagnostic tests to aid in the control and surveillance of PDCoV outbreaks.

Evaluation of the kinetic swine antibody response in serum

Once validated, the assays were used to evaluate the kinetic antibody response over time using serum collected from experimentally infected pigs over weekly intervals. Serological responses detected by the PDCoV ELISA and FMIA following challenge show seroconversion between days 8 and 14 DPI (Fig. 5). Both assays demonstrate a similar dynamic range using the same serum samples and "high" positive standard from which the S/P values are calculated. However, the FMIA appears to detect a slightly higher level of antibody over a longer

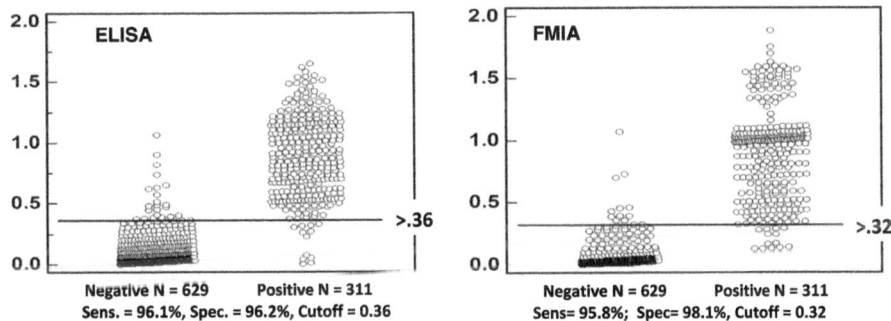

Fig. 4 Receiver operator characteristic (ROC) validation and determination of diagnostic sensitivity (DSe) and specificity (DSp) of the PDCoV-NP ELISA and FMIA assays. DSe and DSp were calculated using serum samples from known PDCoV-infected and known PDCoV-uninfected populations. ROC analysis was performed using MedCalc version 11.1.1.0 (MedCalc software, Mariakerke, Belgium). In each panel, the dot plot on the left represents the negative population, and the plot on the right represents the positive population. The horizontal line bisecting the dot plots for each figure represents the tentative cutoff value that gives the optimal DSe and DSp

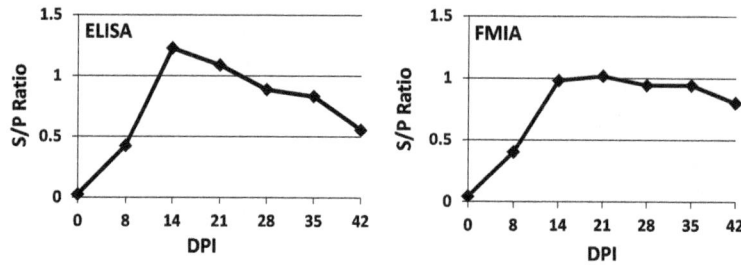

Fig. 5 Serum antibody kinetic time course evaluation. Antibody kinetic responses were calculated for the ELISA and FMIA tests using serum collected weekly over six weeks from experimentally infected pigs. Serological responses detected by PDCoV ELISA and FMIA tests show seroconversion between 8 and 14 DPI

period of time at days 35 and 42 post infection. Additional testing of seroconversion included serum samples collected from a group of 30 piglets near the time of initial field exposure to PDCoV then 28 days later. Figure 6 shows clear seroconversion to the naturally circulating virus within the 28 day time-frame using the same diagnostic cutoff values previously determined by ROC analysis.

Measurement of statistical testing agreement

Multiple comparison, inter-rater (kappa) agreement is a statistical measure of testing agreement, and was calculated among all three tests (ELISA, FMIA & IFA) using 629 positive testing and 311 negative testing serum samples. Statistical comparison calculated kappa values to be 0.940 between FMIA and IFA, 0.902 between ELISA and IFA, and 0.914 between ELISA and FMIA (Table 1). Because all three diagnostic platforms had kappa values above 0.81, it demonstrates that the tests are in "almost perfect" agreement with each other according to the interpretation of kappa by Landis et al. [16].

Cross reactivity

Serological cross reactivity testing was performed between PDCoV and other closely related swine coronaviruses. There was no cross reactivity among the 93 TGEV positive serum samples, 20 PRCV positive serum samples, 167 PEDV field positive serum samples and 84 PEDV experimentally positive serum samples tested via ELISA and FMIA (Table 2). The data show mean OD and MFI readings from ELISA and FMIA tests, respectively. The lack of cross reactivity between PDCoV and aforementioned alphacoronavirus species was also confirmed via western blotting using seropositive, convalescent sera from individual pigs (data not shown).

Development of reagents for detection of PDCoV antigen in diagnostic tests

Both denatured and refolded versions of the NP were used to immunize rabbits for hyperimmune serum and mice for monoclonal antibody production. Rabbit hyperimmune sera specifically recognize the NP and can be used in indirect fluorescent antibody staining at dilutions of 1:1000 to 1:5000. In addition, the polyclonal antisera was used successfully in immunohistochemical staining

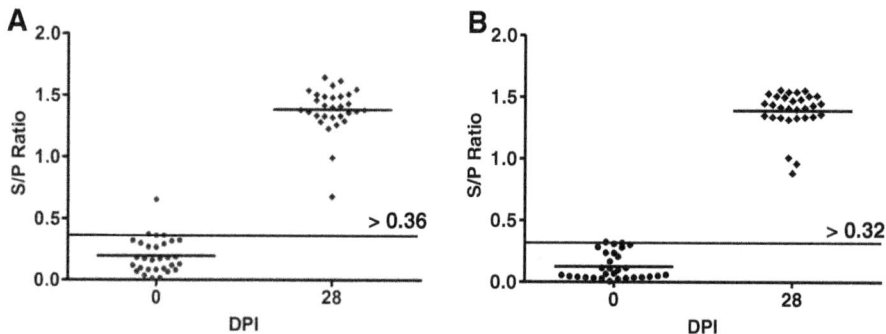

Fig. 6 ELISA (**a**) and FMIA (**b**) results from a group of 30 piglets sampled near the time of initial PDCoV field exposure then 28 days later. Both assays show clear seroconversion to naturally circulating PDCoV. The horizontal line between both positive and negative testing samples shows the diagnostic cutoff value for each test previously determined by ROC analysis

Table 1 Evaluation of statistical testing agreement among three serological testing platforms

	FMIA	Indirect ELISA	IFA
IFA	0.940	0.902	1
Indirect ELISA	0.914	1	0.902
FMIA	1	0.914	0.940
Number positive serum samples	311	311	315
Number negative serum samples	629	629	625
Total	940	940	940

Multiple comparison, inter-rater agreement (kappa association) were calculated among all three tests, IFA, indirect ELISA and FMIA. Kappa values shown represent a statistical measure of test agreement and were calculated using MedCalc version 11.1.1.0

procedures for the detection of PDCoV antigen in intestinal tissues (Fig. 7b). The resulting monoclonal antibodies all recognized native viral protein in infected ST cells demonstrated by bright cytoplasmic immunofluorescent staining (Fig. 7a) and within intestinal enterocytes stained immunohistochemically (Fig. 7c).

Discussion

As a recently identified pathogen, the impact of PDCoV on the swine industry is not yet fully understood. Field observations and recent research studies have suggested that the virus can cause substantial morbidity and mortality in nursing piglets [5–7, 17]. Specific antibody-based reagents and serological tests are essential for the further study and control of PDCoV, as well as the differentiation of PDCoV infection from other related coronaviruses such as PEDV or TGEV.

Therefore, the primary objective of this study was to develop an initial generation of antibody-based diagnostic reagents and serological assays for the further study of PDCoV, including NP-based indirect ELISA and FMIA tests that were developed and validated for diagnostic applications. Specific antibody-based reagents were also not yet available for PDCoV so monoclonal antibodies against selected PDCoV structural proteins were developed. The tools developed during the course of this study can be applied to many ongoing and future studies to better understand and control PDCoV.

The new monoclonal antibody reagents described here should be of substantial value in the detection of PDCoV

Table 2 Serological cross reactivity testing among related swine coronaviruses

	TGEV	PRCV	PEDV
ELISA Mean OD	0.097	0.132	0.074
FMIA Mean FI	0.024	0.016	0.032
Total No. serum samples tested	93	20	251

antigen in a variety of applications including: early verification of virus isolation attempts and virus titrations; immunohistochemistry staining of fixed tissues; development of neutralization assays; fluorescent antibody staining of fresh tissues; development of field-based antigen capture assays such as lateral flow devices; and ELISA applications (competitive ELISA and antigen capture). Through extensive testing via ELISA (over 364 samples) and Western blot analysis, we were not able to demonstrate any cross-reactivity with other major swine coronaviruses including PEDV, PRCV and TGEV. However, since many described deltacoronaviruses of other species have not yet been adapted to cell culture replication or fully characterized, we do not yet know if these reagents may cross-react with other members of the genus.

Several new serological assays for detection of antibody responses to PDCoV were developed during the course of this study. The ELISA and FMIA tests were based on a recombinant nucleoprotein antigen since this protein is highly conserved among PDCoV isolates. In addition, the NP is known to be the most abundant viral protein within host cytoplasmic compartments [18]. The highly immunoreactive PDCoV-NP interacts with itself to form non-covalently linked oligomers and associates with the viral genome to serve as the architectural basis of the ribonucleoprotein complexes during virus assembly [19]. These reasons provided the rationale for its utility as a target antigen for the serodiagnosis of PDCoV in indirect ELISA and FMIA platforms.

Antibody responses to the nucleoprotein of coronaviruses are very robust and have been reported to appear as soon as 7–9 days post infection. We originally hypothesized that antibody reactivity to PDCoV may be conformationally dependent so we designed an experiment to determine whether NP specific immunodominant epitopes tend to be present in greater abundance on conformationally or linearly expressed antigen. A refolded version of the NP was used as an antigen in both assays as it was shown to impart a higher degree of immunoreactivity than its unfolded, linear counterpart. This phenomenon was also observed by Johnson et al. [20] whereby the authors compared reactivity differences among single point serum titrations to provide a surrogate measure of antibody titer. They showed that the enhancement of immunoreactivity of PRRSV-N and nsp1 was completely dependent on refolding, and the reactivity of nsp2P was enhanced by twofold. Furthermore we confirmed their observations that there is a loss of immunoreactivity when the linear protein (solubilized in 8 M urea) was dialyzed in PBS prior to coating on microtiter plates or used for coupling to FMIA microspheres. This may indicate that using an antigen in its more native conformational state may present a higher number of immunoreactive epitopes that are able to

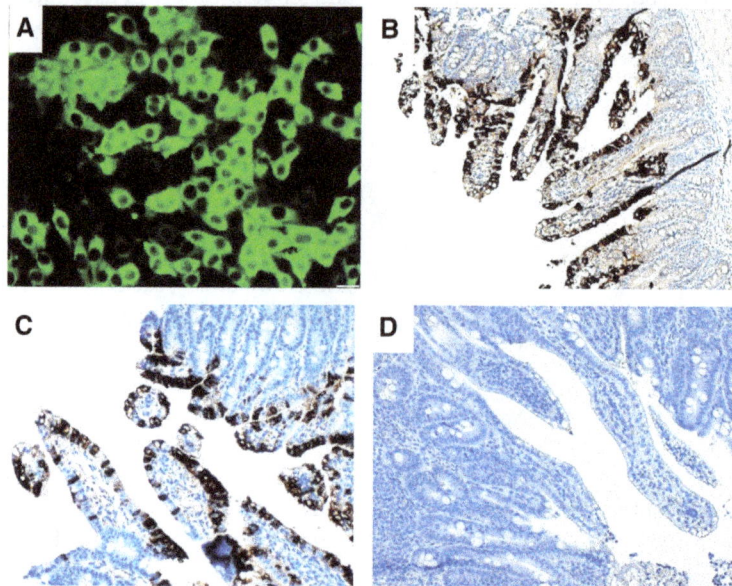

Fig. 7 Development of reagents used for indirect fluorescent antibody (IFA) testing and immunohistochemical (IHC) staining of PDCoV infected cells. Indirect fluorescent antibody staining of PDCoV infected ST cells with PDCoV anti-nucleoprotein monoclonal antibody SD-55-24 [**a**, 100X magnification]. Immunohistochemistry staining of intestinal enterocytes with PDCoV anti-nucleoprotein rabbit polyclonal hyperimmune sera [**b**, 100X magnification]. Immunohistochemistry staining of intestinal enterocytes with PDCoV anti-nucleoprotein monoclonal antibody 55–197 [**c**, 100X magnification]. Uninfected control showing hematoxylin staining of luminal, intestinal brush-border cross section [**d**, 100X magnification]

capture a larger percentage of PDCoV-NP antibodies. Therefore, the production of a well purified, refolded recombinant protein maintained in a near-native conformation was required for the production of an efficacious assay.

Both assays provide the capability of high-throughput testing with reasonable DSe and DSp. ROC analysis was performed for both assays demonstrating DSe of greater than 95 %. The FMIA demonstrated good DSp of 98.1 % while the ELISA showed a slightly lower DSp of 96.2 %. Inter-rater (kappa) agreement, a statistical measure of test agreement demonstrated a significant level of agreement among the IFA, ELISA and FMIA. Furthermore, each of the antibody-capture assays was validated using a large number of well characterized serum samples ($n = 940$) based upon the suggested 5-stage validation methods of Jacobson, which is supported by the office International des Epizooties [21].

Although preliminary DSe and DSp determinations for the first generation serological assays described here were slightly less than ideal, these new assays should provide valuable tools for assessment of PDCoV exposure on a herd level. One explanation for the approximately 95–96 % DSe values determined to date may be related to selection of the presumed positive field populations used for test validation. In this case, the initial stages of validation relied on characterizing the assay using known serum samples from experimentally infected pigs.

Because this was the only sera of known serostatus we had at our disposal, we believe that the resulting initial ROC characteristics were of sufficient value to begin testing samples from other sources believed to not have been exposed to PDCoV and from field sources known to have seroconverted to PDCoV within a specific time frame. By following the methodology outlined by Jacobson [21] in which he recommends the inclusion of these initial ROC cut-off values from experimentally infected pigs, we demonstrated similar assay characteristics on chosen field samples which substantiated our final cut-off value of the assay. These sample sets were collected at approximately 3 weeks after initial diagnosis of PDCoV by PCR. It is possible that PDCoV may not move through a herd at the very rapid rate seen with PEDV. Therefore, some animals in some herds may not have been infected until a week or more after initial detection in the population, resulting in delayed seroconversion in a percentage of the presumed positive population. Likewise, since the initial origination and distribution of PDCoV in the U.S. is not fully understood at this time, it is possible that a small percentage of our presumed seronegative population may have been subject to prior exposure. Many of the samples in our presumed negative population were archived samples collected prior to 2009. However, some originated from recent field submissions from high health, biosecure herds with no clinical or PCR evidence of prior PDCoV or PEDV exposure. The observed DSp values of approximately 98 % for FMIA and 96 % for indirect ELISA are within the expected range for first generation assays using these test formats. Apparent false positive reactions could also be due to an epitope on the expressed antigen having commonality with another low prevalence infectious agent or due to low levels of residual *E. coli* protein contaminants in the purified antigen preparations.

Recently, Thachil et al. [8], developed an indirect anti-PDCoV IgG enzyme-linked immunosorbent assay based on the putative S1 portion of the spike protein and evaluated it using a total of 968 tested serum samples. Although it is a reasonably sensitive (91 %) and specific (95 %) test, there is room for other ELISAs utilizing other target antigens, such as the NP of PDCoV, to serve as primary serological surveillance or confirmatory assays. As noted in Thachil's research, serum samples collected in 2010 were found positive for PDCoV antibody by their ELISA, but not those collected in 2011 and 2012. This was controversial because PDCoV was not thought to be circulating in North America in pigs before its identification in late 2013 [9]. It will be very beneficial to have confirmatory assays to validate this finding. Additional screening of archived historical swine serum samples using multiple serological assays may provide further insight into the origin and epidemiology of PDCoV in North America.

Conclusions

The monoclonal antibody reagents developed here provide important research and diagnostic tools for the industry. They are valuable for fluorescence and immunohistochemical staining methods associated with diagnostic and pathogenesis studies. The serological assays allow the detection of antibodies developed in response to PDCoV infection. The PDCoV indirect ELISA and FMIA will allow high-throughput screening of swine serum samples. These tests should be adequately optimized and validated for serosurveillance on a herd level, but further improvement is needed for full confidence on an individual animal basis. The IFA or other tests may be required for confirmation of individual unexpected results. Work is ongoing to further validate these assays and to adapt them to different sample matrices such as milk or oral fluid samples. Since lactogenic immunity is likely critical for protection of nursing piglets, these assays will be modified for detection of IgA as well.

Abbreviations
ACB, Antigen coating buffer; ADRDL, Animal Disease Research and Diagnostic Laboratory; CAPS, Cyclohexylamino-1-propanesulfonic acid; CPE, Cytopathic effect; CV, Coefficient of variation; DPI, Days post infection; DSe, Diagnostic sensitivity; DSp, Diagnostic specificity; ELISA, Enzyme linked immunosorbent assay; FBS, Fetal bovine serum; FITC, Fluorescein isothiocyanate; FMIA, Fluorescent microsphere immunoassay; HRP, Horseradish peroxidase; IACUC, Institutional Animal Care and Use Committee; IFA, Indirect fluorescent antibody; mAB, Monoclonal antibody; MEM, Eagle's minimum essential medium; MFI, Mean fluorescent intensity; NP, Nucleoprotein; NPB, National Pork Board; OD, Optical density; P/N, Positive to negative; PBS, Phosphate buffered saline; PBST, PBS plus 0.05 % tween 20; PCR, Polymerase chain reaction; PDCoV, Porcine deltacoronavirus; PEDV, Porcine epidemic diarrhea virus; PRCV, Porcine respiratory coronavirus; ROC, Receiver operator characteristic; RT-PCR, Reverse transcriptase polymerase chain reaction; S/P, Sample to positive; SDS-PAGE, Sodium dodecyl sulfate-polyacrylamide gel electrophoresis; SDSU, South Dakota State University; SMD, Sample milk diluent; ST, Swine testicle; TGEV, Transmissible gastroenteritis virus; TMB, 3,3',5,5'-tetramethylbenzidine; TPCK, L-1-Tosylamide-2-phenylethyl chloromethyl ketone; UN-L, University of Nebraska-Lincoln; YT, Yeast extract tryptone

Acknowledgements
The authors thank Drs. Sabrina Swenson and Melinda Jenkins-Moore of the National Veterinary Services Laboratories for providing the cell culture adapted Michigan/8977/2014 isolate of PDCoV; Dr. Richard Hesse of the Kansas State University Veterinary Diagnostic Laboratory and Dr. Sarah Vitosh-Sillman and colleagues at the University of Nebraska-Lincoln for providing serum samples from PDCoV challenge studies; Dr. David Knudsen and Amanda Brock for immunohistochemistry; Dr. Volker Brozel for critical review of the manuscript and the South Dakota State University Animal Resource staff for animal assistance. The authors also thank numerous swine producers and practitioners for providing field samples and herd history information.

Funding
Funding for this work was provided by the National Pork Board through grant #14-184, the South Dakota Agricultural Experiment Station, USDA Hatch funding and the South Dakota Animal Disease Research and Diagnostic Laboratory.

Authors' contributions
FO: Conducted ELISA and FMIA development/validation, statistical analysis and co-wrote paper. SL: Directed reagents and assay development and validation, co-wrote paper. XL: Assisted with ELISA and FMIA development/validation. AS: Assisted with sample acquisition, reagent and assay development and study design. TC: Assisted with sample acquisition and study design. KH: Assisted with assay development and validation. JN: Assisted with reagent development and characterization and manuscript preparation. JH: Assisted with study design and co-wrote paper. EAN: Developed study concept and design, co-wrote paper. All authors read and approved the final manuscript.

Competing interests
The authors are employed by South Dakota State University and the University may receive license fees or royalty payments associated with technologies described in this manuscript.

Consent for publication
Not applicable.

Ethics approval and consent to participate
Procedures involving animals at South Dakota State University (SDSU) were approved by the SDSU Institutional Animal Care and Use Committee (IACUC). Time course swine serum samples provided by Kansas State University and the University of Nebraska-Lincoln (UN-L) were part of a separate PDCoV challenge study approved by the UN-L IACUC. All other samples were obtained as routine diagnostic sample submissions at the South Dakota Animal Disease Research and Diagnostic Laboratory (ADRDL).

Author details
[1]Veterinary & Biomedical Sciences Department, South Dakota State University, Brookings, SD, USA. [2]National Research Center, Giza, Egypt.

References

1. Woo PC, Lau SK, Lam CS, Lau CC, Tsang AK, Lau JH, et al. Discovery of seven novel mammalian and avian coronaviruses in the genus Deltacoronavirus supports bat coronaviruses as the gene source of Alphacoronavirus and Betacoronavirus and avian coronaviruses as the gene source of Gammacoronavirus and Deltacoronavirus. J Virol. 2012;86:3995–4008.

2. Saif L, Pensaert MB, Sestak K, Yeo S-G, Jung K. Coronaviruses. In: Zimmerman JL, Ramirez KA, Schwartz K, Stevenson G, editors. Diseases of Swine. 10th Edition. Chichester, West Sussex, UK: Wiley; 2012. p. 501–524.

3. Wang L, Byrum B, Zhang Y. Detection and genetic characterization of Deltacoronavirus in pigs, Ohio, USA, 2014. Emerg Infect Dis. 2014;20(7): 1227–30. doi:10.3201/eid2007.140296.

4. Marthaler D, Raymond L, Jiang Y, Collins J, Rossow K, Rovira A. Rapid detection, complete genome sequencing, and phylogenetic analysis of Porcine Deltacoronavirus. Emerg Infec Dis. 2014;20(8). doi: 10.3201/eid2008.14-0526.

5. Ma Y, Zhang Y, Liang X, Lou F, Oglesbee M, Krakowka S, Li J. Origin, evolution, and virulence of porcine deltacoronaviruses in the United States. mBio. 2015;6(2):e00064–15. doi:10.1128/mBio.00064-15.

6. Chen Q, Gauger P, Stafne M, Thomas J, Arruda P, Burrough E, et al. Pathogenicity and pathogenesis of a United States porcine deltacoronavirus cell culture isolate in 5-day-old neonatal piglets. Virology. 2015;482:51–9. doi:10.1016/j.virol.2015.03.024.

7. Jung K, Hu H, Eyerly B, Lu Z, Chepngeno J, Saif LJ. Pathogenicity of 2 porcine deltacoronavirus strains in gnotobiotic pigs. Emerg Infect Dis. 2015. Doi: 10.3201/eid2104.141859.

8. Thachil A, Gerber PF, Xiao C-T, Huang Y-W, Opriessnig T. Development and application of an ELISA for the detection of porcine deltacoronavirus IgG antibodies. PLoS One. 2015;10(4):e0124363. doi:10.1371/journal.pone.0124363.

9. APHIS/USDA. Swine enteric coronavirus disease testing summary report. Sept 11, 2014. Available: http://www.aphis.usda.gov/animal_health/animal_dis_spec/swine/downloads/secd_wkly_lab_%20rpt_09_11_14.pdf). Accessed 1 Dec 2015.

10. Langenhorst R, Lawson S, Kittawornrat A, Zimmerman J, Sun Z, Li Y, et al. Development of a fluorescent microsphere immunoassay for detection of antibodies against porcine reproductive and respiratory syndrome virus using oral fluid samples as an alternative to serum-based assays. Clin and Vacc Imm. 2012;19:180–9.

11. Okda F, Liu X, Singrey A, Clement T, Nelson J, Christopher-Hennings J, et al. Development of an indirect ELISA, blocking ELISA, fluorescent microsphere immunoassay and fluorescent focus neutralization assay for serologic evaluation of exposure to North American strains of porcine epidemic diarrhea virus. BMC Vet Res. 2015;11:180. doi:10.1186/s12917-015-0500-z.

12. Nelson EA, Christopher-Hennings J, Drew T, Wensvoort G, Collins JE, Benfield D. Differentiation of U.S. and European isolates of porcine reproductive and respiratory syndrome virus by monoclonal antibodies. J Clin Micro. 1993;31:3184–9.

13. Galfre G, Howe SC, Milstein C, Butcher GW, Howard JC. Antibodies to major histocompatibility antigens produced by hybrid cell lines. Nature (London). 1977;266:550–2.

14. Lawson S, Lunney J, Zuckermann F, Osorio F, Nelson E, Welbon C, et al. Development of an 8-plex Luminex assay to detect swine cytokines for vaccine development: Assessment of immunity after porcine reproductive and respiratory syndrome virus (PRRSV) vaccination. Vaccine. 2010;28:5383–91.

15. Brown E, Lawson S, Welbon C, Gnanandarajah J, Li J, Murtaugh M, et al. Antibody response to porcine reproductive and respiratory syndrome virus (PRRSV) nonstructural proteins and implications for diagnostic detection and differentiation of PRRSV types I and II. Clin and Vacc Imm. 2009;16:628–35.

16. Landis JR, Koch GG. The measurement of observer agreement for categorical data. Biometrics. 1977;33:159–74.

17. Stevenson GW, Hoang H, Schwartz KJ, Burrough ER, Sun D, Madson D, et al. Emergence of porcine epidemic diarrhea virus in the United States: clinical signs, lesions, and viral genomic sequences. J Vet Diagn Invest. 2013;25:649–54.

18. Lee S, Lee C. Functional characterization and proteomic analysis of the nucleocapsid protein of porcine deltacoronavirus. Virus Res. 2015;208: 136–45.

19. McBride R, van Zyl M, Fielding BC. The coronavirus nucleocapsid is a multifunctional protein. Viruses. 2014;6:2991–3018.

20. Johnson CR, Yu W, Murtaugh MP. Cross-reactive antibody responses to nsp1 and nsp2 of porcine reproductive and respiratory syndrome virus. J Gen Virol. 2007;88(4):1184–95.

21. Jacobson RH. Validation of serological assays for diagnosis of infectious diseases. Rev Sci Tech Of Int Epizoot. 1998;17:469–526.

12

Oral administration of inactivated porcine epidemic diarrhea virus activate DCs in porcine Peyer's patches

Chen Yuan, En Zhang, Lulu Huang, Jialu Wang and Qian Yang*

Abstract

Background: Peyer's patches (PPs) can be considered as the immune site of the intestine. Within PPs, Dendritic cells (DCs) can uptake antigens from the gut lumen by extending dendrites into epithelium, and process it and then present to lymphocytes, which effectively antigen produces an immune response. Porcine epidemic diarrhea virus (PEDV) is the causative agent of porcine epidemic diarrhea (PED), an acute and highly contagious enteric viral disease. The interaction between inactivated porcine epidemic diarrhea virus and porcine monocyte-derived dendritic cells (Mo-DCs) has been reported. However, little is known about the interaction between inactivated PEDV and DCs in porcine PPs.

Results: In this study, for the first time we investigated the role of DCs in porcine PPs after oral administration inactivated PEDV. Firstly, a method to isolate DCs from porcine PPs was established, in which the purity of SWC3a$^+$/MHC-II$^+$ DCs was more than 90%. Our findings clearly indicate that DCs in porcine PPs after oral administration of inactivated PEDV not only stimulated the proliferation of allogeneic lymphocytes, but also secreted cytokines (IL-1, IL-4). Furthermore, the number of DCs and IgA$^+$ cells in porcine intestinal mucosal significantly increased and the levels of anti-PEDV specific IgG antibody in the serum and SIgA antibody in the feces increased after oral administration inactivated PEDV.

Conclusions: Our findings indicate that oral administration of inactivated PEDV activate DCs in porcine Peyer's patches and inactivated PEDV may be a useful and safe vaccine to trigger adaptive immunity.

Keywords: Peyer's patches, Dendritic cells, Inactivated porcine epidemic diarrhea virus

Background

Gut is the major immune organ of the body and the intestinal mucosa is thought to be the primary site for performing local-specific immune responses. Gut associated lymphoid tissue (GALT), consists of isolated or aggregated lymphoid follicles forming Peyer's patches (PPs), is considered to be the key inductive tissues for the mucosal immune system [1]. PPs are known as the immune sensors of the intestine because of their ability to transport luminal antigens and bacteria into organized lymphoid tissues within the intestinal mucosa [2]. PPs contain too many immunocompetent cells that are required for the generation of an immune response. Dendritic cells (DCs)

* Correspondence: zxbyq@njau.edu.cn
MOE Joint International Research Laboratory of Animal Health and Food Safety, College of veterinary medicine, Nanjing Agricultural University, Weigang 1, Nanjing, Jiangsu 210095, People's Republic of China

are professional antigen-presenting cells (APC) that possess the unique capacity to trigger primary adaptive immune responses through the antigen-specific activation of naive T cells. DCs in PPs can extend dendrites into the lumen to capture antigens and then present to resting T cells and thus initiate adaptive immune responses [3, 4].

According to the cell lineage, DCs can be divided into two major subsets that include plasmacytoid DCs (pDCs) and myeloid DCs, with the latter commonly referred to as conventional DCs (cDCs) [5]. DCs involved in intestinal immunity were investigated by cDCs induced from bone marrow cells or pDCs induced from blood mononuclear cells so far [6, 7]. But neither pDCs nor cDCs represent the reality of DCs within PPs. Indeed it is difficult to obtain the DCs within PPs in human. Unlike the mouse (isolated PPs), pig had continuous PPs (aggregated PPs), just like human. It is possible to get DCs within PPs in pig. The pig would be an animal model for studying DCs within PPs in human. Therefore, it is important to establish a method that isolate DCs from porcine PPs.

Porcine epidemic diarrhea virus (PEDV) is the causative agent of porcine epidemic diarrhea (PED), an acute and highly contagious enteric viral disease. PEDV infects epithelia in both small and large intestine and cause diarrhea, dehydration, and a high mortality in piglets [8–10]. Currently, PED is globally recognized as an emerging and reemerging disease that has resulted in great economic losses to the swine industry worldwide. PEDV infect piglets mainly through fecal-oral route (digestive tract). Vaccination is a potent tool in the control and prevention of PED [11]. It is important to develop oral vaccines that can elicit effective mucosal immune responses against PEDV infection. Oral vaccines could cut off the route of PEDV invasion. Oral administration of vaccine in pigs have been successfully used to prevent infectious disease [12, 13].

In our previous work, we reported that the interaction between inactivated PEDV and porcine Mo-DCs, which inactivated PEDV enhances the ability of DCs to present, migrate and induce the activation of T lymphocytes in vivo and in vitro [6]. However, little is known about the interaction between inactivated PEDV and DCs in porcine PPs. In this study, for the first time we investigated the role of DCs in porcine PPs after oral inoculation of inactivated PEDV. Firstly, a method to isolate DCs in porcine PPs was established, in which the purity of SWC3a+/MHC-II+ DCs was more than 90%. Our findings clearly indicated that PPs DCs from porcine intestinal mucosal after oral delivery of inactivated PEDV not only stimulated the proliferation of allogeneic lymphocytes, but also secreted cytokines (IL-1,IL-4). Furthermore,the number of DCs and IgA antibody levels in porcine intestinal mucosal significantly increased as compared with the control group and the levels of anti-PEDV specific IgG antibody in the serum and SIgA antibody in the feces increases after oral inactivated PEDV.

Methods
Experimental design
A total of 6 male (cross-bred Duroc/Landrace/Yorkshire) piglets: aged 2-day-old; weight, 1.30–1.50 kg, were obtained from Jiangsu Academy of Agricultural Sciences (Nanjing, China). The piglets were born via natural farrow and were fed with milk. The piglets were housed in Jiangsu Academy of Agricultural Sciences Pig Farm with a constant humidity (60%) and temperature (26 °C) at 12 h light/dark cycle. The PEDV (CV777) was propagated on Vero cells in DMEM (GIBCO, USA) with 5% fetal bovine serum and purified the collection by sucrose gradient centrifugation. Ultraviolet rays inactivated PEDV (UV-PEDV) were produced by exposing the virus to ultraviolet rays for 6 h at an optimal cross linking value. The piglets were randomly divided into two groups (3 piglets each group), respectively fed with

UV-inactivated PEDV (100µg/dose) or equal volume of PBS twice at days 5 and 30 for two months. The next week after the last administration, these pigs were euthanized by intravenous injection of pentobarbital sodium (100 mg/kg) and piglets were sacrificed and ileum samples were collected. All procedures performed on the animals were approved by the Institutional Animal Care and Use Committee of Nanjing Agricultural University and followed the National Institutes of Health guidelines for the performance of animal experiments.

Reagents
FITC-MHCII, PE-SWC3a were from Abcam (New Territories, Hong Kong). Dylight 488-, 594-,-conjugated secondary antibodies were purchased from MultiSciences (Lianke) Biotech Co., Ltd. (China). 4′, 6-diamidino-2-phenylindole (DAPI) solution were obtained from Jackson ImmunoResearch Laboratories (West Grove, PA, USA). CFSE (carboxyfluorescein succinimidyl amino ester) were purchased from Invitrogen (USA). The Cytokine test kits were purchased from Shanghai Huyu Biotechnology (Shanghai, China). Cell Counting Kit-8 were purchased from Beyotime Biotechnology (China).

Isolation of DCs in porcine PPs
After removal of residual mesenteric fat tissue, the ileum was then cut into 1.5 cm pieces. The pieces were incubated in 20 ml of 5 mM EDTA in HBSS for 20 min at 4 °C. Then centrifuged and discarded the supernatant. The ileum was cut in 1 cm pieces and placed in digestion solution containing 4% fetal bovine serum, 2 mg/ml each of Collagenase D (Roche) and DNase I (Sigma), and 100 U/ml Dispase (Fisher) at 37 °C for 20 min with slow rotation. The supernatants were obtained by density gradient centrifugation and then sorted DCs marked SWC3a and MHC-IIfrom porcine PPs by fluorescence activated cell sorting (FACS).

Autologous mixed lymphocyte reaction
DCs had the ability to stimulate lymphocyte proliferation. We examined whether DCs stimulated lymphocyte proliferation in two ways. Firstly, Different groups DCs were incubated with allogeneic lymphocyte labeled CFSE, at a rate of DCs: lymphocyte = 1: 10. Five days later, the proliferation of lymphocyte was detected by FACS.Another way, different groups DCs were incubated with allogeneic lymphocyte, five days later, the proliferation of lymphocyte was detected by CCK8 (Cell Counting Kit-8).

Cytokine assays by enzyme-linked immunosorbent assay
Different groups DCs were incubated with allogeneic lymphocyte for 24 h, at a rate of DCs: lymphocyte = 1:10. The cytokines (IL-2, IL-4, IL-6 and IL-10) in culture mediums were measured using enzyme-linked immunosorbent

assay and performed according to the manufacturer's instructions.

Immunofluorescence (IF) staining of DCs

Tissue sections were permeabilized in 0.4% Triton X-100 in PBS for 5 min. After treating with 5% bovine serum albumin in PBS for 1 h, the tissue sections were incubated with the SWC3a or MHC-IIprimary antibodies overnight at 4 °C, PBS was used in place of the anti-pig antibody for the control. After rinsing in PBS, sections incubated with Alexa Fluor 488 or 647 labeled secondary antibodies were kept at room temperature for 1 h. After staining with DAPI, the cryosections were observed under a confocal laser microscope (LSM-710; Zeiss, Oberkochen, Germany) visualized by CLSM (LSM 710, Zeiss, Oberkochen, Germany).

Immunohistochemistry

After deparaffinization and rehydration, paraffin sections were put in citrate buffer (pH 6) at 90–95 °C for 15 min to retrieve antigen. Then, the sections were put in 0.3% H_2O_2 to quench endogenous peroxidase and washed in PBS. 5% bovine serum albumin were incubated on sections for 30 min to close the non-specific antibody binding sites. After blocking with 5% bovine serum albumin,

sections were incubated with goat anti-pig IgA overnight at 4 °C. Biotinylated secondary antibodies were added to the sections for 1 h at room temperature, and treated with SABC for 60 min. Sections were counterstained with hematoxylin and images were obtained using a light microscope (BH-2; Olympus). Different fields of each tissue in each piglet were counted for the statistical analysis.

Enzyme-linked immunosorbent assay (ELISA)

The PEDV-specific IgA in feces and PEDV-specific IgG in serum were detected by ELISA. ELISA plates were coated 2 µg purified PEDV/well at 4 °C overnight. After antigen removal, ELISA plates were blocked with 3% bovine serum albumin (BSA) in PBS which contains 0.05% Tween (PBST) for 2 h at 37 °C. Then, 100-fold dilutions of serum samples or 2-fold dilutions of lavage fluid from pig were added to the plates and incubated at 37 °C for 2 h. Washed with PBST and added 100 µl of HRP-conjugated goat anti-pig IgA/IgG antibody at 1:2000 dilution and incubated at 37 °C for 1 h. Plates were washed 5 times and incubated with 3, 3′, 5, 5′-tetramethylbenzidine (TMB) for 15 min. Then, the reaction was stopped with sulfuric acid (2 M). Optical densities at 450 nm were measured using an enzyme-linked immunosorbent assay (ELISA) plate reader.

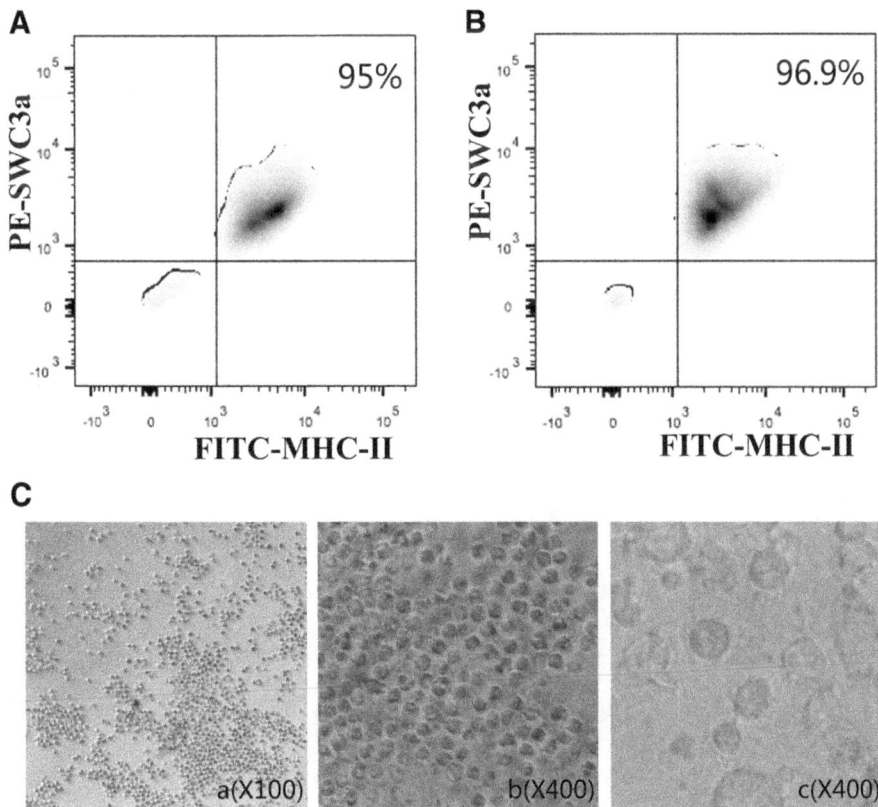

Fig. 1 Sorting DCs from porcine PPSs. two days old piglets were given PBS or inactivated PEDV (100µg/dose) every week for two months. SWC3a$^+$/MHC-II$^+$ DCs from conventional healthy pigs (**a**) and oral inactivated PEDV (**b**) were sorted and analyzed by FACS., $n = 6$. **c** Morphology of DCs from porcine PPs under a light microscope

Statistical analysis

Results were expressed as means ± SD. Analysis of variance and unpaired Student's t-tests were employed to determine statistical differences among multiple groups. P values < 0.05 were considered significant ($*P < 0.05$, $**P < 0.01$).

Results

Sorting DCs from porcine PPs

PPs are important inductive sites for the initiation of innate and adaptive immune responses [2]. To successfully isolate DCs from porcine PPs is an important task, which might contribute to further study on the role of DCs in porcine intestinal mucosal PPs. Cell suspensions were prepared in porcine PPs from conventional healthy pigs and oral inactivated PEDV for two months, and sorted by FACS according to DCs maker SWC3a and MHC-II. Then SWC3a$^+$/MHC-II$^+$ DCs in porcine PPs were analyzed by FACS. The purity of SWC3a$^+$/MHC-II$^+$ DCs from conventional healthy

pigs and oral inactivated PEDV for two months were 95% and 96.9% respectively (Fig. 1a, b). After 5 days of culture, there were many dendrite-like processes on the surface of the DCs from porcine Peyer's patches under the inverted microscope (Fig. 1c).

DCs can promoted the proliferation of allogeneic lymphocytes

DCs can uptake antigens from the gut lumen by extending dendrites into epithelium, then process and present to lymphocytes, which effectively antigen produces an immune response [4, 14, 15]. So in next step, we investigated the potential of DCs to prime lymphocytes responses after its interactions with lymphocytes. As the result showed, DCs in porcine PPs from oral inactivated PEDV pigs for two months had a significantly increased ability to promote lymphocytes proliferation (Fig. 2a, b), while that of DCs from conventional healthy pigs had no significant changes (Fig. 2b).

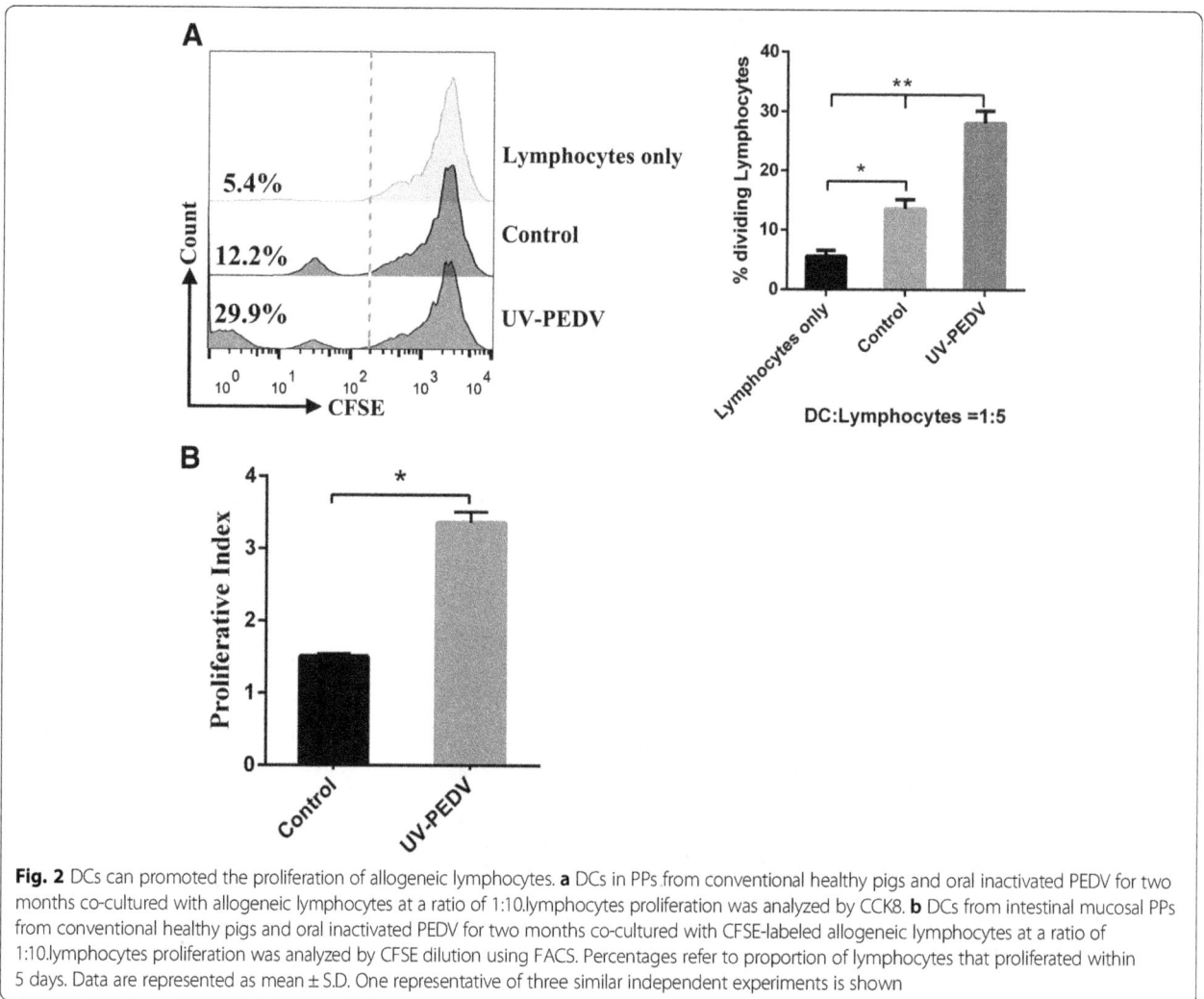

Fig. 2 DCs can promoted the proliferation of allogeneic lymphocytes. a DCs in PPs from conventional healthy pigs and oral inactivated PEDV for two months co-cultured with allogeneic lymphocytes at a ratio of 1:10.lymphocytes proliferation was analyzed by CCK8. b DCs from intestinal mucosal PPs from conventional healthy pigs and oral inactivated PEDV for two months co-cultured with CFSE-labeled allogeneic lymphocytes at a ratio of 1:10.lymphocytes proliferation was analyzed by CFSE dilution using FACS. Percentages refer to proportion of lymphocytes that proliferated within 5 days. Data are represented as mean ± S.D. One representative of three similar independent experiments is shown

DCs can stimulate lymphocytes to secrete cytokines

To examine the effect of DCs stimulate lymphocytes on cytokines, we measured the level of IL2, IL-4, IL-6 and IL-10 in DCs co-cultured with allogeneic lymphocytes by ELISA kit. As the result showed, DCs in porcine PPs from oral inactivated PEDV pigs for two months co-cultured with allogeneic lymphocytes significantly stimulated the secretion of IL2, IL-4 by lymphocytes, as compared with the control group (Fig. 3a, b). However, there was no statistical difference in level of IL-6 and IL-10 between the treated and control group in DCs co-cultured with allogeneic lymphocytes (Fig. 3c, d).

The number of ileum DCs increases after oral inactivated PEDV

DCs are the most potent antigen-presenting cells that bridge innate and adaptive immunity in vivo [3, 4]. The study used IF analysis via dual staining with antibody specific to the DCs markers to detect DCs. SWC3a positive cells were stained red, MHC II positive cells were stained green, double positive cells were SWC3a$^+$/MHC-II$^+$ DCs, which were stained yellow. Our results showed that the number of SWC3a$^+$/MHC-II$^+$ DCs significantly increased in ileum after oral inactivated PEDV (Fig. 4).

The number of ileum IgA$^+$ cells increases after oral inactivated PEDV

IgA favors both maintenance of non-invasive commensal bacteria and neutralization of invasive pathogens [16]. Besides neutralizing pathogens in the intestinal lumen, IgA can intercept microbes and toxins inside intestinal epithelial cells [17]. The distribution patterns of IgA$^+$ cells in ileum were examined by IHC. The positive cells were stained brown. The IgA$^+$ positive cells represented the presence of SIgA molecular adhesion. The number of ileum IgA+ cells have significantly different between oral inoculation of inactivated PEDV group and control group (Fig. 5).

The levels of anti-PEDV specific IgG antibody in the serum and SIgA antibody in the feces increases after oral inactivated PEDV

The levels of anti-PEDV specific IgG antibody in the serum and SIgA antibody in the feces was determined by ELISA. As shown in the Fig. 6, the levels of anti-PEDV specific IgG antibody in the serum increased after oral immunization with UV-inactivated PEDV ($p < 0.01$) as compared to those in the control groups of piglets orally immunized with PBS after the first vaccinations. At 14 and 49 day, anti-PEDV specific IgG antibody in the serum reached a relatively

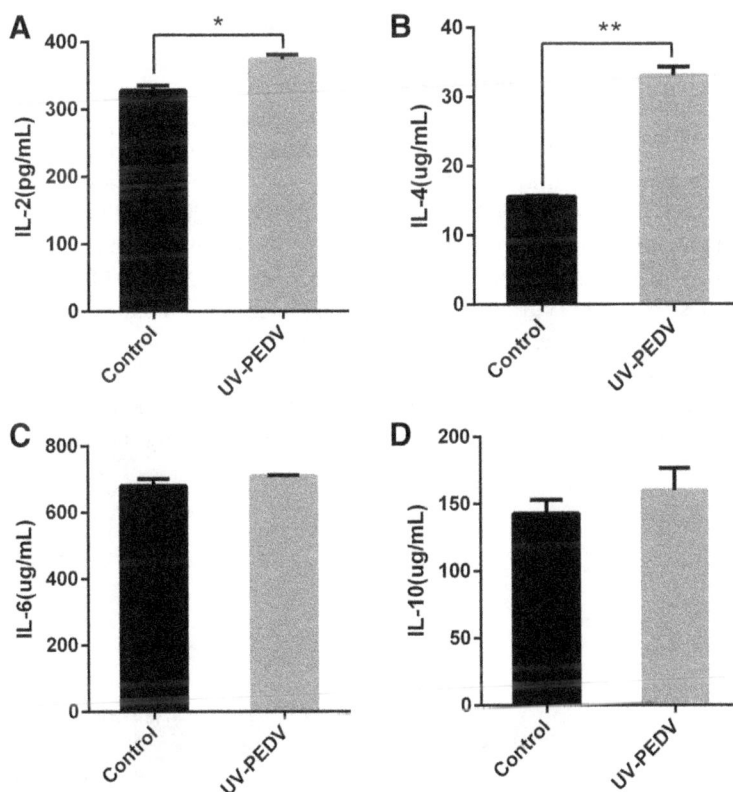

Fig. 3 DCs can stimulate lymphocytes to secrete cytokines. DCs in porcine PPSs from conventional healthy pigs and oral inactivated PEDV for two months co-cultured with allogeneic lymphocytes at a ratio of 1:5 for 24 h respectively, the culture supernatants were collected. IL-2 (**a**), IL-4 (**b**), IL-6 (**c**) and IL-10 (**d**) release in culture supernatants were measured by ELISA. All of the data are presented as means ± SD of three replicates and are representative of three independent experiments. *$P < 0.05$; **$P < 0.01$

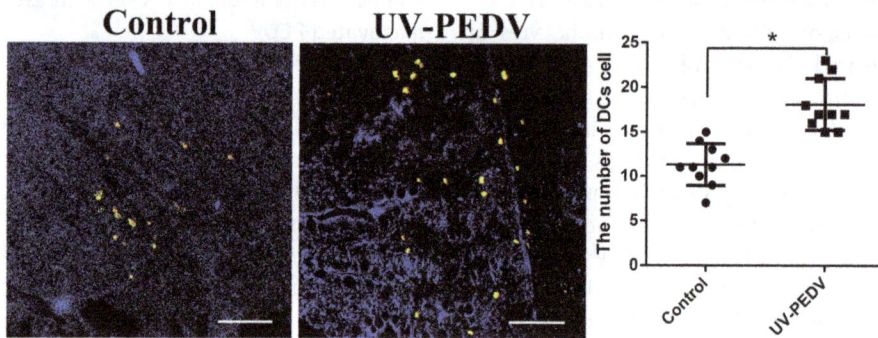

Fig. 4 The number of ileum SWC3a⁺/MHC-II⁺DCs increases after oral inactivated PEDV. SWC3a⁺/MHC-II⁺ DCs were showed by Immunofluorescence staining. The number of SWC3a⁺/MHC-II⁺ DCs in per view was counted and statistical analysis was performed. Significant differences between the treated groups and the control groups are identified as. Scale bar = 100 μm, $n = 10$ *$p < 0.05$

higher level ($p < 0.01$) (Fig. 6a). Similar variation tendency of SIgA titers was detected in the feces (Fig. 6b). No significant difference was detected in the negative control group during the experiment.

Discussion

The small intestinal mucosal immune system is the first line of defense against a wide variety of exogenous molecules [2]. Peyer's Patches (PPs) was named after their detailed description by the Swiss pathologist Johann Conrad Peyer in 1677 [18]. PPs play an important role in distinguishing between potentially harmful agents and common food ingredients within the ingesta as a key component of the gut-associated lymphoid tissue [2, 19]. PPs serve as the primary inductive sites for intestinal immunity [19].

Van Kruiningen et al.noted that the PPs occupied most of the area in the ileum [20, 21]. Within PPs, DCs in the dome and interfollicular areas can uptake antigens from the gut lumen by extending dendrites into epithelium, then and process and present to lymphocytes, which effectively antigen produces an immune response [4, 14, 15]. DCs derived from blood and bone marrow precursors can't represent the real immune status of the body. Mice are the most

commonly used animal models, but because of their small size they have a low amount of DCs in their intestinal mucosal. Moreover, the immune system of mice have less than 10% similarity to humans [22]. In addition, it is difficult to identify and collect DCs from human PPs, study of DCs from human PPs is difficult to achieve [2]. Therefore, the current study is about the isolation of DCs in porcine PPs. Pigs are very similar in anatomy, physiology and genetics to humans. Moreover, the immune system of pigs is over 80% similar to humans [22–24]. A large number of experiments have proved that pigs can be used as an ideal experimental animal model for the establishment of human disease infection and can be used to isolate and characterize the DCs in porcine PPs [25–27]. So, it is important to establish a method to isolate a number of functional DCs in vitro. In this study, we establish a method to isolate DCs in porcine PPs, which could contribute to further study on the role of DCs in porcine PPs.

PEDV infected the host in the mucosal surface of the intestine. Therefore, it is important to develop oral mucosa vaccines that can elicit effective mucosal immune responses against PEDV infection. An oral vaccination is an efficacious strategy owing to induction of potent humoural and

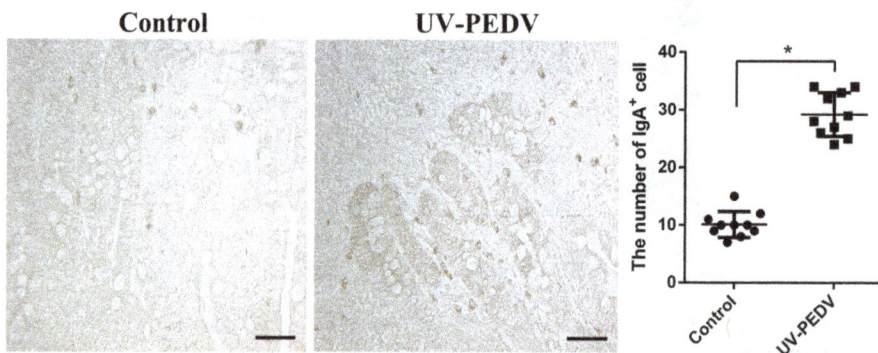

Fig. 5 The number of ileum IgA⁺ cells increases after oral inactivated PEDV. IgA⁺ cells were showed by immumohistochemical staining. The number of IgA⁺ cells in per view was counted and statistical analysis was performed. Scale bar = 100 μm, n = 10, * $P < 0.05$, ** $P < 0.01$

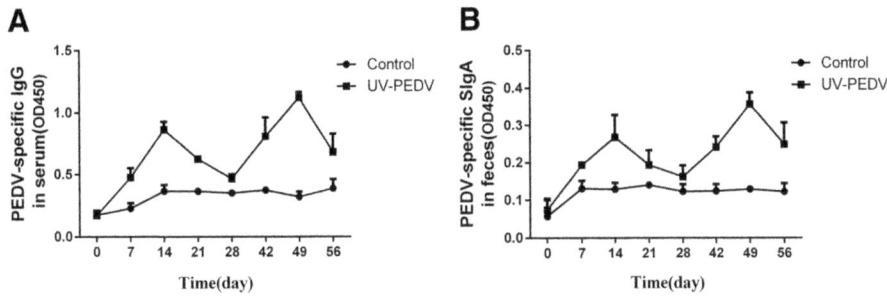

Fig. 6 The levels of anti-PEDV specific IgG antibody in the serum and SIgA antibody in the feces increases after oral inoculation of inactivated PEDV. The levels of anti-PEDV specific IgG antibody in the serum (**a**) and SIgA antibody in the feces (**b**) was determined by ELISA. All data are shown as the mean ± S.D. *$P < 0.05$; **$P < 0.01$

mucosal immune responses and it offers significant advantages such as easy delivery, labor health safety in comparison to other vaccination routes such as intramuscular. Currently live attenuated PEDV vaccines are commercially used [28]. Inactivated whole-virus vaccines and live PEDV can induce significantly protective immune responses respectively. But the problem occurs when live attenuated PEDV mutates into virulent strain or the small intestine mucosal immunity is insufficient, the virus proliferates in the small intestine and can be easily spread in the piglets and cause PED. However, inactivated whole-virus vaccines can solve the secure problem and be effective as well.

PPs is the main site of induction of IgA-producing plasma B cells by DCs loaded with commensal bacteria [29]. Inactivated viruses are usually not sufficiently effective in eliciting local mucosal immune response. DCs play a significant role in the application of inactivated PEDV oral vaccine. After oral administration of antigens, PPs are the first places of T-cell-specific priming and proliferation in the gut. DCs promote lymphocytes proliferation and secreting cytokines against an array of invading pathogens including PEDV. DCs derived from spleen (SP) exhibit strong functional differences as compared to DCs derived from PPs [30]. DCs from PPs are more potent in stimulating allogeneic T-cells proliferation compared with DCs from SP. As well as, the DCs derived from PPs are able to prime the production of IL-4 and IL-10 (Th2 anti-inflammatory cytokines), while DCs from SP lack this ability [30]. Therefore, designing vaccines targeting DCs may be potential agents for the suppression of viral infections. In this study, our results indicated that DCs in porcine PPs after oral administration of inactivated PEDV not only can stimulate the proliferation of allogeneic lymphocytes, but also can stimulate lymphocytes to secrete cytokines (IL-1 and IL-4), after the mixed reaction of DCs and allogeneic lymphocytes. In addition, we also found that the number of ileum IgA$^+$ cells and DCs in porcine ileum significantly increased in piglets in vitro and the levels of anti-PEDV specific IgG antibody in the serum and SIgA antibody in the feces increases after oral administration of inactivated PEDV.

Conclusion

For the first time we investigated the role of DCs in porcine PPs after oral inactivated PEDV. Our findings indicate that oral administration of inactivated PEDV activate DCs in porcine Peyer's patches and inactivated PEDV may be a useful and safe vaccine to trigger adaptive immunity.

Abbreviations

APC: Antigen-presenting cells; cDCs: Conventional DCs; DCs: Dendritic cells; FACS: Fluorescence activated cell sorting; GALT: Gut associated lymphoid tissue; Mo-DCs: Monocyte-derived dendritic cells; pDCs: Plasmacytoid DCs; PED: porcine epidemic diarrhea; PEDV: Porcine epidemic diarrhea virus; PPs: Peyer's patches; SABC: Strept Avidin-Biotin Complex; UV-PEDV: Ultraviolet rays inactivated PEDV. HBSSHank's Balanced Salt Solution

Acknowledgements
We thank Jiangsu Academy of Agricultural Sciences for providing us with the piglets as well as pig farm.

Funding
This work was supported by 31772777 from the National Science Grant of China and a project funded by the Priority Academic Program Development of Jiangsu Higher Education Institutions (PAPD).

Authors' contributions
CY participated in design of the study, analyzed the data and prepared the manuscript. EZ, JW, LH raised piglets and collected the samples, conducted the experiment. QY designed the study and revised the manuscript. All the authors read, revised, and approved the final manuscript.

Consent for publication
Not applicable.

Competing interests
The authors declare that they have no competing interests.

References

1. Brandtzaeg P. Mucosal immunity: induction, dissemination, and effector functions. Scand J Immunol. 2009;70(6):505–15.
2. Jung C, Hugot JP, Barreau F. Peyer's patches: the immune sensors of the intestine. Int J Inflam. 2010;2010:823710.
3. Owen JL, Sahay B, Mohamadzadeh M. New generation of oral mucosal vaccines targeting dendritic cells. Curr Opin Chem Biol. 2013;17(6):918–24.
4. Harman AN, Min K, Nasr N, Sandgren KJ, Cameron PU. Tissue dendritic cells as portals for HIV entry. Rev Med Virol. 2013;23(5):319–33.
5. Sozzani S, Prete AD, Bosisio D. Dendritic cell recruitment and activation in autoimmunity. J Autoimmun. 2017;85:126.
6. Qi G, Zhao S, Tao Q, Yin Y, Yu Q, Qian Y. Effects of inactivated porcine epidemic diarrhea virus on porcine monocyte-derived dendritic cells and intestinal dendritic cells. Res Vet Sci. 2016;106(3–4):149–58.
7. Zhao S, Gao Q, Qin T, Yin Y, Lin J, Yu Q, Yang Q. Effects of virulent and attenuated transmissible gastroenteritis virus on the ability of porcine dendritic cells to sample and present antigen. Vet Microbiol. 2014;171(1–2):74–86.
8. Stevenson GW, Hoang H, Schwartz KJ, Burrough ER, Sun D, Madson D, Cooper VL, Pillatzki A, Gauger P, Schmitt BJ. Emergence of porcine epidemic diarrhea virus in the United States: clinical signs, lesions, and viral genomic sequences. J Vet Diagn Invest. 2013;25(5):649.
9. Jung K, Wang Q, Scheuer KA, Lu Z, Zhang Y, Saif LJ. Pathology of US porcine epidemic diarrhea virus strain PC21A in Gnotobiotic pigs. Emerg Infect Dis. 2014;20(4):668–71.
10. Jung K, Eyerly B, Annamalai T, Lu Z, Saif LJ. Structural alteration of tight and adherens junctions in villous and crypt epithelium of the small and large intestine of conventional nursing piglets infected with porcine epidemic diarrhea virus. Vet Microbiol. 2015;177(3–4):373–8.
11. Jung K, Saif LJ. Porcine epidemic diarrhea virus infection: etiology, epidemiology, pathogenesis and immunoprophylaxis. Vet J. 2015;204(2):134.
12. Monger VR, Stegeman JA, Dukpa K, Gurung RB, Loeffen WL. Evaluation of oral bait vaccine efficacy against classical swine fever in village backyard pig farms in Bhutan. Transbound Emerg Dis. 2016;63(6):e211.
13. Wang X, Wang Z, Xu H, Xiang B, Dang R, Yang Z. Orally Administrated Whole Yeast Vaccine Against Porcine Epidemic Diarrhea Virus Induced High Levels of IgA Response in Mice and Piglets. Viral Immunol. 2016;29(9):526–31.
14. Kelsall BL, Strober W. Distinct populations of dendritic cells are present in the subepithelial dome and T cell regions of the murine Peyer's patch. J Exp Med. 1996;183(1):237.
15. Rescigno M, Sabatino AD. Dendritic cells in intestinal homeostasis and disease. J Clin Investig. 2009;119(9):2441.
16. Strugnell RA, Wijburg OL. The role of secretory antibodies in infection immunity. Nat Rev Microbiol. 2010;8(9):656–67.
17. Macpherson AJ, Köller Y, Mccoy KD. The bilateral responsiveness between intestinal microbes and IgA. Trends Immunol. 2015;36(8):460–70.
18. Reynolds J, Pabst R, Bordmann G. Evidence for the existence of two distinct types of Peyer's patches in sheep.[M] Microenvironments in the Lymphoid System. Springer US, 1985:101.
19. Vazquez-Torres A, Fang FC. Cellular routes of invasion by enteropathogens. Curr Opin Microbiol. 2000;3(1):54–9.
20. Van Kruiningen HJ, Ganley LM, Freda BJ. The role of Peyer's patches in the age-related incidence of Crohn's disease. J Clin Gastroenterol. 1997;25(2):470.
21. Van Kruiningen HJ, West AB, Freda BJ, Holmes KA. Distribution of Peyer's patches in the distal ileum. Inflamm Bowel Dis. 2002;8(3):180–5.
22. .Dawson H. A Comparative Assessment of the Pig, Mouse and Human Genomes[M] The Minipig in Biomedical Research. 2011. p. 323–42.
23. Hart EA, Caccamo M, Harrow JL, Humphray SJ, Gilbert JG, Trevanion S, Hubbard T, Rogers J, Rothschild MF. Lessons learned from the initial sequencing of the pig genome: comparative analysis of an 8 Mb region of pig chromosome 17. Genome Biol. 2007;8(8):R168.
24. Swindle MM. Swine as models in biomedical research. Vet Pathol. 2012;49(2):344.
25. And ERM, Ullrey DE. The pig as a model for human nutrition. Annu Rev Nutr. 1987;7(7):361.
26. Rothkötter HJ, Sowa E, Pabst R. The pig as a model of developmental immunology. Hum Exp Toxicol. 2002;21(9–10):533.
27. Meurens F, Summerfield A, Nauwynck H, Saif L, Gerdts V. The pig: a model for human infectious diseases. Trends Microbiol. 2012;20(1):50–7.
28. Kweon CH, Kwon BJ, Lee JG, Kwon GO, Kang YB. Derivation of attenuated porcine epidemic diarrhea virus (PEDV) as vaccine candidate. Vaccine. 1999;17(20–21):2546.
29. Macpherson AJ, Uhr T. Induction of protective IgA by intestinal dendritic cells carrying commensal bacteria. Science. 2004;303(5664):1662–5.
30. Iwasaki A, Kelsall BL. Freshly isolated Peyer's patch, but not spleen, dendritic cells produce interleukin 10 and induce the differentiation of T helper type 2 cells. J Exp Med. 1999;190(2):229–39.

Rapid and sensitive detection of canine distemper virus by real-time reverse transcription recombinase polymerase amplification

Jianchang Wang[1], Jinfeng Wang[1], Ruiwen Li[2], Libing Liu[1] and Wanzhe Yuan[2*] iD

Abstract

Background: Canine distemper, caused by Canine distemper virus (CDV), is a highly contagious and fatal systemic disease in free-living and captive carnivores worldwide. Recombinase polymerase amplification (RPA), as an isothermal gene amplification technique, has been explored for the molecular detection of diverse pathogens.

Methods: A real-time reverse transcription RPA (RT-RPA) assay for the detection of canine distemper virus (CDV) using primers and exo probe targeting the CDV nucleocapsid protein gene was developed. A series of other viruses were tested by the RT-RPA. Thirty-two field samples were further tested by RT-RPA, and the resuts were compared with those obtained by the real-time RT-PCR.

Results: The RT-RPA assay was performed successfully at 40 °C, and the results were obtained within 3 min–12 min. The assay could detect CDV, but did not show cross-detection of canine parvovirus-2 (CPV-2), canine coronavirus (CCoV), canine parainfluenza virus (CPIV), pseudorabies virus (PRV) or Newcastle disease virus (NDV), demonstrating high specificity. The analytical sensitivity of RT-RPA was 31.8 copies in vitro transcribed CDV RNA, which is 10 times lower than the real-time RT-PCR. The assay performance was validated by testing 32 field samples and compared to real-time RT-PCR. The results indicated an excellent correlation between RT-RPA and a reference real-time RT-PCR method. Both assays provided the same results, and R^2 value of the positive results was 0.947.

Conclusions: The results demonstrated that the RT-RPA assay offers an alternative tool for simple, rapid, and reliable detection of CDV both in the laboratory and point-of-care facility, especially in the resource-limited settings.

Keywords: Canine distemper virus, Nucleocapsid protein gene, Exo probe, Recombinase polymerase amplification, RPA and CDV

Background

Canine distemper, caused by canine distemper virus (CDV), is a highly contagious and fatal systemic disease found worldwide not only in dogs and many other carnivores but also in some non-carnivores [1]. CDV is a non-segmented, negative-stranded, enveloped RNA virus that belongs to the family *Paramyxoviridae* and the genus *Morbillivirus*, and is one of the most lethal infectious agents in both susceptible free-living and captive carnivores [2]. CDV-infected dogs may develop respiratory, gastrointestinal, dermatologic, ophthalmic or neurological disorders, that appear simultaneously or sequentially [3, 4]. The broad spectrum of clinical signs, not dissimilar from the signs observed in other respiratory and enteric diseases of dogs, hampers accurate and early clinical diagnosis of canine distemper [5]. Therefore, rapid and accurate diagnosis of CDV infection would enable veterinarians to implement appropriate

* Correspondence: yuanwanzhe2015@126.com
2College of Veterinary Medicine, Agricultural University of Hebei, No.38 Lingyusi Street, Baoding, Hebei 071001, People's Republic of China
Full list of author information is available at the end of the article

strategies in time to improve disease management and prevent outbreaks, particularly within a shelter environment.

With the advances in molecular detection techniques, a substantial number of assays have been described for CDV diagnosis with a varying degree of sensitivity and specificity, such as reverse transcription polymerase chain reaction (RT-PCR) [6], nested RT-PCR [7], real-time RT-PCR [8], reverse transcription loop-mediated isothermal amplification (RT-LAMP) [9] and insulated isothermal PCR (iiPCR) [1]. The RT-PCR assays however, are cold chain dependent and require relatively expensive equipment with experienced technicians, making these assays difficult to implement in the field and at the point-of-care. Recently, an iiPCR method was reported for rapid and sensitive detection of CDV [1], but the reaction time was about 1 h. A simple, rapid, accurate and user-friendly platform is still needed for early point-of-care (POC) detection of CDV infection.

Recombinase polymerase amplification (RPA) is an isothermal gene amplification technique that has been demonstrated to be a rapid, specific, sensitive, and cost-effective molecular method to identify pathogens [10, 11]. As with PCR, the use of two opposing primers allows exponential amplification of the target sequence in RPA, but the latter is tolerant to 5–9 mismatches in primer and probe showing no influence on the performance of the assay [12–14]. It is thought that RPA possesses superiority in speed, portability and accessibility over PCR [15]. Indeed, RPA has recently been explored to replace PCR for the molecular detection of diverse pathogens with different detection strategies, e.g., bacteria, fungi, parasites and viruses [13, 15–21]. In this study, the development of a real-time RT-RPA assay for simple, rapid, portable and POC detection of CDV was described. The fluorescence in the assay is produced by an exo probe which is detected with a portable, user-friendly tube scanner (Genie III, OptiGene Limited, West Sussex, United Kingdom). The Genie III used in the study weighs only 1.75 kg, measuring 25 cm × 16.5 cm × 8.5 cm, and incorporates a rechargeable battery that can support operation for a whole day, making it suitable for point-of-care testing.

Methods
Virus strains and clinical samples
Canine distemper virus (CDV-FOX-TA strain, genotype: America-2) [22], canine parvovirus (CPV, CPV-b114 strain), canine coronavirus (CCoV, ATCC VR-809 strain), canine parainfluenza virus (CPIV, CPIV/A-20/8 strain), pseudorabies virus (PRV, Barth-K61 strain) were maintained in our laboratory. Newcastle disease virus (NDV, LaSota strain) was from the commercial live vaccine (Weike Biotechnology, Harbin, China). Thirty-two nasal/oropharyngeal swabs were collected from 20 dogs of both

sexes (various breeds and ages) from the animal hospital of Agricultural University of Hebei and 12 raccoon dogs from the farms in Hebei Province, China from 2014 to 2016 and snap-frozen for storage at −80 °C. All the dogs and raccoon dogs clinically were suspected of being CDV infected. Fifteen samples from the dogs and 5 samples from the raccoon dogs had been tested to be CDV positive with the Ct values ranging from 16.36 to 37.03, and the other 12 samples were CDV negative by the real-time RT-PCR [8].

DNA/RNA extraction
CDV, CCoV, CPIV and NDV viral RNA was extracted using Trizol Reagent (Invitrogen, Waltham, USA) according to manufacturer's instructions. CPV and PRV viral DNA was extracted using the TIANamp Virus DNA kit (Tiangen, Beijing, China) according to manufacturer's instructions. Viral DNA and RNA were quantified using a ND-2000c spectrophotometer (NanoDrop, Wilmington, USA). For viral RNA extraction from the nasal/oropharyngeal swabs, the swab was inoculated and vortexed in 1 mL sterile phosphate-buffered saline and centrifuged at 10000 rpm for 10 min at 4 °C. The supernatant was collected and used for viral RNA extraction using the Trizol Reagent. Viral RNA extracted from clinical samples was finally eluted in 20 μL of nuclease-free water. All RNA and DNA templates were stored at −80 °C until tested.

Generation of standard RNA
The 1572 bp RT-PCR product, which covers the nucleocapsid protein gene of CDV, was generated from viral genomic RNA template extracted from CDV-FOX-TA strain using N-forward and N-Reverse primers (Table 1). Primers were synthesized by Sangon (Sangon, Shanghai, China). The 50 μL reaction mixture consisted of 25 μL of 2 × 1 Step Buffer, 2 μL of 1 Step Enzyme Mix (Takara, Dalian, China), 0.5 μL of N-forward primer (20 μmol/L),

Table 1 Sequence of primers and probes for CDV RT-PCR, real-time RT-PCR and RT-RPA assays

Name	Sequence 5'-3'	Amplicon size (bp)
N-Forward	ATGGCCAGCCTTCTTAAG	1572
N-Reverse	TTAATTGAGTAGCTCTCTATCA	
CDV-F	AGCTAGTTTCATCTTAACTATCAAATT	87
CDV-R	TTAACTCTCCAGAAAACTCATGC	
CDV-P	FAM-ACCCAAGAGCCGGATACATAGTTTCA ATGC-BHQ1	
CDV-RPA-F	GCTTACTTCAGACTCGGGCAAGAAATGGTTA	154
CDV-RPA-R	CAGTAGCTCGAATTGTCCGGTCCTCTGTTGT	
CDV-RPA-P	CTTGGCATCACCAAGGAGGAAGCTCAGCTGG (FAM-dT) (THF)(BHQ1- dT)CAGAAATAGCATCCA-C3spacer	

0.5 µL of N-Reverse primer (20 µmol/L), 5 µL of extracted RNA template and 17 µL of ddH$_2$O. The reaction condition was set as follows: 50 °C for 30 min; 94 °C for 2 min; 32 cycles of 94 °C for 30 s, 55 °C for 30 s and 72 °C for 60 s. The resulting fragment was purified with the Gel Extraction Kit (Tiangen, Beijing, China), ligated into a pGEM-T Easy vector (Promega, Madison, WI, USA) and transformed into E.coli DH5α chemically competent cells (Dingguo, Beijing, China) according to standard procedures. Positive clones were identified by sequencing analysis. The recombinant plasmid DNA was linearized by Nde I (Takara, Dalian, China), purified using the Wizard SV Gel and PCR Clean-Up System (Promega, Madison, USA) and transcribed in vitro with the RiboMAX Large Scale RNA Production System-T7 (Promega, Madison, USA). In vitro transcribed CDV RNA was digested with the supplied RQ1 RNase-free DNase and purified. It is 97 nucleotides from the T7 promoter region to the NdeI cut site of the pGEM-T Easy vector. All transcripts generated were about 1669 nucleotides in length and approximate size and RNA integrity were verified by agarose gel electrophoresis. The in vitro transcripts were quantified using a ND-2000c spectrophotometer (NanoDrop, Wilmington, USA), and the copy number of RNA molecules was calculated by the following formula [23]: Amount (copies/µL = [RNA concentration (g/µL) /(transcript length in nucleotides × 340) × 6.02 × 10^{23}.

RPA primers and exo probe
Nucleotide sequence data for CDV strains from GenBank were aligned to identify regions that are conserved in the nucleocapsid gene. According to the reference sequences of different CDV genotypes (accession numbers: AB490678, AF164967, AY386316, GU138403, HQ540292, KF856711, KF914669), three forward primers, three reverse primers, and two exo probes were designed. The RPA primers and probes were tested to select the combination yielding the highest sensitivity (Table 1). Primers and exo probe were synthesized by Sangon (Sangon, Shanghai, China).

Rt-Rpa
RT-RPA reactions were performed in a 50 µL volume using a TwistAmp™ RT exo kit (TwistDX, Cambridge, UK). Other components included 420 nM each RPA primer, 120 nM exo probe, 14 mM magnesium acetate, and 1 µL of viral or sample RNA. All reagents except for the viral template and magnesium acetate were prepared in a master mix, which was distributed into each 0.2 mL freeze-dried reaction tube containing a dried enzyme pellet. One µL of viral RNA was added to the tubes. Subsequently, magnesium acetate was pipetted into the tube lids, then the lids were closed carefully, the magnesium acetate was centrifuged into the rehydrated

material using a minispin centrifuge. The sample was vortexed briefly and spun down once again, and the tubes were immediately placed in the Genie III scanner device to start the reaction at 40 °C for 20 min. The fluorescence signal was collected in real-time and would increase markedly due to successful amplification.

Real-time RT-PCR
Real-time RT-PCR specific for CDV was performed on the ABI 7500 instrument as described previously with some modifications [8]. Sequences for the primers and probe are provided in Table 1. The One Step PrimeScript® RT-PCR Kit (Takara Co., Ltd., Dalian, China) was applied in real-time PCR and the reaction was performed as follows: 42 °C 5 min; 95 °C for 10 s; then 40 cycles of 95 °C for 5 s and 60 °C for 35 s.

Analytical specificity and sensitivity analysis
Ten ng of RNA or DNA was used as template for the specificity analysis of the RT-RPA assay. The assay was evaluated against a panel of pathogens considered important in dogs, CDV, CPV, CCoV, CPIV, PRV, and the virus also belonging to the family Paramyxoviridae, NDV.

The in vitro transcribed RNA was diluted in a 10-fold serial dilutions to achieve RNA concentrations ranging from 9.4 × 10^5 to 9.4 × 10^{-1} copies/µL, which were used as the standard RNA for CDV RT-RPA sensitivity assay. The RT-RPA was tested using the quantitative RNA in eight different times (inter-assay) and in eight replicates in one time (intra-assay) to determine the coefficient of variation (CV). The intra- and inter-assay CVs for the threshold times were calculated. The threshold time in eight different times was plotted against the molecules detected,and a semi-log regression was calculated using Prism software 5.0 (Graphpad Software, SanDiego, USA). For exact determination, a probit regression was performed using the Statistical Product and Service Solutions software (IBM, Armonk, USA).

Validation with clinical samples
RNA extracted from 32 clinical swab samples was tested by RT-RPA, and the results were compared with those obtained using real-time RT-PCR [8].

Results
Analytical specificity and sensitivity
Using 10 ng of viral RNA, DNA or canine genome as template, the results showed that only the CDV was detected by RT-RPA while the other viruses and canine genome templates were not detected (Fig. 1, $n = 5$). No cross detections were observed. The data demonstrated the specificity of RT-RPA assay for the detection of CDV.

Using a dilution range of 9.4 × 10^5 to 9.4 × 10^{-1} copies/µL of in vitro transcribed standard RNA as template, the

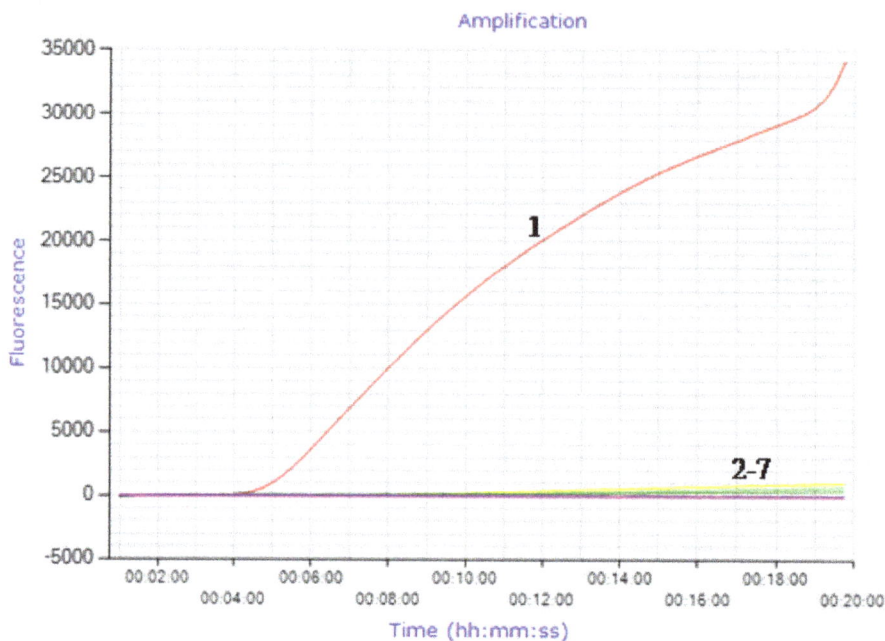

Fig. 1 Analytical specificity of the CDV RT-RPA assay. RT-RPA was carried out at 40 °C for 20 min using 10 ng of viral RNA or DNA as template. The results showed RT-RPA amplified the CDV RNA, but not other viruses tested. 1, CDV; 2, CPV-2; 3, CCoV; 4, CPIV; 5, NDV; 6, PRV; 7, canine genome DNA

RT-RPA and real-time RT-PCR were performed. As shown in Fig. 2, the detection limit of the RT-RPA was 9.4 copies (Fig. 2a and b), which was ten times lower than the real-time RT-PCR (data not shown). The RT-RPA assay was performed eight times on the quantitative RNA, in which 9.4×10^5 to 9.4×10^1 RNA molecules were detected in 8/8 runs, 9.4×10^1, 5/8 and 9.4×10^{-1}, 0/8 (Fig. 2b). Due to the inconsistency in the results, a probit regression analysis was applied, in which the sensitivity in 95% of cases was determined at 31.8 RNA molecules (Fig. 2c). With the data of eight runs on the quantitative RNA standards, a semi-log regression analysis showed the runtime of RT-RPA assay was approximately 3 min– 12 min for 9.4×10^5 to 9.4×10^0 copies (Fig. 2b), while the Ct values of the real-time RT-PCR were 24.43–37.81, which needs approximately 36 min- 57 min.

For the 9.4×10^5 to 9.4×10^0 RNA molecules in the RT-RPA, the CV of the threshold time in the intra-assay ranged from 1.8% to 6.88%, while the CV in the inter-assay ranged from 3.74% to 10.61%.

Evaluation of RT-RPA with clinical samples

The detection results of 32 clinical samples demonstrated that the RT-RPA and real-time RT-PCR showed the same performance (20 positive and 12 negative cases). The further analysis demonstrated the RT-RPA had a diagnostic agreement of 100% with real-time RT-PCR (Table 2). No discrepancy was found in samples (3/20) containing low levels of CDV RNA (Ct > 35, real time RT-PCR),

indicating that the established RT-RPA reliably detected low amounts of CDV in clinical samples. Positive samples had real-time RT-PCR Ct values ranging from 16.36 to 37.03, indicating that the RT-RPA was able to detect CDV RNA across the entire range of the assay. The threshold time (TT) and cycle threshold (Ct) values of RT-RPA and RT-PCR were respectively well at an R^2 value of 0.947 (Fig. 3).

Discussion

In this study, we developed a RT-RPA method based on exo probe for the rapid and sensitive detection of CDV. Specificity analysis revealed that the RT-RPA assay could only detect the CDV, but not other viruses (Fig. 1). Other CDV genotypes were not included in the assay except for the genotype America-2, which is deficiency of the study. The RPA is tolerant to 5–9 mismatches in primer and probe showing no influence on the performance of the assay [12–14], and there were only 2–4 mismatches in the primers and probe in this study with other CDV genotypes. It is assumed the assay would detect all genotypes of CDV, based on targeting a conserved region, but this was not confirmed by testing validated genotypes. The detection limit of the RT-RPA was 9.4 copies, which was 10 times lower than the real-time RT-PCR described previously [8]. We further evaluated this method using clinical samples, and the diagnosis agreement of the RT-RPA and real-time RT-PCR was 100%. Interpretations were limited by the small sample-size, but

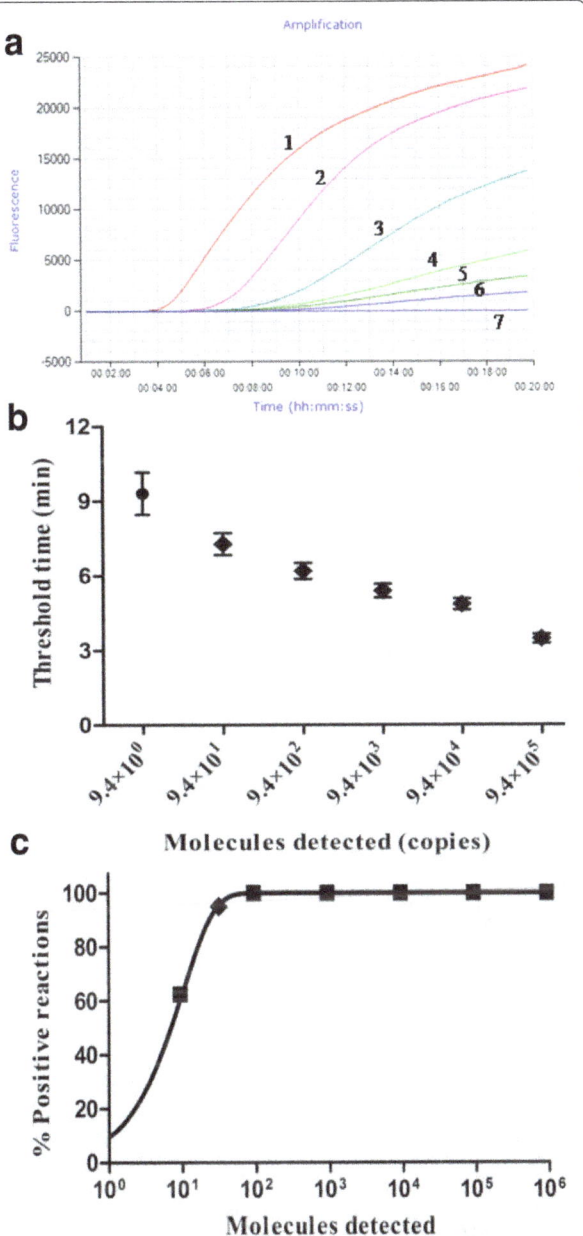

Fig. 2 Performance of the CDV RT-RPA assay. **a** Fluorescence development over time using a dilution range of 9.4×10^5 to 9.4×10^{-1} copies of the CDV standard RNA. Numbers for amplification curves were designated as, 1: 9.4×10^5 copies; 2: 9.4×10^4 copies; 3: 9.4×10^3 copies; 4: 9.4×10^2 copies; 5: 9.4×10^1 copies; 6: 9.4×10^0 copies; 7: 9.4×10^{-1} copies. **b** Semi-logarithmic regression of the data collected from eight CDV RT-RPA tests on the RNA standards using Prism Software 5.0. The run time of the RT-RPA was between 3 min–12 min for CDV RNA from 9.4×10^5 to 9.4×10^0 copies. **c** Probit regression analysis using SPSS software on data of eight runs. The detection limit at 95% probability (31.8 molecules) is depicted by a rhomboid

Table 2 Detection of CDV in clinical samples by RT-RPA and real-time RT-PCR

		real-time RT-PCR		
		Positive	Negative	Total
RT-RPA	Positive	20	0	20
	Negative	0	12	12
	Total	20	12	32

samples from various regions worldwide to evaluate sensitivity of the assay for detecting various strains of the virus circulating.

In recent years, a number of isothermal DNA amplification methods have been developed as a simple, rapid alternative to PCR-based amplification. In the RT-LAMP assay for CDV, four primers were needed and the optimal reaction condition was 60 min at 65 °C [9]. The developed iiPCR could detect as low as 7.6 copies of CDV RNA in approximately 1 h [1]. For the RT-RPA assay developed in this paper, it could detect 9.4 copies of CDV RNA in 12 min, which was more rapid than the above assays. Compared to other isothermal amplification techniques, RPA requires no initial heating for DNA denaturation, and the results could be obtained in less than 20 min; RPA demonstrates a certain tolerance to common PCR inhibitors, and could tolerate a wide range of biological samples [11]; RPA reagents in the lyophilized pellet form could be delivered and stored without cold chain, which could perform satisfactory at 25 °C for up to 12 weeks and at 45 °C for up to 3 weeks [24]. Thus, RPA may be the most applicable approach for the field and point-of-care diagnosis of infectious diseases [11]. A noteworthy feature of the developed RT-RPA assay, i.e., the use of the tube scanner Genie III makes on-site CDV

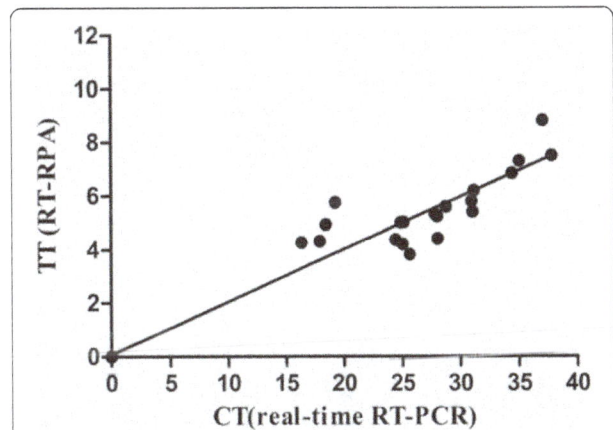

Fig. 3 Comparison between performances of RT-RPA and real-time RT-PCR on clinical samples. Thirty-two RNA extracts of the clinical samples were screened. Linear regression analysis of RT-RPA threshold time (TT) values (y axis) and real-time RT-PCR cycle threshold (Ct) values (x axis) were determined by Prism software. R^2 value was 0.947

the results suggested that the developed RT-RPA performed well in CDV detection. RT-RPA assay should be further tested to more CDV strain RNA extracts or clinical detection feasible, which is especially important for CDV detection and epidemiological surveillance in the field.

Conclusions

An RT-RPA assay with high analytical sensitivity and specificity was successfully developed for the rapid detection of CDV, which could be completed within 20 min. More importantly, the portable feature of the RT-RPA assay makes it applicable at quarantine stations, ports or the site of outbreak. The rapid, sensitive and feasible RT-RPA assay would be a useful tool in CDV control, especially in the resource-limited settings.

Abbreviations

CCoV: Canine coronavirus; CDV: Canine distemper virus; CPIV: Canine parainfluenza virus; CPV-2: Canine parvovirus-2; CT: Cycle threshold; CV: Coefficient of variation; NDV: Newcastle disease virus; POC: Point-of-care; PRV: Pseudorabies virus; RPA: Recombinase polymerase amplification; TT: Threshold time

Acknowledgements

The authors thank the laboratory staff in the animal hospital of Agricultural University of Hebei for sample collection, and we also thank Dr. Yongning Zhang for language review and editing.

Funding

This work was supported by the Natural Science Foundation Youth Project of Hebei Province (C2017325001) and Science and Technology Project Foundation of Animal Husbandry Bureau, Hebei Province, P.R. China (grant number 2014-3-03). The funding agencies had no role in study design, in the collection, analysis and interpretation of data, in the writing of the report, or in the decision to submit the article for publication.

Authors' contributions

JCW and WZY conceived and designed the study. JCW, JFW and RWL performed the experiments and analyzed the data. LBL and WZY helped in the study design and manuscript revision. JCW and WZY wrote the manuscript. All authors read and approved the final manuscript.

Consent for publication

Not applicable.

Competing interests

The authors declare that they have no competing interests.

Author details

[1]Center of Inspection and Quarantine, Hebei Entry-Exit Inspection and Quarantine Bureau, No.318 Hepingxilu Road, Shijiazhuang, Hebei Province 050051, People's Republic of China. [2]College of Veterinary Medicine, Agricultural University of Hebei, No.38 Lingyusi Street, Baoding, Hebei 071001, People's Republic of China.

References

1. Wilkes RP, Tsai YL, Lee PY, Lee FC, Chang HF, Wang HT. Rapid and sensitive detection of canine distemper virus by one-tube reverse transcription-insulated isothermal polymerase chain reaction. BMC Vet Res. 2014;10:213.
2. Lednicky JA, Dubach J, Kinsel MJ, Meehan TP, Bocchetta M, Hungerford LL, Sarich NA, Witecki KE, Braid MD, Pedrak C, et al. Genetically distant American canine distemper virus lineages have recently caused epizootics with somewhat different characteristics in raccoons living around a large suburban zoo in the USA. Virol J. 2004;1:2.
3. Beineke A, Puff C, Seehusen F, Baumgartner W. Pathogenesis and immunopathology of systemic and nervous canine distemper. Vet Immunol Immunopathol. 2009;127(1–2):1–18.
4. Tan B, Wen YJ, Wang FX, Zhang SQ, Wang XD, Hu JX, Shi XC, Yang BC, Chen LZ, Cheng SP, et al. Pathogenesis and phylogenetic analyses of canine distemper virus strain ZJ7 isolate from domestic dogs in China. Virol J. 2011;8:520.
5. Seki F, Ono N, Yamaguchi R, Yanagi Y. Efficient isolation of wild strains of canine distemper virus in Vero cells expressing canine SLAM (CD150) and their adaptability to marmoset B95a cells. J Virol. 2003;77(18):9943–50.
6. Frisk AL, Konig M, Moritz A, Baumgartner W. Detection of canine distemper virus nucleoprotein RNA by reverse transcription-PCR using serum, whole blood, and cerebrospinal fluid from dogs with distemper. J Clin Microbiol. 1999;37(11):3634–43.
7. Shin YJ, Cho KO, Cho HS, Kang SK, Kim HJ, Kim YH, Park HS, Park NY. Comparison of one-step RT-PCR and a nested PCR for the detection of canine distemper virus in clinical samples. Aust Vet J. 2004;82(1–2):83–6.
8. Elia G, Decaro N, Martella V, Cirone F, Lucente MS, Lorusso E, Di Trani L, Buonavoglia C. Detection of canine distemper virus in dogs by real-time RT-PCR. J Virol Methods. 2006;136(1–2):171–6.
9. Cho HS, Park NY. Detection of canine distemper virus in blood samples by reverse transcription loop-mediated isothermal amplification. J Vet Med B Infect Dis Vet Public Health. 2005;52(9):410–3.
10. Piepenburg O, Williams CH, Stemple DL, Armes NA. DNA detection using recombination proteins. PLoS Biol. 2006;4(7):e204.
11. Daher RK, Stewart G, Boissinot M, Bergeron MG. Recombinase polymerase amplification for diagnostic applications. Clin Chem. 2016;62(7):947–58.
12. Daher RK, Stewart G, Boissinot M, Boudreau DK, Bergeron MG. Influence of sequence mismatches on the specificity of recombinase polymerase amplification technology. Mol Cell Probes. 2015;29(2):116–21.
13. Abd El Wahed A, El-Deeb A, El-Tholoth M, Abd El Kader H, Ahmed A, Hassan S, Hoffmann B, Haas B, Shalaby MA, Hufert FT, et al. A portable reverse transcription recombinase polymerase amplification assay for rapid detection of foot-and-mouth disease virus. PLoS One. 2013;8(8):e71642.
14. Boyle DS, Lehman DA, Lillis L, Peterson D, Singhal M, Armes N, Parker M, Piepenburg O, Overbaugh J. Rapid detection of HIV-1 proviral DNA for early infant diagnosis using recombinase polymerase amplification. MBio. 2013; 4(2) doi:https://doi.org/10.1128/mBio.00135-13.
15. Crannell ZA, Cabada MM, Castellanos-Gonzalez A, Irani A, White AC, Richards-Kortum R. Recombinase polymerase amplification-based assay to diagnose Giardia in stool samples. Am J Trop Med Hyg. 2015;92(3):583–7.
16. Yang Y, Qin X, Zhang W, Li Y, Zhang Z. Rapid and specific detection of porcine parvovirus by isothermal recombinase polymerase amplification assays. Mol Cell Probes. 2016;30(5):300–5.
17. Boyle DS, McNerney R, Teng Low H, Leader BT, Perez-Osorio AC, Meyer JC, O'Sullivan DM, Brooks DG, Piepenburg O, Forrest MS. Rapid detection of mycobacterium tuberculosis by recombinase polymerase amplification. PLoS One. 2014;9(8):e103091.
18. Wang J, Liu L, Li R, Yuan W. Rapid detection of porcine circovirus 2 by recombinase polymerase amplification. J Vet Diagn Investig. 2016;28(5):574–8.
19. Murinda SE, Ibekwe AM, Zulkaffly S, Cruz A, Park S, Razak N, Paudzai FM, Ab Samad L, Baquir K, Muthaiyah K, et al. Real-time isothermal detection of Shiga toxin-producing Escherichia Coli using recombinase polymerase amplification. Foodborne Pathog Dis. 2014;11(7):529–36.
20. Zaghloul H, El-Shahat M. Recombinase polymerase amplification as a promising tool in hepatitis C virus diagnosis. World J Hepatol. 2014;6(12):916–22.
21. Krolov K, Frolova J, Tudoran O, Suhorutsenko J, Lehto T, Sibul H, Mager I, Laanpere M, Tulp I, Langel U. Sensitive and rapid detection of Chlamydia trachomatis by recombinase polymerase amplification directly from urine samples. J Mol Diagn. 2014;16(1):127–35.
22. Yang D, Lv P, Hu C, Jia Y, Xie Z, Cao D, Ren Y, Liu H. Isolation and identification of canine distemper virus from fox. Southwest China Journal of Agricultural Science. 2008;21(1):204–7.

23. Yun JJ, Heisler LE, Hwang II, Wilkins O, Lau SK, Hyrcza M, Jayabalasingham B, Jin J, McLaurin J, Tsao MS, et al. Genomic DNA functions as a universal external standard in quantitative real-time PCR. Nucleic Acids Res. 2006; 34(12):e85.

24. Lillis L, Siverson J, Lee A, Cantera J, Parker M, Piepenburg O, Lehman DA, Boyle DS. Factors influencing Recombinase polymerase amplification (RPA) assay outcomes at point of care. Mol Cell Probes. 2016;30(2):74–8.

14

Isolation and characterization of porcine epidemic diarrhea virus associated with the 2014 disease outbreak in Mexico

María Elena Trujillo-Ortega[1], Rolando Beltrán-Figueroa[1], Montserrat Elemi García-Hernández[2],
Mireya Juárez-Ramírez[3], Alicia Sotomayor-González[1], Erika N. Hernández-Villegas[2], José F. Becerra-Hernández[2]
and Rosa Elena Sarmiento-Silva[2]*

Abstract

Background: Interest in porcine epidemic diarrhea has grown since the 2013 outbreak in the United States caused major losses, with mortality rates up to 100 % in suckling piglets. In Mexico, an outbreak of porcine epidemic diarrhea, characterized by 100 % mortality in piglets, began in March 2014 in the State of Mexico.

Methods: The aim of this study was to confirm and identify porcine epidemic diarrhea virus (PEDV) in samples from piglets with suggestive clinical signs using virological, histological, and molecular techniques. Necropsy was performed on 13 piglets from two litters with initial and advanced clinical signs. Suggestive lesions of acute infection with PEDV were detected in histological sections of the small and large bowels; specifically, multiple virus particles with visible crown-shaped projections were observed using electron microscopy and negative staining. Viral isolation was performed in Vero cells with trypsin. Infection was monitored by observation of cytopathic effect, and titration was determined by $TCID_{50}$/ml. The presence of the PEDV in cultures and clinical samples was confirmed by RT-PCR amplification and sequencing of a 651-bp segment of the S glycoprotein gene, as well as a 681-bp matrix protein gene.

Results: The nucleotide sequence analysis of the Mexican isolates showed marked homology to viruses that circulated in 2013 in Colorado, USA.

Conclusions: In this paper we confirm the isolation and characterization of PEDV from animals with early and advanced clinical signs.

Keywords: Porcine epidemic diarrhea virus, Mexico, Outbreak, Characterization

Background

Recently, there has been a growing interest in porcine epidemic diarrhea (PED), stemming from the major losses caused by the 2013 outbreak in the United States, with mortality rates up to 100 % in suckling piglets [1–5]. The infectious agent causing the outbreak was identified on May 10, 2013, at the Veterinary Diagnostic Laboratory of Iowa State University in Ames. Infection by a coronavirus-like virus, known as porcine epidemic diarrhea virus (PEDV), was confirmed [6, 7].

The disease first obtained recognition as a devastating enteric disorder affecting pigs in the UK in 1971. However, it wasn't until 1978, in Belgium, that the etiologic agent was identified as a coronavirus and given the name PEDV (PEDV strain CV777). While the disease was first reported in the United Kingdom in 1970, it has since then spread to Belgium, Hungary, Korea, Italy, Thailand, Japan, and China. In Asia, PED has been considered endemic since 1982, causing substantial economic losses to the swine industry [6, 8–10].

* Correspondence: rosass@unam.mx
[2]Departamento de Microbiología e Inmunología, Facultad de Medicina Veterinaria y Zootecnia, Universidad Nacional Autónoma de México, Mexico City 04510, Mexico

PEDV is an Alphacoronavirus classified in the subfamily *Coronavirinae* of the family *Coronaviridae*. PEDV is an enveloped, single-stranded, positive sense RNA virus that infects swine, usually causing respiratory and gastrointestinal disease [2, 8, 9, 11]. The complete genomic sequence of PEDV has a length of 28,038 nucleotides (nt) starting with a 292 nt untranslated region (UTR), followed by six genes—replicase (Rep), spike (S), ORF3, envelope (E), membrane (M), and nucleoprotein (N)— and ending with a 3'untranslated region from 27 706 to 28 038 nt [9].

PEDV infects the epithelium of the small intestine, an environment rich in proteases, and causes atrophy of the villi resulting in diarrhea and dehydration. Therefore, this disease is characterized, as its name implies, by diarrhea, often watery, as well as some systemic signs such as vomiting, fever, anorexia, and lethargy. The disease is more severe in suckling piglets because of their increased susceptibility to dehydration, but outbreaks are also observed in growing pigs and occasionally in adults [12].

In terms of diagnosis of PEDV infection, there are reports of veterinary diagnostic laboratories that have developed molecular detection techniques. However, viral isolation in cell culture remains the confirmatory test. This procedure is considered difficult to perform due to specific conditions required by the virus, such as trypsin supplementation [6, 13].

Although there have been studies describing the disease in the United States and Canada, to our knowledge, this is the first report of PEDV in Mexico [1, 10, 14]. The aim of this study was to isolate, identify, and characterize PEDV from samples collected during an outbreak that occurred in Mexico in 2014.

Methods

The outbreak began on March 22, 2014, and samples were taken two days after. Directed sampling was performed in pigs in the weaning phase experiencing diarrhea, vomiting, and dehydration that caused 100 % mortality in piglets with permission of the owner. On April 2, samples of lung, gastric contents, stomach, intestine (duodenum), and intestinal contents (feces) were taken from 5 piglets in a litter with early clinical signs (litter 1, ID C1, 6 h of age), and five piglets in a litter with advanced clinical signs (litter 2, ID C2, 24–36 h of age). Euthanasia of both litters was performed by electrocution and subsequent exsanguination (AVMA Guidelines for the Euthanasia of Animals: 2013 Edition). In addition, two samples (lung, gastric contents, stomach, and duodenum) from dead piglets (RIP1 and RIP2) that were 36–48 h old at the time of death, as well as feces from two finishing pigs with diarrhea, were taken.

Histopathology

Samples of lung, stomach, small intestine (duodenum, jejunum and ileum), large intestine (cecum and colon), and mesenteric lymph nodes were collected and preserved in 10 % formalin buffered at pH 7.2. These samples were subsequently processed by routine paraffin embedding technique and staining with hematoxylin and eosin [15]. Tissues were evaluated using Leica DM500 optical microscopy.

Transmission electron microscopy

Fragments of small intestine (jejunum) were fixed in 2.5 % glutaraldehyde for 24 h, washed with a cacodylate solution buffered to pH 7.2. Then were treated with 1 % osmium tetroxide, and washed with collected cacodylate buffer. Subsequently, they were dehydrated with increasing concentrations of acetone and embedded in epoxy resin. Semi-thin sections (200 nm) were cut and mounted on slides and contrasted with toluidine blue. Finally, fine cuts of 60 nm were mounted on copper grids and contrasted with uranyl acetate and lead citrate for posterior observation and evaluation in a Zeiss EM 900 electron microscope.

Viral isolation

Pools of samples including stool and bowel from litter 1 (advanced clinical signs), and litter 2 (initial clinical signs) were macerated by adding 5 ml of D-MEM (Gibco Cat. 10313-021 Lot. 1374740) in the presence of antibiotics (Penicillin 10.00 U/ml, streptomycin 10,000 mcg/ml Gibco Cat. 15140-122 and 200 mM L-glutamine, Gibco 25030-081), and centrifuged at 1500 rpm for 10 min. The supernatant was filtered using a Millex GP filter unit, 0.22 μm (Millipore Express PES membrane, Cat. SLGP033RB).

Vero cells were grown in six-well plates (Corning Inc. COSTAR 3527) previously washed with D-MEM (5 times) to remove the fetal bovine serum. These plates were then inoculated with filtered supernatants in D-MEM supplemented with different concentrations of trypsin (2.5, 5, 10, and 20 μg/ml, and 2 mg/ml) (DifcoTripsine 250 Cat. 215240, Lot. 4181462). After 2 h, this material was removed and the plates were filled with fresh complete medium containing trypsin.

At 48 h post-infection the cells were resuspended, collected, sonicated for 5 min at 37 °C (BransonicSonifier, 5510 Ultrasonic cleaner), and centrifuged at 5000 rpm. An aliquot of 500 μl was collected for posterior RNA extraction. The remaining supernatant was used to infect serial passages in Vero cells in the above conditions (Passage 2).

Supernatant titrations were performed with $TCID_{50}$/ml, using Reed & Münch formula [16].

RT-PCR

RNA extraction

RNA extraction from different passages of the supernatants and samples (intestine, feces, gastric contents, stomach) was achieved with Trizol reagent according to the manufacturer's instructions (Gibco Life Technologies Cat. 15596-018). Table 1 shows the conditions used in the one step RT-PCR reaction.

Table 1 Conditions used for the RT-PCR

ID	Gene	Sequence	Position	Tm °C	Amplicon size
PEDVF (1)	S[a]	TTCTGAGTCACGAACAGCCA	1,466	55	651
PEDVR (1)	S[a]	CATATGCAGCCTGCTCTGAA	2,097	55	651
MPED2F (2)	M[b]	AGTCTTACATGCGAATTGACC	2,565	55	681
MPED2R (2)	M[b]	AGCTGACAGAAGCCATAAAGT	2,398	55	681

[a][20], [b][21]

RT-PCR conditions

RT-PCR reaction was performed using Onestep Kit (Qiagen Cat. 210212) with some variations for a total 20 µl reaction mixture: 4 µl 5× Buffer, dNTPs 0.8 µl, 0.8 µl Primer, 0.8 µl Primer R, 0.8 µl enzyme mixture, 0.4 µl RNase inhibitor, 7.4 l of H_2O, and 5.0 µl of RNA.

The thermocycler was preheated to 50 °C for 2 min, after which the reaction tubes were placed, reverse transcription was carried out at 50 °C for 30 min, and reverse transcriptase was inactivated for 15 min at 95 °C. The reaction was carried out under the following conditions: 94 °C 2 min, (94 °C 1 min, 55 °C 1 min, 72 °C 90 min) for 40 cycles, with a final period of 10 min at 72 °C. Selected primers and their specifications are shown in Table 1.

In order to confirm the presence of viral antigens and to discard the possibility of TGEV and rotavirus infection, Immunochromatography in Sandwich PEDV Ag (Bionote PED Ag Cat. RG14-01), TGE Ag (Bionote TEG/PED Ag Cat. RG14-03), and Rotavirus Ag test (Bionote PED/ROTA Ag Cat. RG14-05) were performed according to manufacturer instructions in gastric content, feces, lung, and intestine samples of one litter as well as in cell culture supernatant of VERO infected cells.

Sequencing

The amplification products were purified by SE-Gel® 2 % (Size Select Agarose Gels™, Invitrogene Cat. G6610-02) according to the manufacturer's protocol and sequenced by the Sanger method using ABI 3130 platform Sequencing. The sequences were edited, assembled, and aligned using Bio-Edit, Clone Manager Version 9, MultAlin 5.4.1, and MEGA6 programs.

Results

Gross lesions

Necropsy was performed on piglets less than one week old from two litters, all of which exhibited a body condition of 1–3/5, anorexia, depression, dehydration, and watery yellow diarrhea. Postmortem examination in 20 % of the piglets showed small areas of lung consolidation in the cranial lobes. The stomach was markedly dilated and full of coagulated milk. 100 % of piglets exhibited small bowel dilatation and in the intestinal lumen, abundant water content and yellow coagulated milk was present (Fig. 1).

Microscopic lesions

Histological sections of lung, stomach, small intestine (duodenum, jejunum and ileum), large intestine (cecum and colon), and mesenteric lymph nodes were evaluated. In the lung, mild interstitial lymphohistiocytic infiltrate was observed; as well as neutrophilic infiltrate intra-alveolar bronchiolar moderate. In the gastric mucosa, parietal cell necrosis, neutrophilic and lymphocytic infiltrate,

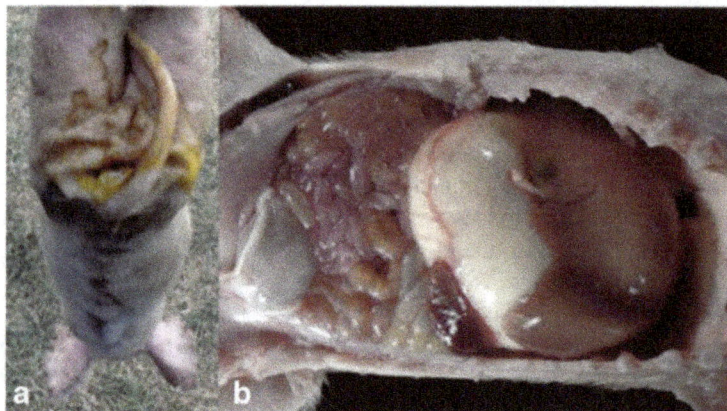

Fig. 1 a Suckling pig, 6 days old. The perianal region exhibits abundant yellow watery stools. **b** Abdominal cavity. The stomach is markedly dilated and full of coagulated milk. Small bowel dilatation and thinning, through which abundant yellow water content and undigested food remains can be observed

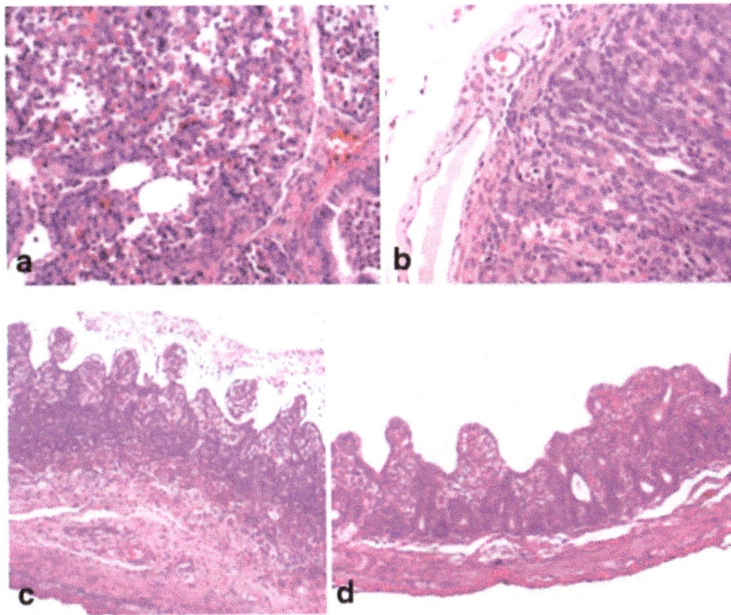

Fig. 2 a Lung. Slight lymphohistiocytic interstitial infiltrate is observed, as well as moderate neutrophilic infiltrate intra-alveolar bronchiolar. **b** Stomach. The gastric mucosa parietal cells exhibited necrosis and multifocal neutrophilic infiltrate with mild dilatation of lymphatic vessels. **c** and **d** Jejunum. Degeneration and necrosis of intestinal epithelial cells, severe villous atrophy, and mild to moderate lymphocytic infiltrate can be observed

and mild multifocal and dilated lymphatic vessels were seen. In histological sections of small and large intestine, degeneration and necrosis of intestinal epithelial cells, severe villous atrophy, lymphocytic infiltrate, mild to moderate congestion, and dilatation of lymphatic vessels were detected. The most significant changes were noted in the jejunum and ileum. Finally, the mesenteric lymph nodes exhibited moderate lymphoid hyperplasia (Fig. 2).

Transmission electron microscopy

Ultrastructural evaluation of the small intestine of piglets from both litters was accomplished. Enterocytes exhibited shortening and degeneration of microvilli. In the cytoplasm, the rough endoplasmic reticulum exhibited different degrees of expansion and loosening of ribosomes, and mitochondrial cristae were lost due to swelling. Numerous spherical viral particles were observed, with a size and morphology compatible with the described characteristics of coronavirus structure (Fig. 3).

Viral isolation and characterization

In the first passage, Vero infected cells with 2 mg/ml and 20 µg/ml of trypsin were detached, as well as uninfected cells with the same concentrations of trypsin. Infected cells treated with 2.5, 5, and 10 µg of trypsin showed cytopathic effect (CPE), which consists of rounded cells, small plaques, intracytoplasmic vacuoles, and detachment not observed in uninfected cells. The effect was observed 24 h post-infection in cells infected with samples from litter 2.

For subsequent passages, trypsin concentrated at 10 µg/ml was selected, as the cells did not detach, CPE was observed, and identification by RT-PCR was positive (Fig. 4).

A total of three passages of virus isolates in Vero cells were carried out to determine whether the isolated virus

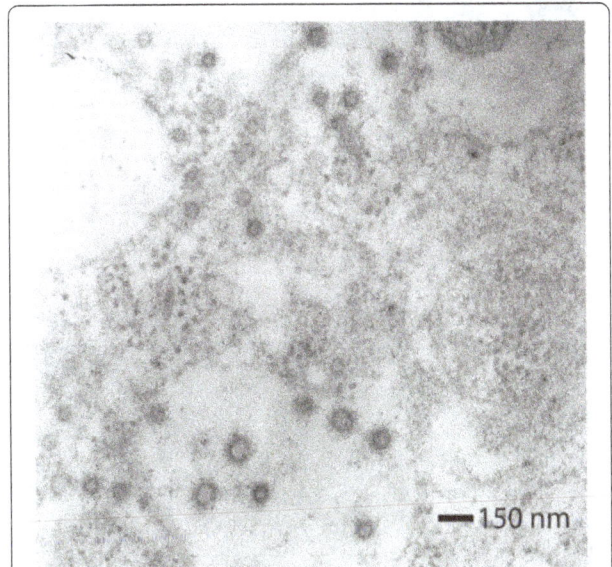

Fig. 3 Electronic transmission photography of an enterocyte's cytoplasm. Numerous viral particles measuring 75–83 nm in diameter are observed. These viral particles possess a membrane with numerous slightly electrodense projections that are 20 nm in length (*arrow*). Adjacent to these viral particles, clusters of ribosomes (*inset*) are appreciated. Contrast technique with uranyl acetate and lead citrate. 50,000× magnification

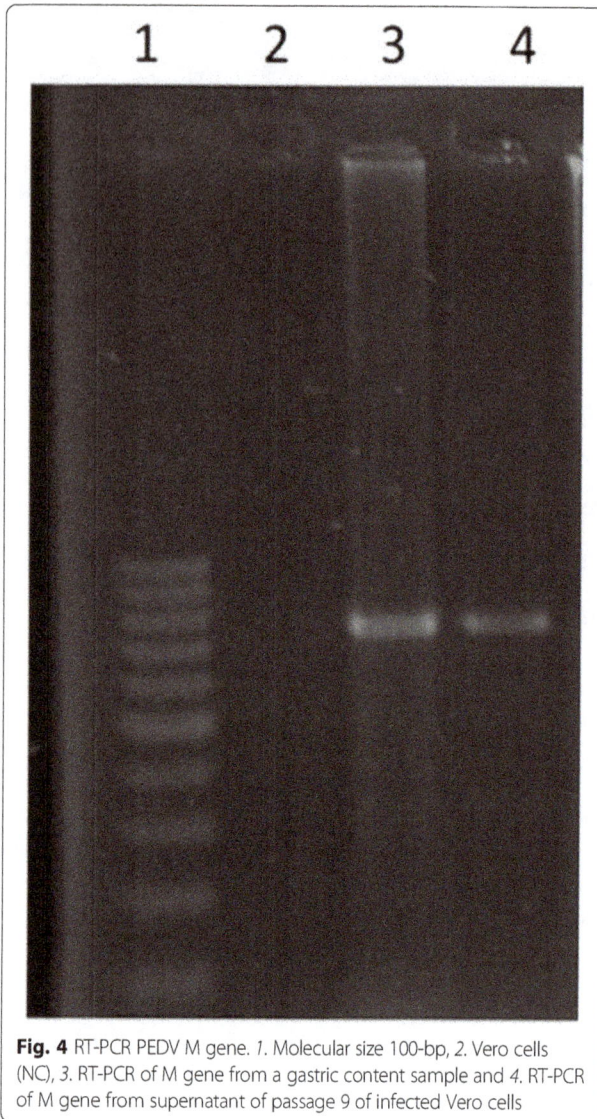

Fig. 4 RT-PCR PEDV M gene. *1.* Molecular size 100-bp, *2.* Vero cells (NC), *3.* RT-PCR of M gene from a gastric content sample and *4.* RT-PCR of M gene from supernatant of passage 9 of infected Vero cells

could propagate and survive efficiently in cell culture. A clear CPE was observed after 24 h post-infection. Figure 5 shows a representative effect from passage nine after 24 h post-infection. During these passages the title of infectious virus fluctuated between 2.32E +03 and 2.81E + 06 $TCID_{50}$/ml.

Molecular diagnosis
RT-PCR in clinical samples
The S gene fragment of 651-bp was amplified with PEDVF and PEDVR primers, (Table 1) while the primers MPED2F and MPED2R amplified a M gene fragment of 681-bp. Table 2 summarizes RT-PCR results in different types of samples, as well as direct Immunochromatography in Sandwich results of PED, TGE, and Rotavirus Ag tests, confirming the presence of PEDV and the absence of TGEV and porcine rotavirus antigens.

Nucleotide sequence accession numbers
The 4 partial sequences of each gene (M and S) were deposited in GenBank under accession numbers KM044328, KM044329, KM044330, KM044331, KM044332, KM044333, KM044334, and KM044335, respectively.

Molecular analysis was completed with partial amplifications of S and M viral genes; the sequences obtained for S gene (593-bp) showed 99 % similarity with the same region of the strains NPL-PEDV/2013/P10 (KJ778616, Lawrence PK unpublished date May 2014), K13JA11-4 (KJ539153, Cho, YY. 2014), and OH14 (access No. KJ408801). For M gene (681-bp) 95 % similarity was found with strains NPL-PEDV/2013/P10 (KJ778616 [17], OH14 (access No. KJ408801), and USA/Indiana/17846/2013 (Access No. KF452323) (Fig. 6) [18].

Discussion
Enteric diseases in piglets cause severe losses in the swine industry. Infectious agents commonly involved are *Escherichia coli*, *Clostridium perfringens* type C, *Isospora suis*, rotavirus, and coronavirus (TGEV and PEDV).

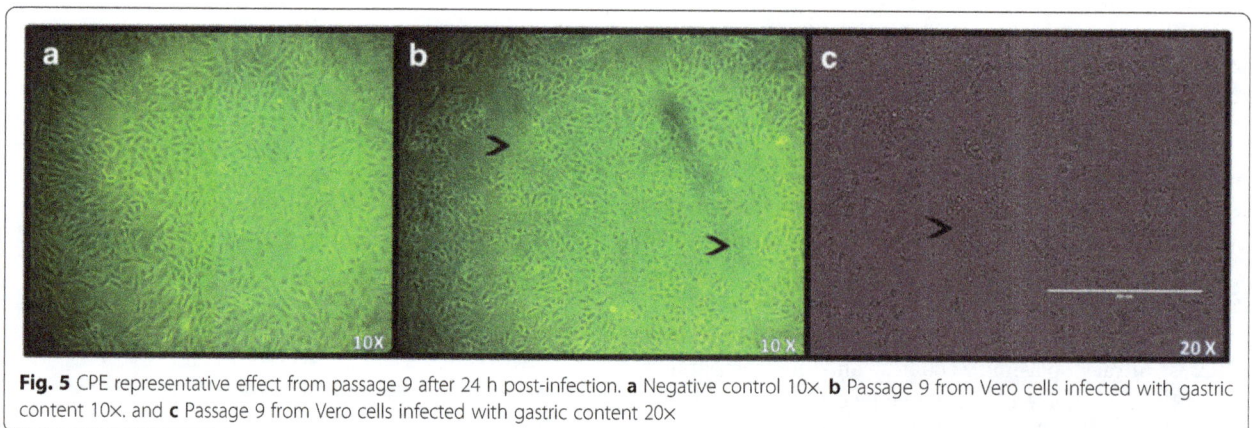

Fig. 5 CPE representative effect from passage 9 after 24 h post-infection. **a** Negative control 10x. **b** Passage 9 from Vero cells infected with gastric content 10x. and **c** Passage 9 from Vero cells infected with gastric content 20x

Table 2 Summary of RT-PCR results for the amplification of fragments M and S from different tissues and Immunochromatography in Sandwich PEDV, TGEV, and porcine rotavirus antigen (Ag)

SAMPLE	RT-PCR M			RT-PCR S			ISOLATE			Ag test		
										Rotavirus	PEDV	TGEV
	C1	C2	RIP	C1	C2	RIP	C1	C2	RIP			
STOMACH CONTENTS	+	+	+	+	+	+				-	+	-
INTESTINE	+	+	+	+	+	+	+	+	+	-	+	-
STOMACH	+	+	+	+	+	+				-		-
FECES	+	+	N.C.	+	+	N.C.	+	+		-	+	-
LUNG	-	-	-	-	-	-				-	-	-

NC not collected

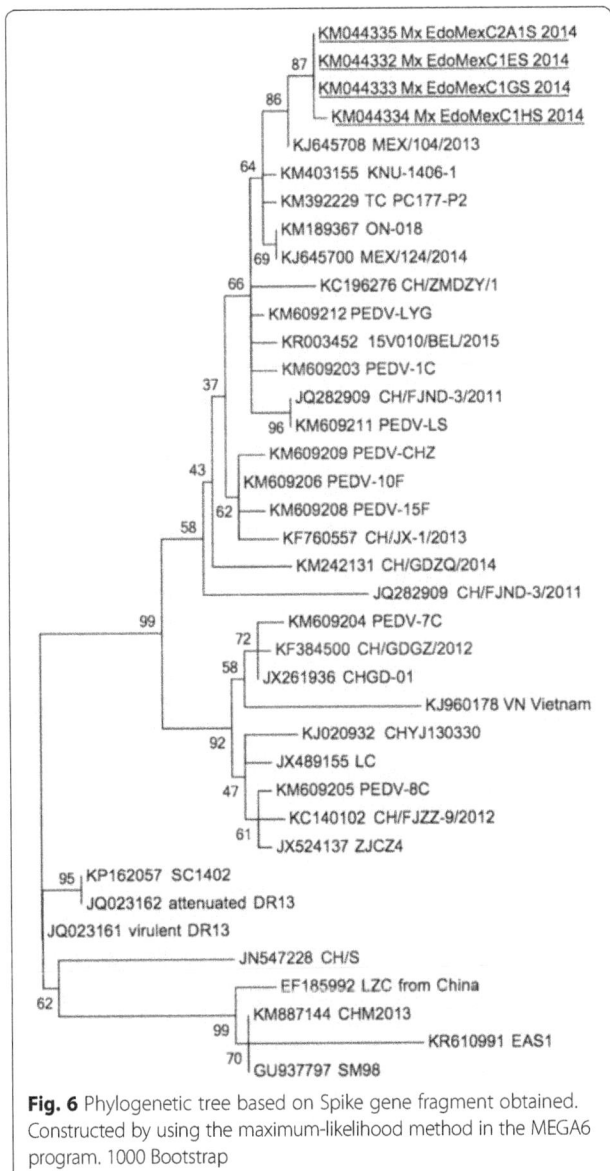

Fig. 6 Phylogenetic tree based on Spike gene fragment obtained. Constructed by using the maximum-likelihood method in the MEGA6 program. 1000 Bootstrap

Recently, the porcine epidemic diarrhea virus (PEDV) has become highly relevant as a result of the outbreak detected in the United States in 2013 [19].

In Mexico, the National Health, Food Safety, and Food Quality Service (SENASICA), sent a notification in May 2014 to the OIE alerting the existence of suggestive clinical evidence indicating the presence of PEDV. Official veterinary services, in coordination with farmers, developed a diagnostic protocol for the disease, including epidemiological sampling of finishing, fattening, breeding stock, and backyard farm animals exhibiting clinical signs. In the research conducted from August 2013 to May 2014, 2309 samples were analyzed by Real Time PCR, of which 30 % were positive. It should be noted that, until that report, authorities had not been able to isolate PEDV, so participation of other infectious agents cannot be ruled out.

In the present study, the clinical symptoms, morbidity, mortality, as well as macroscopic and microscopic lesions observed, coincided with field and experimental infections with PEDV [2, 5, 7, 8, 12]. Moreover, we detected the presence of viral particles compatible with coronavirus in jejunal enterocytes using transmission electron microscopy. These findings, along with the absence of histopathological evidence of bacterial and parasitic pathogens such as *Escherichia coli, Clostridium perfringens* type C, and *Isospora suis*, allowed us to discard possible differential diagnoses to TGEV and porcine rotavirus, which were ruled out using the antigen detection test.

This is the first study reporting isolation of PEDV from the outbreak in Mexico. Isolation from intestine and fecal samples of two litters of animals with initial and advanced clinical signs was achieved using several concentrations of trypsin: 2.5, 5, 10, 20 µg/ml, and 2 mg/ml. Uninfected cells treated with 20 µg/ml and 2 mg/ml detached from all culture bottles; 5 and 10 µg/ml treated cells remained attached allowing observation of cytopathic effect after 24 h post-infection. These results show that using a lower quantity of trypsin (15 µg/ml) than reported in other studies still allows the virus to infect the target cells [13].

RT-PCR was performed to confirm the presence of the virus in cell culture with the amplification of viral genes S and M as reported by Kim et al. (2001) and Li et al. (2012), with observation of expected amplification products throughout all passages [20, 21]. Likewise, titration of each passage was obtained via $TCID_{50}$/ml. Titers ranging between $2.32x10^3$ and $2.81x10^6$ $TCID_{50}$/ml demonstrate progressive replication throughout serial passages.

PEDV isolation represents an important tool for the investigation of its pathogenesis, as well as for the development of serological and molecular diagnostic techniques. A standardized technique for PEDV propagation is essential for potential development of an attenuated vaccine that could be used in nursery pigs and pregnant sows to mitigate the negative impact caused by the disease.

Conclusions
The pigs described in this study already showed clinical signs consistent with the PEDV infection as we demonstrated by differential diagnosis and in this paper we confirm the isolation and characterization of PEDV from animals with early and advanced clinical signs.

Abbreviations
AVMA, American Veterinary Medical Association; CPE, cytopathic effect; OIE, World Organization for Animal Health; PCR, polymerase chain reaction; PED, porcine epidemic diarrhea; RT, reverse transcriptase; TGE, transmissible gastroenteritis

Acknowledgments
Viral isolation advice was given by Eric Nelson, Ph.D. (Veterinary & Biomedical Sciences, South Dakota State University, EE. UU.). Phylogenetic study advice was given by Fernando González Candelas, PhD. (University of Valencia, Spain). Samples were obtained with the help of the Teaching and Research Center in Swine Production (CEIEPP, Centro de Enseñanza e Invesigación en Producción Porcina), FMVZ, UNAM, and Javier Díaz Castorena for help with references assistance.

Funding
This study was financially supported by PAPIIT project No. IN220515; Modulation of the Concentration of Acute Phase Proteins in pigs immunized with Mexican isolates of Porcine Epidemic Diarrhea virus and its association with antigenic variation.

Authors' contributions
RESS directed the research, reviewed the data and manuscript; and directed revisions. METO conducted research and compiled data. RBF was involved in sampling and phylogenetic analysis and participated in drafting the manuscript. MJR provided pathological analysis and participated in drafting the manuscript. MEGH was involved in the development of molecular techniques and phylogenetic analysis and participated in drafting the manuscript; ASG, ENHV and JFBH performed viral isolation and the manuscript draft. All authors have read and approved the manuscript.

Competing interests
The authors declare that they have no competing interests.

Consent for publication
Not applicable.

Author details
[1]Departamento de Medicina y Zootecnia de Cerdos, Facultad de Medicina Veterinaria y Zootecnia, Universidad Nacional Autónoma de México, Mexico City 04510, Mexico. [2]Departamento de Microbiología e Inmunología, Facultad de Medicina Veterinaria y Zootecnia, Universidad Nacional Autónoma de México, Mexico City 04510, Mexico. [3]Departamento de Patología, Facultad de Medicina Veterinaria y Zootecnia, Universidad Nacional Autónoma de México, Mexico City 04510, Mexico.

References
1. Zhao J, Shi BJ, Huang XG, Peng MY, Zhang XM, He DN, Pang R, Zhou B, Chen PY. A multiplex RT-PCR assay for rapid and differential diagnosis of four porcine diarrhea associated viruses in field samples from pig farms in East China from 2010 to 2012. J Virol Methods. 2013;194(1-2):107–12.
2. Jung K, Saif LJ. Porcine epidemic diarrhea virus infection: Etiology, epidemiology, pathogenesis and immunoprophylaxis. Vet J. 2015;204(2):134–43.
3. Bowman AS, Krogwold RA, Price T, Davis M, Moeller SJ. Investigating the introduction of porcine epidemic diarrhea virus into an Ohio swine operation. BMC Vet Res. 2015;11:38.
4. Marthaler D, Bruner L, Collins J, Rossow K. Third strain of porcine epidemic diarrhea virus. United States Emerg Infect Dis. 2014;20(12):2162–3.
5. Opriessnig T. Re-emergence of porcine epidemic diarrhea virus in the global pig population. Vet J. 2015;204(2):131.
6. Mole B. Deadly pig virus slips through US borders. Nature. 2013;499(7459):388.
7. Song D, Huang D, Peng Q, Huang T, Chen Y, Zhang T, Nie X, He H, Wang P, Liu Q, et al. Molecular characterization and phylogenetic analysis of porcine epidemic diarrhea viruses associated with outbreaks of severe diarrhea in piglets in Jiangxi, China 2013. PLoS One. 2015;10(3):e0120310.
8. Huang YW, Dickerman AW, Pineyro P, Li L, Fang L, Kiehne R, Opriessnig T, Meng XJ. Origin, evolution, and genotyping of emergent porcine epidemic diarrhea virus strains in the United States. MBio. 2013;4(5):e00737–00713.
9. Marthaler D, Jiang Y, Otterson T, Goyal S, Rossow K, Collins J. Complete Genome Sequence of Porcine Epidemic Diarrhea Virus Strain USA/Colorado/ 2013 from the United States. Genome Announc. 2013;1(4):e00555–13.
10. Chen Q, Li G, Stasko J, Thomas JT, Stensland WR, Pillatzki AE, Gauger PC, Schwartz KJ, Madson D, Yoon KJ, et al. Isolation and characterization of porcine epidemic diarrhea viruses associated with the 2013 disease outbreak among swine in the United States. J Clin Microbiol. 2014;52(1):234–43.
11. Su S, Wong G, Shi W, Liu J, Lai AC, Zhou J, Liu W, Bi Y, Gao GF. Epidemiology, Genetic Recombination, and Pathogenesis of Coronaviruses. Trends Microbiol. 2016;24(6):490–502.
12. Madson DM, Magstadt DR, Arruda PH, Hoang H, Sun D, Bower LP, Bhandari M, Burrough ER, Gauger PC, Pillatzki AE, et al. Pathogenesis of porcine epidemic diarrhea virus isolate (US/Iowa/18984/2013) in 3-week-old weaned pigs. Vet Microbiol. 2014;174(1-2):60–8.
13. Wicht O, Li W, Willems L, Meuleman TJ, Wubbolts RW, van Kuppeveld FJ, Rottier PJ, Bosch BJ. Proteolytic activation of the porcine epidemic diarrhea coronavirus spike fusion protein by trypsin in cell culture. J Virol. 2014;88(14):7952–61.
14. Ojkic D, Hazlett M, Fairles J, Marom A, Slavic D, Maxie G, Alexandersen S, Pasick J, Alsop J, Burlatschenko S. The first case of porcine epidemic diarrhea in Canada. Can Vet J. 2015;56(2):149–52.
15. Prophet EB, Mills B, Arrington JB, Sobin LH. Hematoxilina y Eosina. Métodos histotecnológicos. In: Métodos histotecnológicos. Washington: Instituto de Patología de las fuerzas armadas de los Estados Unidos de América; 1995.
16. Payment P, Trudel M. Methods and Techniques in Virology. New York: Marcel Dekker, INC New York; 1993.
17. Lawrence PK, Bumgardner E, Bey RF, Stine D, Bumgarner RE. Genome sequences of porcine epidemic diarrhea virus: in vivo and in vitro phenotypes. Genome Announc. 2014;2(3):e00503–14.

18. Stevenson GW, Hoang H, Schwartz KJ, Burrough ER, Sun D, Madson D, Cooper VL, Pillatzki A, Gauger P, Schmitt BJ, et al. Emergence of Porcine epidemic diarrhea virus in the United States: clinical signs, lesions, and viral genomic sequences. J Vet Diagn Invest. 2013;25(5):649–54.
19. Crawford K, Lager K, Miller L, Opriessnig T, Gerber P, Hesse R. Evaluation of porcine epidemic diarrhea virus transmission and the immune response in growing pigs. Vet Res. 2015;46(1):49.
20. Kim SY, Song DS, Park BK. Differential detection of transmissible gastroenteritis virus and porcine epidemic diarrhea virus by duplex RT-PCR. J Vet Diagn Invest. 2001;13(6):516–20.
21. Li ZL, Zhu L, Ma JY, Zhou QF, Song YH, Sun BL, Chen RA, Xie QM, Bee YZ. Molecular characterization and phylogenetic analysis of porcine epidemic diarrhea virus (PEDV) field strains in south China. Virus Genes. 2012;45(1):181–5.

15

Coronavirus and paramyxovirus in bats from Northwest Italy

Francesca Rizzo[1]* (iD), Kathryn M. Edenborough[2], Roberto Toffoli[3], Paola Culasso[3], Simona Zoppi[1], Alessandro Dondo[1], Serena Robetto[1], Sergio Rosati[4], Angelika Lander[2], Andreas Kurth[2], Riccardo Orusa[1], Luigi Bertolotti[4†] and Maria Lucia Mandola[1†]

Abstract

Background: Bat-borne virus surveillance is necessary for determining inter-species transmission risks and is important due to the wide-range of bat species which may harbour potential pathogens. This study aimed to monitor coronaviruses (CoVs) and paramyxoviruses (PMVs) in bats roosting in northwest Italian regions. Our investigation was focused on CoVs and PMVs due to their proven ability to switch host and their zoonotic potential. Here we provide the phylogenetic characterization of the highly conserved polymerase gene fragments.

Results: Family-wide PCR screenings were used to test 302 bats belonging to 19 different bat species. Thirty-eight animals from 12 locations were confirmed as PCR positive, with an overall detection rate of 12.6% [95% CI: 9.3–16.8]. CoV RNA was found in 36 bats belonging to eight species, while PMV RNA in three *Pipistrellus* spp. Phylogenetic characterization have been obtained for 15 alpha- CoVs, 5 beta-CoVs and three PMVs; moreover one *P. pipistrellus* resulted co-infected with both CoV and PMV. A divergent alpha-CoV clade from *Myotis nattereri SpA* is also described. The compact cluster of beta-CoVs from *R. ferrumequinum* roosts expands the current viral sequence database, specifically for this species in Europe. To our knowledge this is the first report of CoVs in *Plecotus auritus* and *M. oxygnathus*, and of PMVs in *P. kuhlii*.

Conclusions: This study identified alpha and beta-CoVs in new bat species and in previously unsurveyed Italian regions. To our knowledge this represents the first and unique report of PMVs in Italy. The 23 new bat genetic sequences presented will expand the current molecular bat-borne virus databases. Considering the amount of novel bat-borne PMVs associated with the emergence of zoonotic infections in animals and humans in the last years, the definition of viral diversity within European bat species is needed. Performing surveillance studies within a specific geographic area can provide awareness of viral burden where bats roost in close proximity to spillover hosts, and form the basis for the appropriate control measures against potential threats for public health and optimal management of bats and their habitats.

Keywords: Bat-borne viruses, Coronavirus, Emerging viruses, Genetic characterization, Paramyxovirus, Surveillance

Background

Bats (order Chiroptera) represent at least one-fifth of existing mammals, consisting of over 1300 known species of which at least 44 are present in Europe [1] and 34 in Italy [2]. Species diversity is expected to increase as some taxa, i.e. *Myotis nattereri* complex, are in the processes of being defined as cryptic species using molecular approaches rather than using morphological characteristics [3]. Bats are grouped into two suborders: the fruit-eating megabats (Megachiroptera), or flying foxes consisting of the single family *Pteropodidae*, and the echolocating insectivorous microbats (Microchiroptera) comprising 16 bat families [4].

Bat borne viruses are arousing increased interest since viral infections in bats have been associated with zoonotic disease outbreaks in humans and domestic animals, including livestock. Rabies virus, Hendra and Nipah viruses, Severe Acute Respiratory Syndrome

* Correspondence: francesca.rizzo@izsto.it
†Equal contributors
[1]Istituto zooprofilattico sperimentale del Piemonte, Liguria e Valle d'Aosta, Via Bologna 148, 10148 Torino, Italy
Full list of author information is available at the end of the article

(SARS) and Middle East Respiratory Syndrome (MERS) coronaviruses, as well as Filoviruses exemplify the role of bats in spreading viruses [5–7].

In the last fifteen years, at least two widespread outbreaks have been caused by novel coronaviruses jumping the species barrier, SARS in 2002–2003 and MERS starting from the Arabian Peninsula since 2012 [6, 7]. Genetic similarities between the viral sequences detected during outbreaks and CoV sequences in bats suggest the viruses originated in flying mammals and presumably passed to humans through a previous adaptation in intermediate hosts, i.e. civet cats and dromedaries [8]. Coronaviruses (family *Coronaviridae*, subfamily *Coronavirinae*) are divided into four main genera: Alphacoronavirus (alpha-CoV) and Betacoronavirus (beta-CoV) found mainly in mammals, Gammacoronavirus detected in birds and marine mammals and Deltacoronavirus found mainly in birds. Several alpha and beta-CoVs have been described worldwide in different bat species (e.g. [9–17]). From the first report in China, *Rhinolophus* species have been specifically associated with SARS-like CoVs [18–20], belonging to the lineage b of beta-CoV genus. Further investigations are needed to clarify the origin of all mammalian coronaviruses, assumed to be from viral ancestors residing in bats [21], untill the recent discovery of a new and highly divergent CoV (i.e. WESV) from house shrews in China [22].

As of 2010, the circulation of CoV in Italian bat population has been notified in only few published studies: SARS-like beta-CoVs have been identified in *Rhinolophus* species [23] and CoVs sequences are available only for Italian *Pipistrellus kuhlii, Hypsugo savii, Nyctalus noctula, Epseticus serotinus, Myotis blythii* and *R. hipposideros* species from fecal samples [24, 25]. Despite the rapid accumulation of bat CoV sequences in the last decade, any viral isolation trial, on different mammalian and bat cell lines failed till 2013, when the first isolation of SARS-like CoV from bat fecal samples succeeded in China [26].

On the list of emerging zoonoses there is a broad diversity of bat-borne paramyxoviruses (PMV), belonging to the wide *Paramyxoviridae* family, as the emergent Nipah virus and Hendra virus (Henipaviruses) and rubulaviruses (e.g. Menangle virus, Tioman virus and Tuhoko virus 1, 2 and 3) (e.g. [27–29] and references therein). Detection and isolation of paramyxoviruses from tissues and urine have been obtained mainly from flying foxes of the genus Pteropus in Africa, Asia, and South America (e.g. [27, 30, 31]) and in Australia (e.g. [32–34]), but also microbat species not previously indicated as PMV reservoirs tested positive for PMV RNA in Africa and Europe [27, 35–37]. Moreover, the ever-increasing attention paid to bat-associated pathogens, has led to the discovery of numerous novel and yet unclassified PMV, revealing an unexpected genetic diversity in the *Paramyxovirinae* subfamily [36]. PMV identification has been reported in only

few studies in insectivorous bats in Europe from Germany, Bulgaria, Romania and Luxembourg, with none of the novel viruses closely related with highly or human pathogenic paramyxoviruses [16, 17, 27, 36].

Following the increasing need of surveillance for bat-borne viruses and the wide range of bat species potentially representing reservoirs for known or unknown pathogens, this study aimed to estimate the viral diversity and distribution in the bat population resident in Northwest Italy. Our investigation was focused on coronaviruses and paramyxoviruses due to their proved ability to switch host and their zoonotic potential. Here we provide the phylogenetic characterization of viral polymerase gene fragments, which are highly conserved within the viral families under investigation.

Methods

Sites and sample collection

Since all bat species in Europe are protected under the Habitats Directive of the European Union [38] and the Agreement on the Conservation of Populations of European Bats [39], samples collection and bat species identification were performed by expert chiropterologists authorized by the Italian Ministry of Environment (authorization number DPN/2010/0011879 and 000882/PNM/08052014).

Bats were captured, during the three years of surveillance (2013–2016) in the Northwestern Italian regions of Piedmont and Liguria, following ethical and safety recommendations [40]. Samplings were conducted from mid-June to October, a period that approximately corresponds to the pregnancy, lactation, dispersion and mating activity of European bats. To minimize animal disturbance, bats were caught soon after parturition with nylon mist-nets of mesh size of 16 to 19 mm positioned at 10–20 m from the reproductive and temporary roost along flight paths towards foraging and drinking areas. During autumn catches were focused particularly at swarming sites in caves where individuals from different colonies meet to mate [41]. All nets were checked every 10 min and captured bats were removed carefully from nets as soon as possible to minimize injury, drowning, strangulation, or stress and individually placed into disposable cloth bags awaiting species identification, collection of biometric data and biological samples.

Species identification was carried out according to Dietz & Kiefer [1] and individual details such as age, class, sex, reproductive status, forearm length, and body mass were recorded. Saliva and urine drops, when present, were collected directly on the animal by swabbing, while feces were recovered, when present, from the cotton bag. All bats were released in the same place of capture after minimal manipulations and were not tagged.

Based on the results of the first two years of surveillance, an increase in feces collection was performed in

2016 setting up random, non-invasive feces samplings underneath single- species reproductive roosts. Briefly, plastic films were left on the ground under different areas of each reproductive colony, then 15 min later single fresh droppings were collected with clean disposable forks, placed in 1 ml of buffered peptone water and kept at 4 °C till analyses. Dead animals in good post-mortem conditions were also collected and stored at −20 °C for further analyses.

RNA extraction and cDNA synthesis

Swabs and feces were maintained in 1 ml of UTM™ Viral Transport Medium (Catalog Number: 360C; Copan Diagnostics, Corona, California) and stored at −20 °C. Before any further analyses took place, the presence of the rabies virus antigen was investigated on dead animals by direct immunofluorescent staining in a BSL3 Laboratory, after necropsy. Once rabies infection has been excluded, samples underwent a pre-treatment before being submitted to automatic nucleic acid purification with magnetic beads.

Pre-treatment for tissues involved the preparation of a tissues pool composed by heart, lung, spleen and intestine from individual animals. The pools were homogenized at a ratio of 1:10 w/V in 1 ml of DEPC-treated PBS in a TissueLyser (Qiagen, Hilden, Germany). Tissue homogenates were then clarified at 13,000×g for 10 min at 4 °C, then 200 μl of tissues pool supernatant were incubated at 56 °C for 10 min with 180 μl of ATL buffer and 20 μl of Qiagen protease provided by the EZ1 Virus Mini Kit v2.0 (Qiagen, Hilden, Germany).To avoid any biosafety risk, the pre-treatment for swabs (saliva and urine) and feces suspensions involved the direct inactivation of 200 μl of each suspension in 200 μl of ATL buffer under a BSL3 hood. Nucleic acid purification (RNA/DNA) was finally accomplished on the EZ1 Advanced XL Instrument using an amount of 400 μl as sample input and a final elution volume of 60 μl of RNase-DNase free water, following the manufacturer's guidelines. RNA was stored at −80 °C until amplification protocols were performed.

cDNA was synthetized from 5 μl of each RNA/DNA sample with the Transcriptor First Strand cDNA Synthesis Kit (Roche Diagnostics, Mannheim, Germany), according to manifacturer's instructions.

Coronavirus detection

For coronavirus detection, 2 μl of cDNA were amplified with an end-point PCR assay targeting a conserved RNA-dependent RNA polymerase (RdRp) gene fragment (537 bp), as described by Poon et al. [42]. The amplification was set up in a 25 μl reaction mixture containing 0.2 mM deoxynucleoside triphosphates, 1.5 mM MgCl2, 0.2 μM of IN-6 and IN-7 primer and 1 U of Platinum Taq Polymerase (Invitrogen, Carlsbad, CA). The

cycling conditions were 94 °C for 2 min, 40 cycles at 94 °C for 1 min, 48 °C for 1 min, 72 °C for 1 min and final elongation step at 72 °C for 7 min.The annealing temperature of primer was modified from 58 °C to 48 °C.

Upon amplification, 20 μl of PCR products were run in 1.5% agarose gel electrophoresis and visualized by GelGreen Nucleic Acid Gel Stain (Biotium) staining; bands of the expected size were excised from the gel for sequencing.

Paramyxovirus detection

For paramyxovirus detection, a broadly reactive seminested PCR assay specific for the RNA polymerase (L)-gene (538 bp) of the *Paramyxovirinae* subfamily was applied. 0 2 μl of cDNAs were amplified using the PAR primers designed by Tong et al. [43] and the protocol optimized with Taguchi method by Kurth et al. (36). Briefly for first round, the final concentration of the 25 μl reaction mix was: 0.1 mM deoxynucleoside triphosphates, 10 mM MgCl2, 0.12 μM of PAR F1 and PAR R primers and 1.25 U of Platinum Taq Polymerase (Invitrogen, Carlsbad, CA). The cycling conditions were 94 °C for 2 min, 40 cycles at 94 °C for 15 s, 50 °C for 30 s, 72 °C for 30 s and a final elongation step at 72 °C for 7 min. Then 1 μl first round PCR product was used in the second round with the same concentrations except for the MgCl2, set up at 1 mM and the use of PAR F2 and PAR R primers, cycling parameters were identical to the first round.

PCR products (20 μl) were run and recovered from a 1.5% agarose gel, as described before.

Sequencing and phylogenetic analysis

Amplicons were purified by gel extraction with the QIAquick Gel Extraction kit (Qiagen, Hilden, Germany), according to the manufacturer's instructions. After elution, nucleic acid quantification of the recovered DNA was done using Thermo Scientific Nanodrop spectrofotometer and submitted for direct sequencing to BMR Genomics, Padua, Italy. The obtained chromatograms were manually checked for unclear base calls and edited using Geneious R7.1.7 software (Geneious, Auckland, New Zealand).

The sequences were aligned using Muscle (implemented in Geneious software) and the alignment was used to evaluate the best evolutionary model (Modeltest ver 3.7) and to draw a bayesian phylogenetic tree (MrBayes ver. 3.1.2). Consensus tree was created after at least 1 million of heuristic search generations and after eliminating the first 25% of evaluated tree topologies (burnin = 25%).

Biomolecular species identification

A total genomic DNA extraction was performed only for PCR positive individuals starting from the original swab suspensions using the QIAmp DNA Mini kit (Qiagen, Hilden, Germany) and following the manufacturer protocol. To confirm species identification by genetic determinations, the complete mitochondrial Cytochrome b gene (Cytb) was amplified as in Puechmaille et al. [3]. PCR products were submitted for direct sequencing to BMR Genomics, Padua, Italy. The obtained chromatograms were manually checked for unclear base calls and edited using Geneious R7.1.7 software (Geneious, Auckland, New Zealand). Species identification was conducted by comparing the obtained sequences to on-line available reference sequences (BLAST alignment, NCBI web site).

Results

Samples collection

Starting from June 2013 till October 2016 a total of 302 animals (35 dead; 267 live) belonging to 19 bat species were collected during 49 capture sessions in 38 locations of Piedmont and five of Liguria regions. Collection of saliva, urine and feces from the same animal was not possible for each of the 267 live bats handled, leading to the final collection of 123 oral swabs (37%), 49 urine swabs (15%) and 158 fecal drops (48%). Sex definition was determined for 195 bats: 117 males and 78 females; the additional 107 single fecal droppings collected in 2016 under 4 different monospecific colonies were considered

as non-assigned individual samples. All captured species are listed in Table 1.

No animal captured during the active surveillance showed signs of disease. During necropsies, no macroscopic lesions referring to infectious diseases were observed, and all the examined bats were negative in the rabies virus antigen IF test.

Coronavirus and paramyxovirus detection

CoV and PMV positive sample types included feces (33/158; 21%) and urine swabs (6/49; 12.2%). None of the tissue pools from dead bats or oral swabs were PCR positive. A significantly greater percentage of female bats, 11.5% (9/78), were PCR-positive than males, 4.3% (5/117), ($p = 0.05$).

Coronavirus and/or paramyxovirus RNA was found in 38 animals belonging to eight bat species (Table 1). Specifically, CoV RNA was detected in 36 bats from 12 sampling sites in Piedmont and one in Liguria, while PMV RNA in three animals from three sampling sites in Piedmont; a map showing the positive sites is presented in Fig. 1. In our sample set, the detection rate of CoV was 12% (36/302; 95% confidence interval [CI] = 9.6–17) ranging between 3.6% for *P. kuhlii*, despite representing the most abundant species in our sample, and 47.4% for *R. ferrumequinum*.

Table 1 Sampled bat species and CoV and PMV prevalences detected

Genus	Species	n°sampled (n° pos)	CoV detection; n/N (%)	PMV detection; n/N (%)
Pipistrellus	*Pipistrellus kuhlii*	56 (4)	2/56; 3.6%	2/56; 3.6%
	Pipistrellus pipistrellus	20 (5)	4/20; 20%	1/20; 5%
	Pipistrellus nathusii	2		
Myotis	*Myotis myotis*	43 (4)	4/43; 9.3%	
	Myotis brandtii	1		
	Myotis bechsteinii	1		
	Myotis nattereri	22 (3)	3/22; 13,6%	
	Myotis daubentonii	24 (2)	2/24; 8.3%	
	Myotis emarginatus	29		
	Myotis oxygnathus	23 (2)	2/23; 8.7%	
	Myotis mistacinus	3		
Hypsugo	*Hypsugo savii*	5		
Plecotus	*Plecotus auritus*	14 (1)	1/14; 7.1%	
	Plecotus austriacus	1		
	Plecotus macrobullaris	1		
	Barbastella barbastellus	17		
Nyctalus	*Nyctalus leisleri*	1		
Rhinolophus	*Rhinolophus ferrumequinum*	38 (18)	18/38; 47.4%	
	Rhinolophus hipposideros	1		
	Total	302 (39)	36/302; 12% [95% CI: 9.6–17]	3/302; 1% [95% CI: 0.3–3,1]

95% Confidence Interval (95% CI) is expressed only for CoV and PMV overall rates

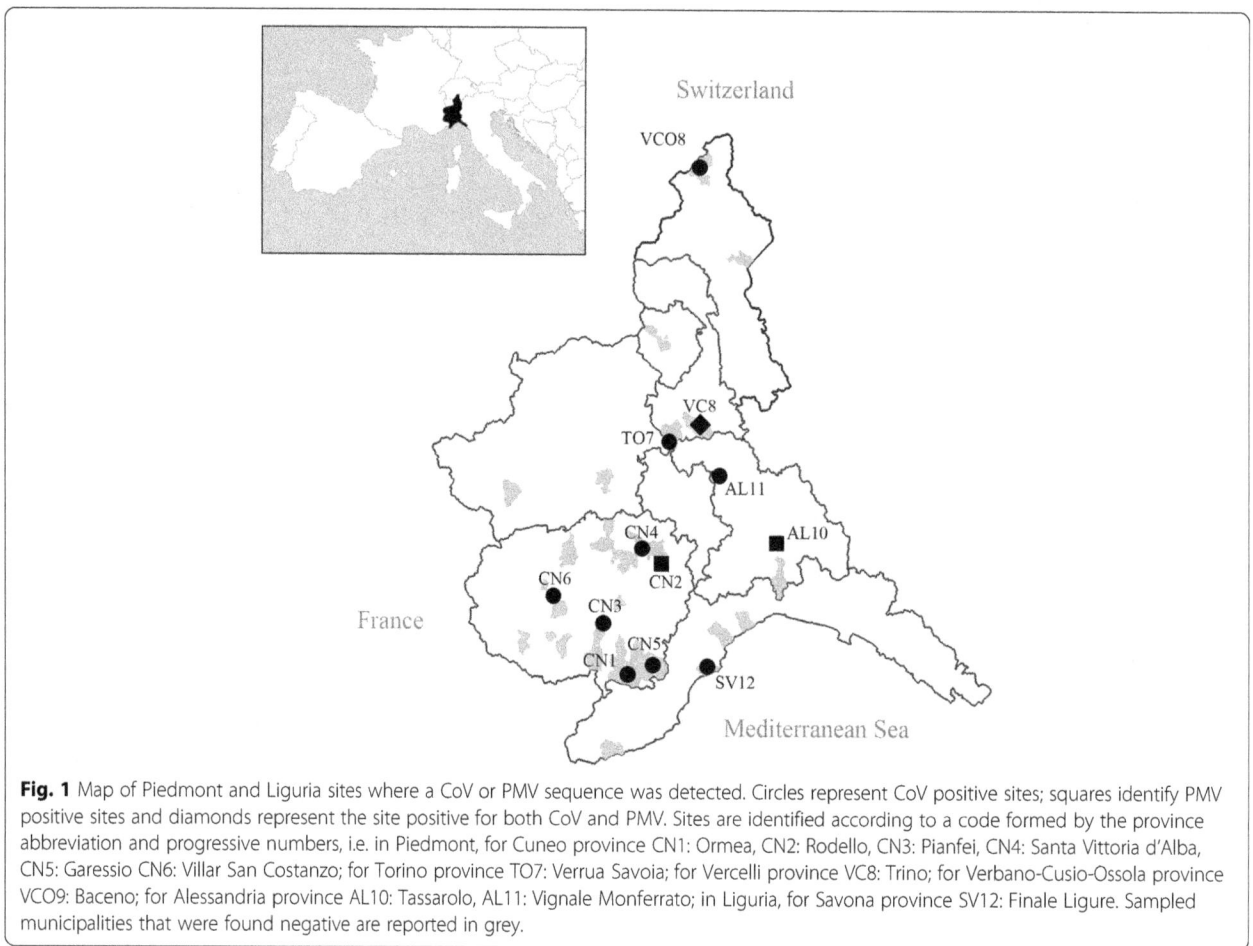

Fig. 1 Map of Piedmont and Liguria sites where a CoV or PMV sequence was detected. Circles represent CoV positive sites; squares identify PMV positive sites and diamonds represent the site positive for both CoV and PMV. Sites are identified according to a code formed by the province abbreviation and progressive numbers, i.e. in Piedmont, for Cuneo province CN1: Ormea, CN2: Rodello, CN3: Pianfei, CN4: Santa Vittoria d'Alba, CN5: Garessio CN6: Villar San Costanzo; for Torino province TO7: Verrua Savoia; for Vercelli province VC8: Trino; for Verbano-Cusio-Ossola province VCO9: Baceno; for Alessandria province AL10: Tassarolo, AL11: Vignale Monferrato; in Liguria, for Savona province SV12: Finale Ligure. Sampled municipalities that were found negative are reported in grey.

Phylogenetic analysis was performed on 20 unique sequences obtained from 36 samples that yielded a PCR product of the expected size after the CoV PCR screening. The positive samples were collected from: *M. nattereri* (*n* = 3), *M. myotis* (*n* = 2), *M. oxygnathus* (*n* = 1), *P. kuhlii* (*n* = 1), *P. pipistrellus* (*n* = 3), *P. auritus* (*n* = 1) and *R. ferrumequinum* (*n* = 9). Any new sequences identified were submitted to GenBank and the accession numbers assigned are given in Table 2. The PMV strains were detected in three different provinces from two *P. kuhlii* at CN2 and AL10 sites and one *P. pipistrellus* at VC8 site; moreover, phylogenetic analysis based on the L-gene fragment was possible for all the three strains retrieved in this study. Interestingly, one *P. pipistrellus* from VC8 site was coinfected by both CoV and PMV as PCR positive results were obtained from the same urine sample. Details of positive sequenced samples are displayed in Table 2.

CoV phylogeny

RdRp phylogeny is presented in Fig. 2 and shows that 15 CoV strains from this study clustered in the alphacoronavirus genus and 5 in the beta-coronavirus genus.

As shown in Fig. 2, the three *M. nattereri* alpha-CoV strains (560, 562 site CN1 and 1021 site TO7) cluster with nucleotide similarities ranging from 94 to 96% within a CoV clade composed of three *M. nattereri* and one *M. bechstenii* from Germany (AN: KT94921–924) and another *M. nattereri* from Hungary (AN: KJ652333), but show only an 86% identity with *M. nattereri* CoVs strains from UK 2009.

Genetic species determination based on the Cyt B gene fragment of 837 bp for these *M. nattereri* species showed a 99% sequence identity with a French *M. nattereri* isolate (AN: JF412408) named "MspA Mnat22 cytochrome b gene" was highlighted.

Based on this finding, our new CoVs strains belong to the *M. nattereri* SpA, a putative new species within the *M. nattereri* species complex.

Three alpha-CoV strains found in feces samples of three bats belonging to the *Myotis* genus show 100% identity to each other (4235 from *M. oxygnathus*, site SV12 and 4658 and 4663 from *M. myotis*, site CN4) and form a divergent clade. When compared to other CoVs, this clade showed the highest identity (~97%) with two

Table 2 CoV and PMV positive samples for which a sequence is available

Species	ID	Sample type	Capture date	Site	Setting[a]	Sex/age[b]	CoV sequence (AN)/CoV genus	PMV sequence (AN)
Myotis nattereri	560	Feces	31/08/13	CN1	T roost	M/ad	Mnat560_IT_13 (KY780381)/alpha	
	562	Urine				F/juv	Mnat562_IT_13 (KY780382)/alpha	
	1021	Feces	16/08/14	TO7	R roost	F/juv	Mnat1021_IT_14 (KY780387)/alpha	
Pipistrellus pipistrellus	1015	Urine	05/08/14	VC8	R roost	F/ad	Ppip1015C_IT_14 (KY780385)/alpha	Ppip1015P_IT_14 (KY780403)
	1016	Feces				F/juv	Ppip1016_IT_14 (KY780386)/alpha	
	1000	Feces	11/08/14	VC9	Fora-ging	M/ad	Ppip1000_IT_14 (KY780384)/alpha	
Pipistrellus kuhlii	600	Feces	19/08/14	CN2	R roost	F/ad		Pkuh600_IT_14 (KY780401)
	605	Feces				F/ad	Pkuh605_IT_14 (KY780383)/alpha	
	621	Urine	06/08/14	AL10	R roost	F/ad		Pkuh621_IT_14 (KY780402)
Myotis myotis	4658	Feces	15/08/16	CN4	R roost		Mmyo4658_IT_16 (KY780397)/alpha	
	4663	Feces					Mmyo4663_IT_16 (KY780398)/alpha	
Myotis oxygnathus	4235	Feces	06/07/16	SV12	R roost		Moxy4235_IT_16 (KY780395)/alpha	
Plecotus auritus	4241	Feces	20/09/16	CN5	Swar-ming	M/ad	Paur4241_IT_16 (KY780396)/beta	
Rhinolophus ferrumequinum	4009	Feces	04/07/16	CN6	R roost		Rfer4009_IT_16 (KY780388)/alpha	
	4011	Feces					Rfer4011_IT_16 (KY780389)/alpha	
	4015	Feces					Rfer4015_IT_16 (KY780390)/alpha	
	4019	Feces					Rfer4019_IT_16 (KY780391)/beta	
	4024	Feces					Rfer4024_IT_16 (KY780392)/alpha	
	4025	Feces					Rfer4025_IT_16 (KY780393)/alpha	
	4027	Feces					Rfer4027_IT_16 (KY780394)/beta	
	4674	Feces	13/07/16	AL11	R roost		Rfer4674_IT_16 (KY780399)/beta	
	4675	Feces					Rfer4675_IT_16 (KY780400)/beta	

ID: Identification number corresponds to the progressive and unique number assigned to each analyzed sample. Site codes are displayed in Fig. 1

[a]the setting where bats were caught, R roost: reproductive roost; T roost: temporary roost

[b]age definitions are juv: juvenile and ad: adult

M. myotis CoV strains, from Germany and Hungary (AN: HM368166 and KJ652331).

Two *P. pipistrellus* CoV sequences (1000 site VCO9 and 1015 site VC8) cluster together with two *P. pipistrellus* strains (Pip1, Pip2) from the same species detected in France in 2014 (AN: KT345294–95) and one *P. pipistrellus* strain from Italy (AN: KF500945); interestingly the third *P. pipistrellus* CoV (1016 site VC8) is ~27% divergent from the others and clusters near Pip3 CoV strain from France (AN: KT345296).

The *P. kuhlii* CoV sequence clusters (605 site CN2) with a similarity of ~97%, within a clade of two *P. kuhlii* strains from Italy 2007 (AN: KF500949) and Spain (AN: HQ184058).

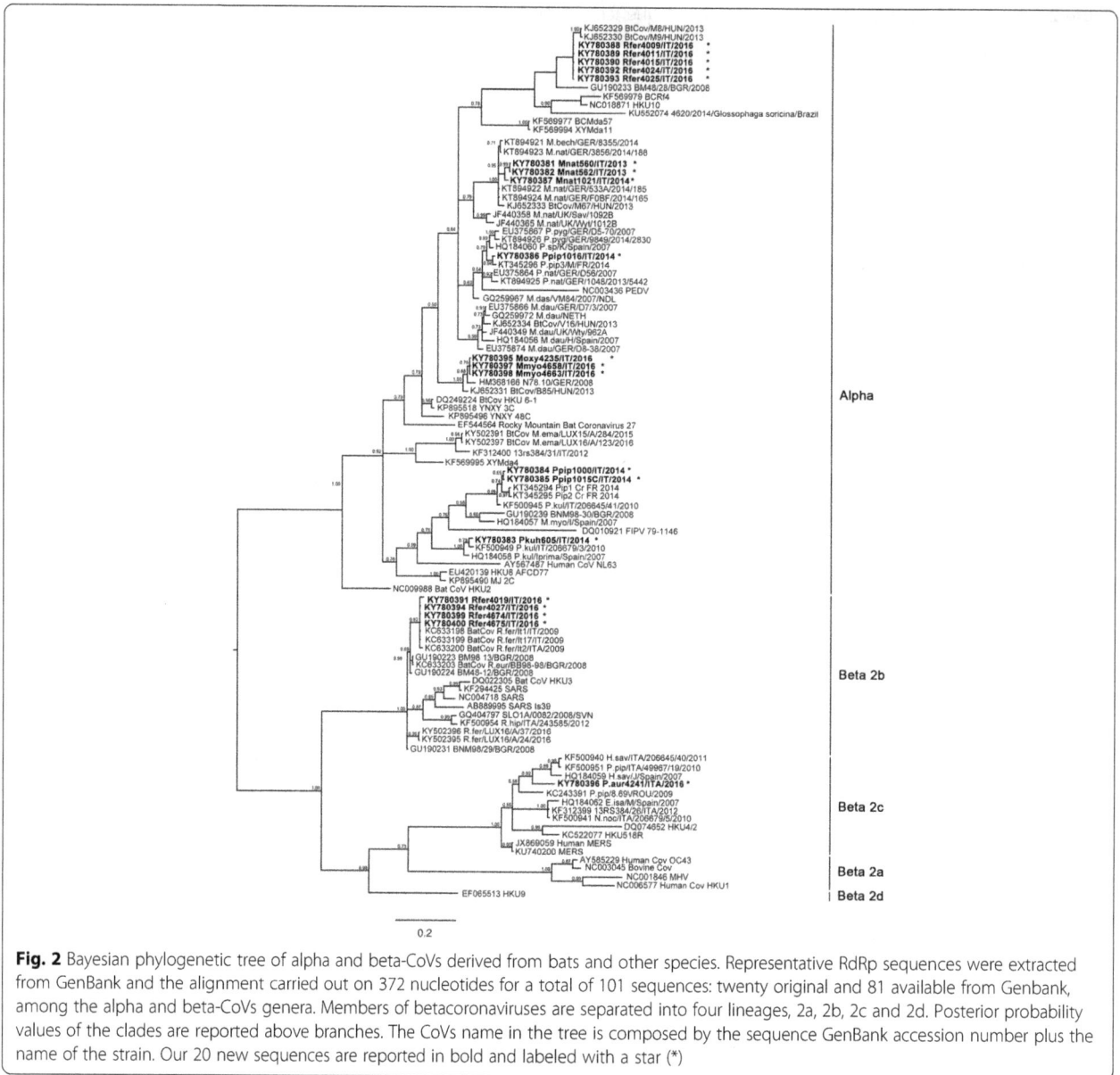

Fig. 2 Bayesian phylogenetic tree of alpha and beta-CoVs derived from bats and other species. Representative RdRp sequences were extracted from GenBank and the alignment carried out on 372 nucleotides for a total of 101 sequences: twenty original and 81 available from Genbank, among the alpha and beta-CoVs genera. Members of betacoronaviruses are separated into four lineages, 2a, 2b, 2c and 2d. Posterior probability values of the clades are reported above branches. The CoVs name in the tree is composed by the sequence GenBank accession number plus the name of the strain. Our 20 new sequences are reported in bold and labeled with a star (*)

Five *R. ferrumequinum* alpha-CoV sequences (4009, 4011, 4015, 4024, 4025 site CN6) found in fecal droppings from the same monospecific roost, showed 100% identity with each other clustering within the clade formed by the only three *R. ferrumequinum* alpha-CoV sequences detected in Europe so far, 3% divergent from the ones from Hungary (AN: KJ652329–30) and 13% from the Bulgarian one (AN:GU190233).

Among the beta-CoV group (lineage b) four *R. ferrumequinum* CoV strains (4019, 4027 site CN6 and 4674, 4675 site AL11) cluster together with other three Italian beta-CoV sequences from the same species (AN: KC33198–200). Interestingly, the 4027 sequence is 100% identical with 4674 and 4675, although originating from two *R. ferrumequinum* roosts located at 130 km distance.

One novel beta-CoV sequence from *Plecotus auritus* (AN: KY780396) clusters separately in the beta-CoV group (lineage c) showing only a ~88% similarity with two *H. savii* CoV strains one from Spain (AN: HQ184059) and one from Italy (AN: KF500940) and a *P. pipistrellus* strain from Italy (AN: KF500951). It's divergence from a MERS CoV strain isolated in 2014 from a camel (AN: KU740200) is 14%. Phylogenetic analyses of this short fragment show that CoVs cluster based on the relatedness of host species.

PMV phylogeny

PMV phylogeny based on representative L-gene sequences available from GenBank is presented in Fig. 3.

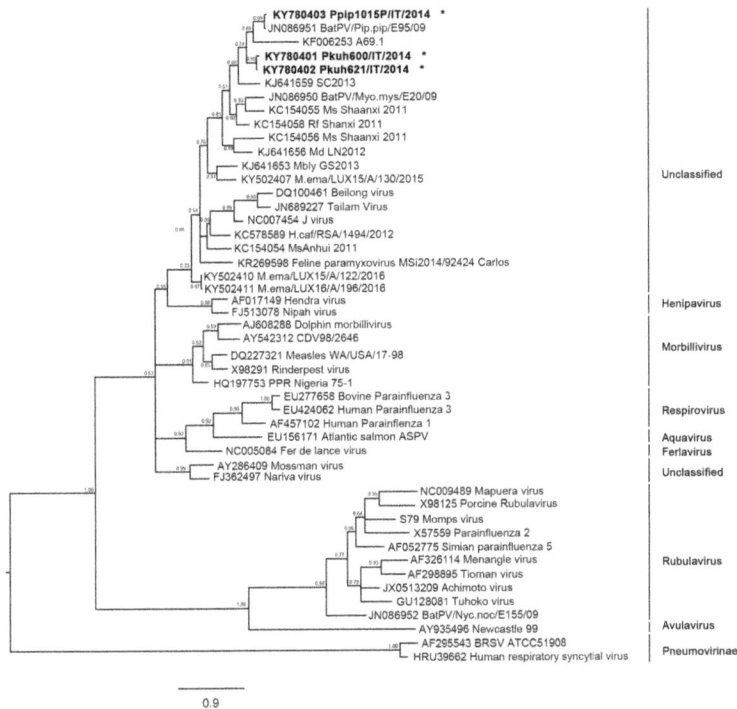

Fig. 3 Bayesian phylogenetic tree of the *Paramixoviridae* family. The tree is built on a L-gene fragment of 393 nucleotides on a total of 48 taxa: three original sequences and 45 sequences. Available L-gene sequences, representative of the seven currently known and unclassified genera of the *Paramixovirinae* sub-family, together with two strains from the *Pneumovirinae* sub-familiy were extracted from GenBank. Posterior probability values of the clades are reported above branches. The samples name in the tree is composed by the GenBank accession number plus the name of the strain. New obtained sequences are sequences are reported in bold and labeled with a star (*)

The new PMV strains were detected in three different locations from one *P. pipistrellus* at VC8 site and two *P. kuhlii* bats at CN2 and AL10 sites (80 km distance).

Our 1015 *P. pipistrellus* strain revealed a 97% nucleotide identity with the E95 PMV strain (AN: JN086951) detected in the same bat species in Germany in 2009, but is more than 23% divergent from any other known PMV sequence.

The two PMV sequences from *P. kuhlii* (600 site CN2 and 621 site AL10) are 97% similar to each other and cluster separately from previously known PMV sequences (18–20% divergence) in the L-gene fragment phylogenetic tree.

To our knowledge, the *P. kuhlii* species was never previously implicated as paramyxovirus host.

Discussion

Recently, emerging disease surveillance programs have intensified to investigate the role of bats in the evolution and spillover of zoonotic pathogens from wildlife. Our study involved three years of active and passive surveillance to characterize the viral diversity of the Northwestern Italian bat population. Using viral family-wide PCRs we identified and phylogenetically characterized 20 new CoVs and 3 PMVs strains. To date, studies on bat CoVs

phylogeny are mainly based on datasets of short sequences (i.e. 440 bp) (e.g. [9, 10, 13, 14, 24, 25, 44]) due to the difficulties of obtaining isolates and good quality viral RNA from bats, but ideally long sequence fragments would be beneficial to infer more reliable phylogenies.

The high prevalence of positive fecal samples (21%) in our study is in concordance with other studies, which identified feces as the best sample type for CoVs detection in bats [9, 18]. Rather than collecting samples from individually caught bats, which is time consuming and labor intensive, collecting single fecal droppings under mono-species roosts turned out to be a reliable and non-invasive method for virological surveillance of bat roosts during their reproductive period. Moreover, urine is confirmed as the most suitable and appropriate sample types for detection of paramyxoviruses in bat populations [33], considering that 2 out of the 3 PMV positive samples from our study were urine swabs.

In 2013–2014 coronavirus circulation was identified in at least four species-specific reproductive roosts of Piedmont: TO7 site for *M. nattereri*, VC8 site for *P. pipistrellus*, CN2 and AL10 for *P. kuhlii*. Unfortunately, attempts to re-test the same roosts in 2016 failed since the VC8 colony moved due to the effect of human disturbance (i.e. robbery of the copper roof cover used as refuge by *P. pipistrellus* bats), and the other three colonies located in private buildings were inaccessible due to logistical reasons. The likelihood of roost disturbance should be taken into account when putting in place bat surveillance plans to enable a steady follow up of the colonies over time.

Bats social behavior could explain the significantly higher infection rate detected in our study for female bats all sampled in August near maternity roosts. Previous studies documented higher virus detection rates in females and juveniles captured near maternity roosts in summer, supporting the hypothesis that virus amplification occurs mainly in reproductive roosts [11, 45].

The identification of the same CoV strains (100% identical) in different roosts of the same bat species (i.e. *R. ferrumequinum* and *P. pipistrellus*) located also at over 100 km distance, seems to confirm that most bat-CoVs appear species-specific and thus more closely associate with the host species than the sampling location [11, 15, 20]. Interestingly, we identified a divergent alpha-CoV lineage in *M. nattereri SpA*, representing a cryptic lineage within the *Myotis nattereri* species complex in the Mediterranean region. The lineage is known to be present in Italy, however no information is available for Germany and Hungary [46]. Following the host-virus coevolution theory based on their close phylogenetic concordance [47], the small divergence (from 3.5 to 5%) between our *M. nattereri SpA* CoV strains and the German or Hungarian *M. nattereri* ones could indicate that they all reside in the *M. nattereri SpA* host, considering that molecular species identification for those specimens is lacking.

The detection of identical alpha-CoV sequences in two different species belonging to the *Myotis* genus (*M. oxygnathus* and *M. myotis*) from two distinct roosts (sites SV12 and CN4) 90 km apart could be due to the expansion and overlapping of habitats and foraging areas of *Myotis* spp. through the Maritime Alpine chain and valleys. To our knowledge this is the first report of CoV in the *M. oxygnathus* species.

The compact cluster of almost identical beta-CoV (lineage b) strains from two separate *R. ferrumequinum* roosts gives further indications that the *Rhinolophus* genus may represent the specific host for SARS-like CoVs and gives an important contribution in terms of available beta-CoV sequences from this species in Europe. To our knowledge, this is the first report of CoV in the *P. auritus* species. This sequence clusters separately within the beta-CoV (lineage c) showing a 14% divergence with a MERS strain identified from a camel in Egypt.

The detection of highly divergent alpha-CoV strains within one *P. pipistrellus* reproductive roost, the circulation of both alpha and beta-CoVs within one *R. ferrumequinum* roost and the co-infection of *P. pipistrellus* with both CoV and PMV provide further evidence that bats are able to carry more than one virus. While infection

with multiple CoVs in the same species/bat/colony is well known, and has been previously reported in China [26, 48, 49] and Europe [19], apart from metagenomic studies notably biased towards the identification of sequences from dsDNA viruses, to our knowledge the coinfection of different ssRNA viral families in the same animal was so far reported only in one study in Europe from *P. pygmaeus* in Hungary [44]. In the specific, the co-infection with two ssRNA viral families within the same host may be explicable in the light of the IFN inhibition used by paramyxoviruses to circumvent host' innate immune response [50]. This mechanism, known as IFN antagonism, may be exploited by other viruses able to escape the adaptive immunity, e.g. CoVs, to be introduced and proliferate in the same host, as observed in mallards [51].

By the increased viral surveillance, a considerable number of novel paramyxoviruses has been discovered in pteropoid and non-pteropoid species, but to date the number of bat PMV sequences for Europe is very scarce and only from few bat species [16, 17, 27, 36]. The three new PMV strains, two in *P. kuhlii* and one in *P. pipistrellus* species, couldn't be classified within any of the current seven known PMV genera, but cluster in the crowded, unassigned PMV clade, which comprises several bat derived strains. Our report represents the first identification of PMVs in the *P. kuhlii* species worldwide. The two sequences, retrieved from two roosts located 90 km apart, are divergent from previously known PMV clusters, which may indicate a stronger association to the host species rather than the geographic area also for paramyxoviruses. This viral tropism is also strongly supported by the high similarity of our *P. pipistrellus* sequence to that of the one other E95 PMV sequence retrieved in Germany from the same bat species. In support of this hypothesis, a study on renal tissues from African bats underlined how paramyxovirus divergence in pteroid and non-pteroid bats correlates with bat taxonomy, suggesting a strong association with bat genera [37]. Because the L-gene fragment used as genetic marker in the aforementioned study is not overlapping with the sequence we used, we couldn't phylogenetically compare them. Nevertheless, given the high similarity our *P. pipistrellus* sequence shows with the E95 PMV strain, our findings support this association. Moreover, an extensive collection of urine samples from the colony would be necessary to facilitate PMVs isolation, which remains a critical requirement for full genome and pathogenic characterization of the strains detected.

Conclusions

Compared to previous studies published in Italy [24, 25], we detected alpha and beta-CoVs in not previously surveyed Italian regions and in new bat species; moreover, this report represents the first and novel identification of PMVs in Italy. The 23 new bat genetic sequences will fill gaps and expand the current molecular bat-borne virus databases.

Considering the amount of novel bat-borne PMVs associated with the emergence of zoonotic infections in animals and humans in the last years define the virus diversity within European bat species is needed. Performing surveillance studies within a specific geographic area can provide awareness of viral burden where bat roosts are in close proximity to spillover hosts, and can form the basis for the appropriate control measures to curb potential threats for public health and optimal management of bats and their habitats.

Acknowledgements
We thank Carla Lo Vecchio for the assistance with the laboratory work and Mara Calvini for dead animals' collection.

Funding
Financial support for this study and its publication was provided by the Italian Ministry of Health in the context of Ricerca Sanitaria Corrente 2013 (Code: IZS PLV 09/13 RC).

Authors' contributions
MLM, SR1. and RO participated in the conception and coordination of the study; RT and PC performed all the captured and bat species identification in the field; FR and AL performed the experiments and FR wrote the paper; LB and KE analyzed the genetic data; SZ and AD performed the necropsies; AK and SR2 contributed reagents, materials and analysis tools. All authors read and approved the final manuscript.

Consent for publication
Not applicable

Competing interests
The authors declare no competing interests. The founding sponsors had no role in the design of the study; in the collection, analyses, or interpretation of data; in the writing of the manuscript, and in the decision to publish the results.

Author details
[1]Istituto zooprofilattico sperimentale del Piemonte, Liguria e Valle d'Aosta, Via Bologna 148, 10148 Torino, Italy. [2]Robert Koch Institute, Seestraße 10, 13353 Berlin, Germany. [3]Chirosphera, via Tetti Barbiere 11, 10026 Santena, TO, Italy. [4]Department of Veterinary Science, Largo Paolo Braccini 2, 10095 Grugliasco, TO, Italy.

References
1. Dietz C, Kiefer A. Bats of Britain and Europe. London: Bloomsbury Publishing; 2016.
2. Lanza B. Mammalia V Chiroptera. In: Calderini editors. Fauna d'Italia, vol XLVII. Milano: Calderini de Il Sole 24; 2012.
3. Puechmaille SJ, Allegrini B, Boston ES, Dubourg-Savage MJ, Evin A, Knochel A, et al. Genetic analyses reveal further cryptic lineages within the Myotis Nattereri species complex. Mammalian Biology-Zeitschrift für Säugetierkunde. 2012;77:224–8.
4. Calisher CH, Childs JE, Field HE, Holmes KV, Schountz T. Bats: important reservoir hosts of emerging viruses. Clin Microbiol Rev. 2006;19:531–45.
5. Wong S, Lau S, Woo P, Yuen KY. Bats as a continuing source of emerging infections in humans. Rev Med Virol. 2007;17:67–91.
6. Peiris JS, Lai ST, Poon LLM, Guan Y, Yam LYC, Lim W. Coronavirus as a possible cause of severe acute respiratory syndrome. Lancet. 2003;361:1319–25.
7. Memish ZA, Mishra N, Olival KJ, Fagbo SF, Kapoor V, Epstein JH, et al. Middle East respiratory syndrome coronavirus in bats. Saudi Arabia Emerg Infect Dis. 2013;19:1819–23.
8. Hu B, Ge X, Wang LF, Shi Z. Bat origin of human coronaviruses. Virol J. 2015;12:221.
9. Tang XC, Zhang JX, Zhang SY, Wang P, Fan XH, Li LF, et al. Prevalence and genetic diversity of coronavirus in bats from China. J Virol. 2006;80:7481–90.
10. Dominguez SR, O'Shea TJ, Oko LM, Holmes KV. Detection of group 1 coronaviruses in bats in North America. Emerg Infect Dis. 2007;13:1295–300.
11. Gloza-Rausch F, Ipsen A, Seebens A, Gottsche M, Panning M, Drexler JF, et al. Detection and prevalence patterns of group I coronaviruses in bats northern Germany. Emerg Infect Dis. 2008;14:626–31.
12. Falcon A, Vazquez-Moron S, Casas I, Aznar C, Ruiz G, Pozo F, et al. Detection of alpha and betacoronaviruses in multiple Iberian bat species. Arch Virol. 2011;156:1883–90.
13. Anthony SJ, Ojeda-Flores R, Rico-Chávez O, Navarrete-Macias I, Zambrana-Torrelio CM, Rostal MK, et al. Coronaviruses in bats from Mexico. J Gen Virol. 2013;94:1028–38.
14. Goffard A, Demanche C, Arthur L, Pinçon C, Michaux J, Dubuisson J. Alphacoronaviruses detected in french bats are phylogeographically linked to coronaviruses of european bats. Viruses. 2015;7:6279–90.
15. Asano KM, Hora AS, Scheffer KC, Fahl WO, Iamamoto K, Mori E, et al. Alphacoronavirus in urban Molossidae and Phyllostomidae bats Brazil. Virol J. 2016;13:110.
16. Fischer K, Zeus V, Kwasnitschka L, Kerth G, Haase M, Groschup MH, et al. Insectivorous bats carry host specific astroviruses and coronaviruses across different regions in Germany. Infect Genet Evol. 2016;37:108–16.
17. Pauly M, Pir JB, Loesch C, Sausy A, Snoeck CJ, Hübschen JM, Muller CP. Novel Alphacoronaviruses and paramyxoviruses Cocirculate with type 1 and severe acute respiratory system (SARS)-related Betacoronaviruses in Synanthropic bats of Luxembourg. Appl Environ Microbiol. 2017;83:e01326–17.
18. Lau SK, Woo PC, Li KS, Huang Y, Tsoi HW, Wong BH, et al. Severe acute respiratory syndrome coronavirus-like virus in Chinese horseshoe bats. Proc Natl Acad Sci U S A. 2005;102:14040–5.
19. Drexler JF, Gloza-Rausch F, Glende J, Corman VM, Muth D, Goettsche M, et al. Genomic characterization of severe acute respiratory syndrome-related coronavirus in European bats and classification of coronaviruses based on partial RNA-dependent RNA polymerase gene sequences. J Virol. 2010;84: 11336–49.
20. Balboni A, Battilani M, Prosperi S. The SARS-like coronaviruses: the role of bats and evolutionary relationships with SARS coronavirus. Microbiologica-quarterly journal of. Microbiol Sci. 2012;35:1.
21. Woo PC, Lau SK, Lam CS, Lau CC, Tsang AK, Lau JH, et al. Discovery of seven novel mammalian and avian coronaviruses in the genus deltacoronavirus supports bat coronaviruses as the gene source of alphacoronavirus and betacoronavirus and avian coronaviruses as the gene source of gammacoronavirus and deltacoronavirus. J Virol. 2012;86:3995–4008.
22. Wang W, Lin X-D, Liao Y, Guan X-Q, Guo W-P, Xing J-G, et al. Discovery of a highly divergent coronavirus in the Asian house shrew from China illuminates the origin of the alphacoronaviruses. J Virol. 2017;91:e00764–17.
23. Balboni A, Palladini A, Bogliani G, Battilani M. Detection of a virus related to betacoronaviruses in Italian greater horseshoe bats. Epidemiol Infect. 2011;139:216–9.
24. Lelli D, Papetti A, Sabelli C, Rosti E, Moreno A, Boniotti MB. Detection of coronaviruses in bats of various species in Italy. Viruses. 2013;5:2679–89.
25. De Benedictis P, Marciano S, Scaravelli D, Priori P, Zecchin B, Capua I, et al. Alpha and lineage C betaCoV infections in Italian bats. Virus Genes. 2014;48:366–71.
26. Ge XY, Li JL, Yang XL, Chmura AA, Zhu G, Epstein JH, et al. Isolation and characterization of a bat SARS-like coronavirus that uses the ACE2 receptor. Nature. 2013;503:535–8.

27. Drexler JF, Corman VM, Müller MA, Maganga GD, Vallo P, Binger T, et al. Bats host major mammalian paramyxoviruses. Nat Commun. 2012;3:796.

28. Edwards S, Marsh GA. Henipaviruses: bat-borne paramyxoviruses. Microbiol Aust. 2017;38:4–7.

29. Baker KS, Todd S, Marsh GA, Crameri G, Barr J, Kamins AO, et al. Novel, potentially zoonotic paramyxoviruses from the African straw-colored fruit bat Eidolon Helvum. J Virol. 2013;87:1348–58.

30. Drexler JF. CormanVM, Gloza-Rausch F, Seebens a, Annan a, Ipsen a, et al. Henipavirus RNA in African bats. PLoS One. 2009;4:e6367.

31. Chua KB, Koh CL, Hooi PS, Wee KF, Khong JH, Chua BH, et al. Isolation of Nipah virus from Malaysian island flying-foxes. Microbes Infect. 2002;4:145–51.

32. Halpin K, Young PL, Field HE, Mackenzie JS. Isolation of Hendra virus from pteropid bats: a natural reservoir of Hendra virus. J Gen Virol. 2000;81:1927–32.

33. Barr JA, Smith C, Marsh GA, Field H, Wang LF. Evidence of bat origin for Menangle virus a zoonotic paramyxovirus first isolated from diseased pigs. J Gen Virol. 2012;93:2590–4.

34. Barr J, Smith C, Smith I, de Jong C, Todd S, Melville D, et al. Isolation of multiple novel paramyxoviruses from pteropid bat urine. J Gen Virol. 2015;96:24–9.

35. Wilkinson DA, Temmam S, Lebarbenchon C, Lagadec E, Chotte J, Guillebaud J, et al. Identification of novel paramyxoviruses in insectivorous bats of the Southwest Indian Ocean. Virus Res. 2012;170:159–63.

36. Kurth A, Kohl C, Brinkmann A, Ebinger A, Harper JA, Wang LF, et al. Novel paramyxoviruses in free-ranging European bats. PLoS One. 2012;7:e38688.

37. Mortlock M, Kuzmin IV, Weyer J, Gilbert AT, Agwanda B, Rupprecht CE, et al. Novel paramyxoviruses in bats from sub-Saharan Africa 2007–2012. Emerg Infect Dis. 2015;21:1840–3.

38. The Council of the European Communities. Council directive 92/43/EEC of 21 may 1992 on the conservation of natural habitats and of wild fauna and flora. Off J Eur Union. 1992;206:7–50.

39. UNEP/EUROBATS Agreement on the Conservation of Populations of European Bats. EUROBATS, London. 1991. http://www.eurobats.org/official_documents/agreement_text. Accessed on 25 February 2013.

40. Kunz TH, Parsons S. Ecological and behavioural methods for the study of bats. 2nd ed. Baltimore: Johns Hopkins University Press; 2009.

41. Veith M, Beer N, Kiefer A, Johannesen J, Seitz A. The role of swarming sites for maintaining gene flow in the brown long-eared bat (Plecotus Auritus). Heredity. 2004;93:342–9.

42. Poon LL, Chu DK, Chan KH, Wong OK, Ellis TM, Leung YH, et al. Identification of a novel coronavirus in bats. J Virol. 2005;79:2001–9.

43. Tong S, Chern SW, Li Y, Pallansch MA, Anderson LJ. Sensitive and broadly reactive reverse transcription-PCR assays to detect novel paramyxoviruses. J Clin Microbiol. 2008;46:2652–8.

44. Kemenesi G, Dallos B, Görföl T, Boldogh S, Estók P, Kurucz K, et al. Molecular survey of RNA viruses in Hungarian bats: discovering novel astroviruses coronaviruses and caliciviruses. Vector Borne Zoonotic Dis. 2014;14:846–55.

45. Plowright RK, Field HE, Smith C, Divljan A, Palmer C, Tabor G, et al. Reproduction and nutritional stress are risk factors for Hendra virus infection in little red flying foxes (Pteropus Scapulatus). Proc R Soc Lond B Biol Sci. 2008;275:861–9.

46. Salicini I, Ibáñez C, Juste J. Deep differentiation between and within Mediterranean glacial refugia in a flying mammal the Myotis Nattereri bat complex. J Biogeogr. 2013;40:1182–93.

47. Cui J, Han N, Streicker D, Li G, Tang X, Shi Z, et al. Evolutionary relationships between bat coronaviruses and their hosts. Emerg Infect Dis. 2007;13:1526–32.

48. Lau SK, Poon RW, Wong BH, Wang M, Huang Y, Xu H, et al. Coexistence of different genotypes in the same bat and serological characterization of Rousettus bat coronavirus HKU9 belonging to a novel Betacoronavirus subgroup. J Virol. 2010;84:11385–94.

49. Yuan J, Hon CC, Li Y, Wang D, Xu G, Zhang H, et al. Intraspecies diversity of SARS-like coronaviruses in Rhinolophus Sinicus and its implications for the origin of SARS coronaviruses in humans. J Gen Virol. 2010;91:1058–62.

50. Parks GD, Alexander-Miller MA. Paramyxovirus activation and inhibition of innate immune responses. J Mol Biol. 2013;425:4872–92.

51. Wille M, Avril A, Tolf C, Schager A, Larsson S, Borg O, et al. Temporal dynamics diversity and interplay in three components of the virodiversity of a mallard population: influenza a virus avian paramyxovirus and avian coronavirus. Infect Genet Evol. 2015;29:129–37.

16

Evolutionary phylodynamics of foot-and-mouth disease virus serotypes O and A circulating in Vietnam

Van Phan Le[1*], Thi Thu Hang Vu[2], Hong-Quan Duong[3], Van Thai Than[4*] and Daesub Song[5]

Abstract

Background: Foot-and-mouth disease virus (FMDV) is one of the highest risk factors that affects the animal industry of the country. The virus causes production loss and high ratio mortality in young cloven-hoofed animals in Vietnam. The VP1 coding gene of 80 FMDV samples (66 samples of the serotype O and 14 samples of the serotype A) collected from endemic outbreaks during 2006–2014 were analyzed to investigate their phylogeny and genetic relationship with other available FMDVs globally.

Results: Phylogenetic analysis indicated that the serotype O strains were clustered into two distinct viral topotypes (the SEA and ME-SA), while the serotype A strains were all clustered into the genotype IX. Among the study strains, the amino acid sequence identities were shared at a level of 90.1–100, 92.9–100, and 92.8–100% for the topotypes SEA, ME-SA, and genotype IX, respectively. Substitutions leading to changes in the amino acid sequence, which are critical for the VP1 antigenic sites were also identified. Our results showed that the studied strains are most closely related to the recent FMDV isolates from Southeast Asian countries (Myanmar, Thailand, Cambodia, Malaysia, and Laos), but are distinct from the earlier FMDV isolates within the genotypes.

Conclusions: This study provides important evidence of recent movement of FMDVs serotype O and A into Vietnam within the last decade and their genetic accumulation to be closely related to strains causing FMD in surrounding countries.

Keywords: Foot-and-mouth disease virus, Serotype O, Serotype A, Vietnam

Background

Foot-and-mouth disease virus (FMDV), causing foot-and-mouth disease (FMD), is a contagious virus affecting cloven-hoofed domestic (pig, cattle, goat, and sheep) and wild animals. FMDV has been detected in >100 countries worldwide, mostly in Asia, Africa, and the Middle East [1]. The FMD causes economic losses to the livestock population and reduces food security and economic development. For this reason, FAO and OIE have launched a necessary strategy for global FMD control.

FMDV, a picornavirus, is the prototypical member of the *Aphthovirus* genus within the *Picornaviridae* family.

The virus particle is about 25–30 nm in diameter and roughly spherical in shape [2]. Similar to that of other picornaviruses, the FMDV genome organization consists of a large single open reading frame that encodes for the structure proteins, VP4, VP2, VP3, and VP1 (also known as 1A, 1B, 1C, and 1D, respectively), in which VP1, VP2, and VP3 are surface proteins; while VP4 is located internally.

FMDV is well identified as having seven immunological distinct serotypes, including serotype O, A, C, Asia 1, and the South African Territories (SAT) serotypes (including SAT1, SAT2, and SAT3) subsequently with numerous identified subtypes [3]. The infection of a single viral serotype does not confer, in consequence, the full protection against the infection of other viral serotypes [4]. The FMDV serotype A has been considered to be one of the most antigenically diverse among the seven serotypes [5, 6]. The FMDV serotype A has been classified into 10 major genotypes

* Correspondence: letranphan@vnua.edu.vn; thai@cau.ac.kr
[1]Department of Microbiology and Infectious Disease, Faculty of Veterinary Medicine, Vietnam National University of Agriculture, Hanoi, Vietnam
[4]Department of Microbiology, Chung-Ang University College of Medicine, Seoul, South Korea
Full list of author information is available at the end of the article

(designated as I to X) based on the VP1 phylogenetic trees [5, 6]. FMDV serotype O is classified into 11 topotypes, designated as Europe-South America (Euro-SA), Middle East-South Asia (ME-SA), Southeast Asia (SEA), Cathay (CHY), West Africa (WA), East Africa 1 (EA-1), East Africa 2 (EA-2), East Africa 3 (EA-3), East Africa 4 (EA-4), Indonesia-1 (ISA-1), and Indonesia-2 (ISA-2) [7, 8]; and FMDV serotype Asia 1 is classified into seven genotypes (designated as I to VII) [9].

FMD is an endemics and widespread disease in African, Asia, and its further spread into the FMD-free areas like American, Europe, and Australia is a direct threat [1]. Among seven serotypes of FMDV, serotype O and A have been distributed extensively and are responsible for outbreaks in Asia and Africa; the three SAT serotypes have been generally restricted in their distribution to Africa, and the serotype Asia 1 has never been found outside of Asia [10, 11].

Agriculture plays an important role in the national economy of Vietnam, in which animal production contributes about approximately 32% to the total GDP. FMD is considered the most economically important infectious disease affecting domestic cattle, buffalo, and pigs. The concurrent circulation of FMDV serotypes O, A, and Asia 1 are detected in which the serotype O remains the most prevalent and is responsible for the highest numbers of outbreaks [12, 13]. The FMDV serotype O and its outbreak in Vietnam were first described in academic research between 1996 and 2001 [14]. The FMDV serotype Asia 1 and A were subsequently identified in 2005 and 2009, respectively [12, 15]. In 2008 and the first 2 months of 2009, the FMDV serotypes O and A were reported to be the prevalent serotype and caused approximately 166 FMD outbreaks in 128 communes in 47 districts of 14 provinces throughout the country [16].

Currently, limited information is available regarding the genetic characteristics and geographical distribution of the FMDV serotype O and A causing sporadic outbreaks in Vietnam. In this study, the VP1 coding gene of 80 FMDV samples (66 samples of the serotype O and 14 samples of the serotype A) collected from endemic outbreaks during 2006–2014 were analyzed to investigate their phylogeny and genetic relationship with other available FMDVs globally. These data will provide important evidence of recent movement of FMDVs serotype O and A into Vietnam within the last decade and their genetic characterization with strains causing FMD in neighboring countries.

Methods
Sample collection and virus isolation
A total of 80 FMD-positive samples were collected in a passive surveillance program from 19 provinces located in north and northern central Vietnam during 2006–2014

(Table 1, Fig. 1). All the virus isolates were initially confirmed by FMDV antigen ELISA (WRL Pirbright, UK) and then isolated from the BHK-21 cell culture system with subsequence passages. Briefly, BHK-21 cells were cultured in minimum essential medium (MEM; Gibco BRL, Grand Island, NY, USA) supplemented with 5% fetal bovine serum (FBS; Gibco BRL) and 0.1% Gentamicin (Gentamicin Reagent Solution, Gibco BRL) at 37 °C in a humidified atmosphere containing 5% CO_2. The epithelial homogenate was centrifuged at 10,000 g for 10 min, and the supernatant was then filtered by using a 0.45-μm sterile syringe filter (Corning Costar, Corning, NY, USA). The filtered samples were inoculated into a monolayer of BHK-21 cells at 37 °C for 1 h, followed by two washes with MEM media. The infected cells were maintained in MEM media containing 2% FBS. Infected cells were harvested after 2–4 days post-infection and were subsequently passed into the BHK-21 cells until cytopathic effects appeared. The names of the isolates were assigned as follow: serotype, country code, laboratory record number, and the year of sample collection, and stored at −80 °C for further exam (Additional file 1).

RNA extraction and reverse transcription-polymerase chain reaction (RT-PCR)
The viral RNA was extracted from infected cell culture supernatants using the QIAamp viral RNA mini kit (Qiagen, Valencia, CA, USA), according to the manufacturer's instructions. The cDNA step was performed using the Superscript™ III First-Strand Synthesis System for RT-PCR (Invitrogen, Carlsbad, CA, USA) according to the manufacturer's instructions. The primer set of the VN-VP1F/VN-VP1R (VN-VP1F: 5′-AGYGCYGGYAARGAYTTTGA-3′, VP1R: 5′-CATGTCYTCYTGCATCTGGTT-3′) was used for the PCR-amplification of DNA fragments containing the 639 nt length of the VP1 coding region [17]. Briefly, the reaction was carried out at 42 °C for 60 min (reverse transcription), 35 cycles of 95 °C for 1 min (for denaturation), 52 °C for 1 min (for annealing), and 72 °C for 1 min (for extension), followed by 72 °C for 10 min (for final extension). The PCR products were separated on 1.2% SeaKem LE agarose gel and viewed on a BioRad Gel Doc XR image-analysis system.

Nucleotide sequencing and sequence analysis
Capsid VP1 is the most studied FMDV protein because of its significance for virus attachment and entry, protective immunity, and serotype specificity [3, 18, 19]. The amplified capsid VP1 PCR products were either purified with QIAquick PCR purification kit or QIAquick gel extraction kit according to the manufacturer's instructions (Qiagen). RT-PCR primers were used for the direct sequencing of internal gene segments by using

Table 1 Origin of the serotypes O and A FMDVs sisolated in this study

Year ofisolation	Province	Host	Number of sample	Type	Topotype
2006	Vinh Phuc	Buffalo	1	O	SEA
	Thai Nguyen	Buffalo	1	O	SEA
		Cattle	2	O	SEA
	Ha Noi	Cattle	1	O	SEA
		Pig	1	O	SEA
	Lang Son	Cattle	1	O	SEA
		Pig	1	O	SEA
	Son La	Cattle	5	O	SEA
		Pig	1	O	SEA
		Buffalo	2	O	SEA
2007	Lai Chau	Buffalo	1	O	SEA
	Thai Nguyen	Cattle	2	O	SEA
	Ha Tinh	Cattle	1	O	SEA
2008	Nghe An	Buffalo	3	A	Genotype IX
		Cattle	1	A	Genotype IX
2009	Yen Bai	Buffalo	2	O	SEA
	Son La	Buffalo	6	O	SEA
		Pig	1	O	SEA
	Ha Giang	Cattle	1	O	SEA
	Tuyen Quang	Buffalo	1	O	SEA
	Nghe An	Cattle	1	O	SEA
	Quang Ninh	Buffalo	2	O	SEA
	Hoa Binh	Buffalo	3	O	SEA
		Cattle	2	O	SEA
		Buffalo	1	A	Genotype IX
	Lang Son	Buffalo	1	O	SEA
	Lao Cai	Cattle	1	O	SEA
	Phu Tho	Cattle	2	O	SEA
		Buffalo	1	A	Genotype IX
	Bac Can	Buffalo	1	A	Genotype IX
		Cattle	1	A	Genotype IX
	Ha Giang	Cattle	2	A	Genotype IX
	Quang Tri	Buffalo	1	A	Genotype IX
2010	Son La	Buffalo	2	O	SEA
	Lao Cai	Buffalo	1	O	SEA
	Dien Bien	Cattle	1	O	SEA
	Tuyen Quang	Buffalo	1	O	SEA
	Yen Bai	Cattle	2	O	SEA
		Buffalo	1	O	SEA
2013	Ha Noi	Pig	1	O	ME-SA
		Bovine	2	A	Genotype IX
	Quang Tri	Bovine	4	O	ME-SA
	Ha Tinh	Bovine	1	A	Genotype IX
2014	Ha Noi	Pig	6	O	ME-SA
	Ha Nam	Pig	3	O	SEA
		Bovine	1	O	SEA

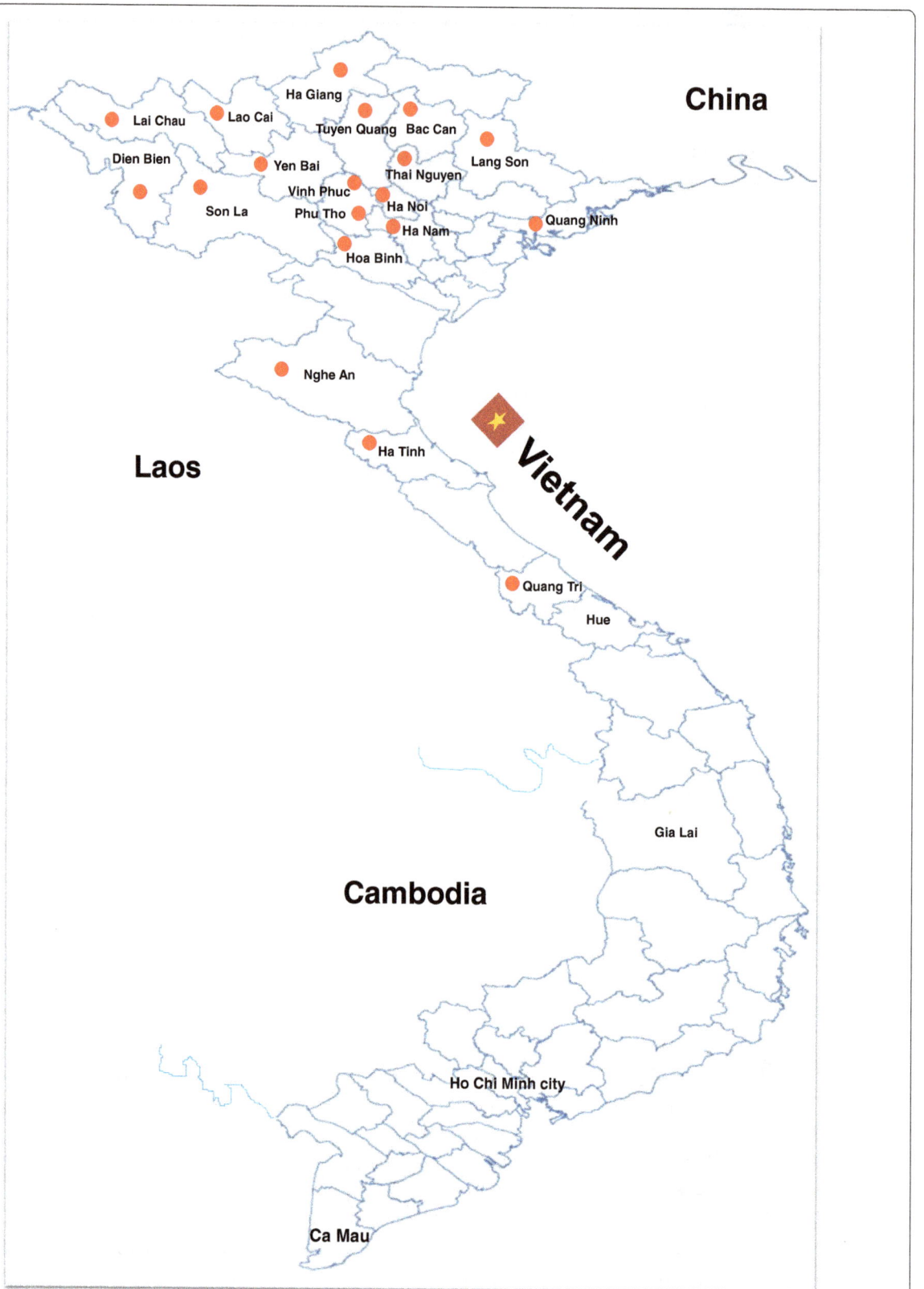

Fig. 1 Map of Vietnam showing the provinces (*red*) from which the FMDV isolates were collected during outbreaks

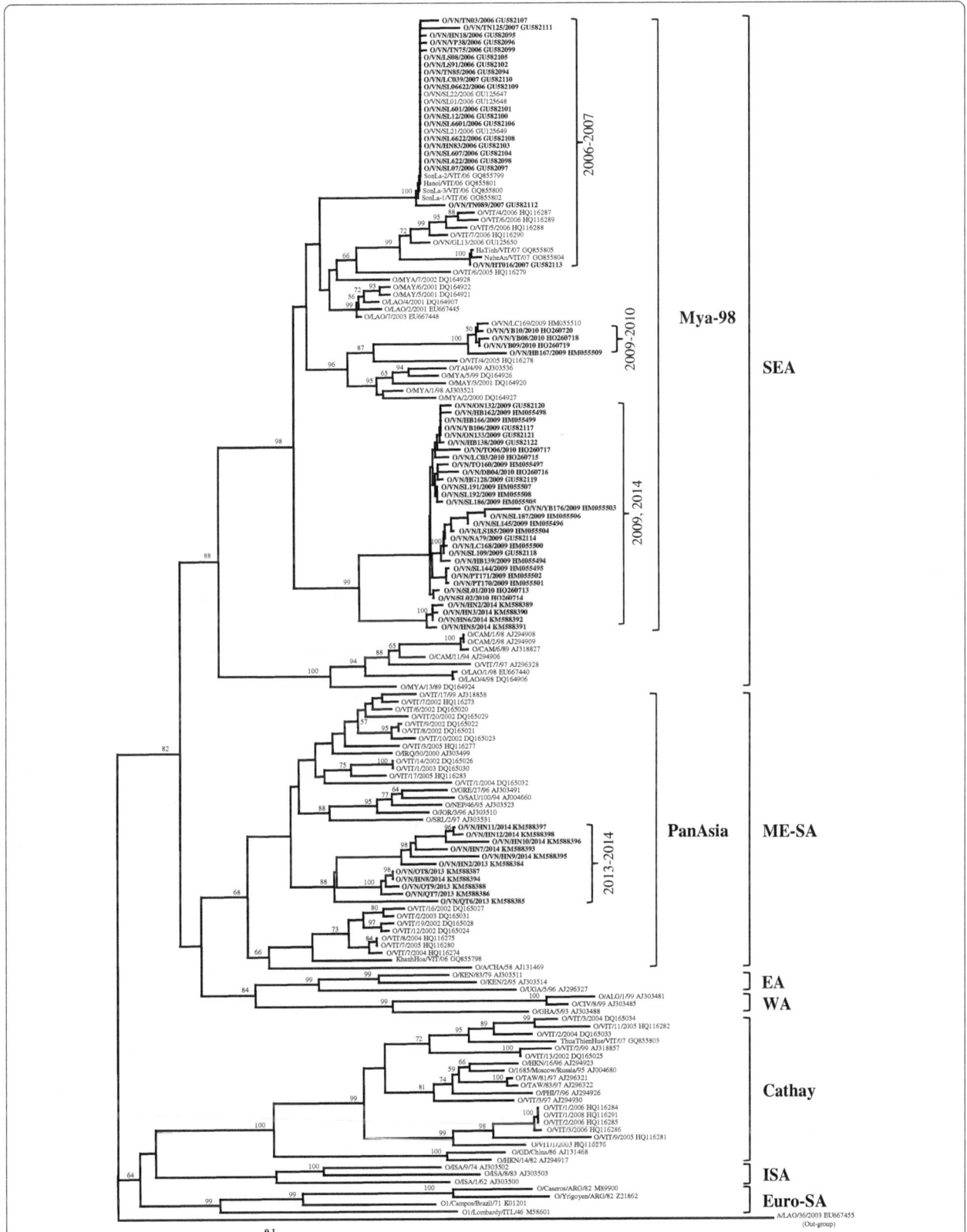

Fig. 2 (See legend on next page.)

a BigDye terminator cycle sequencing kit and an automatic DNA sequencer (Model 3730, Applied Biosystems, Foster City, CA, USA). The obtained nucleotide and deduced amino acid sequences in this study were aligned using the ClustalX 2.1 program [20] and Lasergene software (DNASTAR; Madison, WI, USA) by using the parameters set against the corresponding FMD viral sequences from the NCBI GenBank.

Phylogenetic analysis

The complete nucleotide sequences of the VP1 coding gene from the 80 FMDV samples examined in this study were compared against a representative VP1 coding gene from the available FMDV sequences in the GenBank database. Phylogenetic trees were constructed by the neighbor-joining algorithm put into practiced with MEGA 6.06 software suite [21]. The bootstrap resampling method with 1000 replicates was used to evaluate the topology of the phylogenetic tree. The pairwise distance was calculated using MEGA 6.06 software package [21].

Results

Detection of FMDV genome and genome sequencing

The viral RNA was sufficiently extracted from 80 FMDV samples (66 samples of the serotype O and 14 samples of the serotype A) (Table 1). The VP1 coding gene was successfully amplified by RT-PCR using the primer set as described above. The full length sequence of the VP1 coding gene was determined by direct sequencing of the PCR amplicons. Thereafter, the obtained nucleotide sequences have been deposited in the NCBI GenBank database under the accession numbers described in Additional file 1.

Genetic diversity and phylogenetic analyses of the VP1 coding gene

For the FMDV serotype O, the nucleotide sequence identity among 66 FMDV serotype O isolates showed diversity at a level of 79.1–100%. These strains shared nucleotide sequence identity at 100, 91.5–97.8, 86.2–100, 89.9–99.8, 94.2–99.3 and 83.8–99.8% in 2006, 2007, 2009, 2010, 2013 and 2014, respectively. Moreover, phylogenetic analysis demonstrated that the VP1 coding gene of the FMDV serotype O was clustered into two distinct viral topotypes, the SEA (lineage Mya-98) and ME-SA (lineage PanAsia) (Fig. 2). Most of the strains were clustered into the SEA topotype, while only the strains isolated between 2013 and

2014 were clustered into the ME-SA topotype (Fig. 2). The study strains shared nucleotide identities at a level of 89.0–92.6 and 84.1–86.6% compared to the O/MYA/2/2000/SEA (SEA) and O/A/CHA/58 (ME-SA) prototype strains, respectively [7, 22].

For FMDV serotype A, the VP1 nucleotide sequence identity among the study strains was 89.3–100%. Phylogenetic analysis indicated that the VP1 coding gene of the 14 FMDV serotype A strains was classified within the genotype IX (topotype Asia), together with other Vietnamese strains isolated during 2004–2009 (Fig. 3). The study strains showed nucleotide identity at a level of 87.9–91.1% compared to the A/TAI/118/87 prototype strain. Moreover, these FMDV serotype A showed genetic diversity among the strains circulating in Vietnam in the 2004–2005, 2008–2009, and 2013 seasons by grouping into three distinct sub-clusters (Fig. 3).

Comparison of VP1 amino acid sequences

The deduced amino acid sequences obtained from the VP1 gene segments were aligned and compared to investigate the consequences of the observed genetic characterization of FMDV serotype O and A in Vietnam during 2006–2014. For the FMDV serotype O, the O/MYA/2/2000/SEA and O/A/CHA/58 prototype strains were used as references. Among the SEA topotype (lineage Mya-98), amino acid sequence identities significantly showed genetic variation of 90.1–100% within the Vietnamese strains and 89.6–93.4% to the O/MYA/2/2000 prototype strain. Compared to the O/MYA/2/2000/SEA prototype strain, seven amino acid substitutions were detected at positions 14, 24, 51, 93, 111, 131, and 184 (Figs. 4 and 5). Among the ME-SA topotype (lineage PanAsia), the study strains shared the amino acid identity at a level of 92.9–100% and shared an identity with O/A/CHA/58 prototype strain at a level of 86.8–93.4%. Eight amino acid substitutions were detected at positions 34, 56, 87, 127, 138, 139, 143, and 158 (Figs. 4 and 5). Notably, the change of the two amino acids at positions 45^{K-Q} and 154^{K-R} occurring at residues among the SEA topotypes, as well as the change of two amino acids at positions 43^{T-V} and 208^{P-H} occurring at residues among the ME-SA topotypes (Figs. 4 and 5).

For the FMDV serotype A, the A/TAI/118/87 prototype strain was used as a reference strain for genotype IX. Within the genotype IX (topotype Asia), the study strains shared 92.8–100% at the amino acid level and shared an identity with the A/TAI/118/87 prototype strain at level of 91.1–93.4%. Compared to the A/TAI/

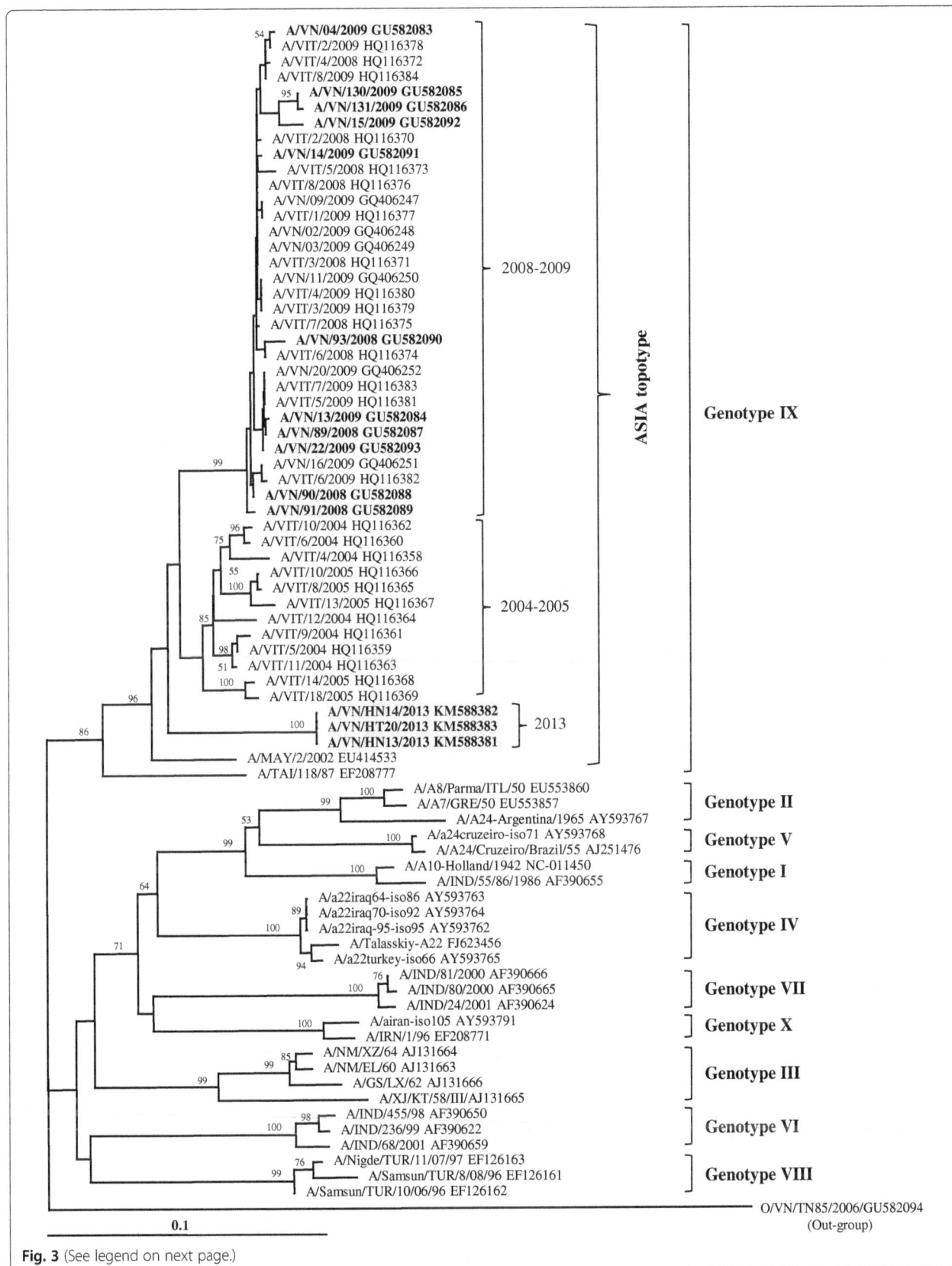

Fig. 3 (See legend on next page.)

(See figure on previous page.)
Fig. 3 Phylogenetic tree based on the complete nucleotide sequence of the VP1 coding region of type A FMDVs showing relationships between the study strains and other type A representatives worldwide. The Vietnamese strains are marked in *bold*. *Numbers* at nodes indicate the level of bootstrap support based on the neighbor-joining analysis of 1000 re-sampled datasets. Only values above 50% are given. A *bar* represents 0.1 substitutions per nucleotide position

118/87 prototype strain, four amino acid substitutions were detected at positions 138, 139, 153, and 195 (Fig. 6). An amino acid changes in the antigenic sites noted at position 148^{S-P} of the A/VN/130/2009 and A/VN/131/2009 strains (Fig. 6).

Discussion

The sequences of the VP1 coding gene are widely used to identify and characterize FMDV lineages and sub-lineages [23]. In addition, the VP1 capsid protein is the most helpful protein to investigate the relationship between different isolates of the FMDV because of its significance for viral attachment and entry, protective immunity, and serotype specificity [18, 19]. In this study, we used the VP1 coding region of the FMDV serotype O and type A isolated from the north and northern central regions of Vietnam during

2006–2014 for the determination of their phylogeny and genetic relationships with other available Vietnamese and global FMDV strains in the NCBI GeneBank database. Importantly, the 19 provinces enrolled in this study have historically been highly affected by FMDVs and share both a border and a trade of animals with the highly affected FMDV countries of China, Laos, Cambodia, and Thailand [9, 14, 16].

The FMDV serotype O is the most prevalent out of the seven serotypes that circulate in many parts of the world. The FMD outbreaks of type O viruses were first identified in 1987 from FMD outbreaks in Europe by analysis of its nucleotide sequence [24]. Based on the accumulation of VP1 genome sequences, 10 topotypes of FMDV serotype O were designated as the Euro-SA, ME-SA, SEA, CHY, WA, EA-1, EA-2, EA-3, ISA-1, and ISA-2. The topotypes

```
          Amino acid    10        20        30        40        50        60        70        80        90        100
Strain name            ....|....|....|....|....|....|....|....|....|....|....|....|....|....|....|....|....|....|....|....|
O/MYA/2/2000 (ref)     TTSTGESADPVTATVENYGGETQVQRRHHTDISFILDRFVKVTPKDQINVLDLMQTPPHTLVGALLRTATYYFADLEVAVKHEGDLTWVPNGAPEAALDN
O/VN/SL07/2006, n=13   ...T........TT.........V...Q.......................L.....S.....................N........A..T..E.
O/VN/TN75/2006, n=1    ----------------------------------------------.....L.....S.....................N........A..T..E.
O/VN/VP38/2006, n=1    ----------------------------------------------.....L.....S.....................N........A..T..E.
O/VN/TN03/2006, n=1    ----------------------------------------------.....L.....S.....................N........A..T..E.
O/VN/LC039/2007, n=1   ----------------------------------------------.....L.....S.....................N........A..T..E.
O/VN/TN089/2007, n=1   ...T........TT.........V...Q.......................L.....S.....................N........A..T..E.
O/VN/TN125/2007, n=1   ...T........TT.........V...Q...D...................L.....S.....................N........A..T..E.
O/VN/HT016/2007, n=1   ...T........T.........V..........Q....L.....................................N............E.
O/VN/HB138/2009, n=1   ...TD.........T.........V..........Q....L...................................A.....G.
O/VN/HB139/2009, n=1   ...T...G......T.........V..........Q....L...................................A.....G.
O/VN/HB162/2009, n=1   ...T..........T.........V..........Q....L...................................A.....G.
O/VN/HB167/2009, n=1   ...T........APT..S.V.Q..V...........S....L..................................A.....D.
O/VN/QN132/2009, n=1   ...T..........T.........V..........Q....L...................................A.....G.
O/VN/YB176/2009, n=1   ...T........S.T.........V..........Q....LNQ..AA.N....S......................A.....G.
O/VN/TQ160/2009, n=1   ...T..........T.........V..........Q....L...................................A.....G.
O/VN/SL144/2009, n=1   ...T..........T.........V....V.....Q....L...................................A.....G.
O/VN/SL191/2009, n=9   ...T..........T.........V..........Q....L...................................A.....G.
O/VN/PT170/2009, n=1   ...T..........T.........V....V.....Q....L...................................A.....G.
O/VN/LC168/2009, n=1   --------------------------I..Q....L...................................A.....G.
O/VN/LC169/2009, n=1   ...T..........T.........V..........S...L...................................A.....D.
O/VN/TQ06/2010, n=1    ...T..........T.........V..........Q....L...................................A..T..G.
O/VN/LC03/2010, n=1    ...T..........T.........V..........Q....L...................................A..T..G.
O/VN/DB04/2010, n=1    ...T..........T.........V..........QG....L...................................A.....G.
O/VN/SL01/2010, n=1    ...T..........T.........V..........Q....L...................................A.....G.
O/VN/SL02/2010, n=1    ...T..........T.........V..........Q....L...................................A.....G.
O/VN/YB08/2010, n=3    ...T..........T.........V..........S...L...................................A.....D.
O/VN/HN2/2014, n=1     ...A..........T.........V..........LN.....S...............K.........A..T..E.
O/VN/HN3/2014, n=1     ...A..........T.........V..........L.....S.....................A..T..E.
O/VN/HN5/2014, n=1     ...RS.........T.........V..........L.....S.....................A..T..E.
O/VN/HN6/2014, n=1     ...T..........T.........V..........L.....S.....................A..T..E.
O/A/CHA/58 (ref)       TTSPGESADPVTATVENYGGETQVQRRQHTDVSYILDRFVKVTPKDQINVLDLMQIPAHTLVGALLRTATYYFADLEVAVKHKGNLKWVPNGAPETALDN
O/VN/HN2/2013, n=1     ...T...................R.........F..................T...................E..T.............
O/VN/QT6/2013, n=1     ...T..........R..........F.....T.V.............T...................E..T.............
O/VN/QT7/2013, n=1     ...A...................R.........F..................T...................E..T.............
O/VN/QT8/2013, n=1     ...T....................F..................T...................E..T.............
O/VN/QT9/2013, n=1     ...TD....................F..................T...................E..T.............
O/VN/HN7/2014, n=1     C..TD.........R..........F..................T...................E..T.............
O/VN/HN8/2014, n=1     ...T....................F..................T...................E..T.............
O/VN/HN9/2014, n=1     ...T..........R..........F..................T...................E..T.............
O/VN/HN10/2014, n=1    D..T....................F..................T.........G...E..T.....GR..G.E.
O/VN/HN11/2014, n=1    ...T..........R..........F..................T...................E..T.............
O/VN/HN12/2014, n=1    ...T.......I..T.......R..........F..................T...................E..T.............
```

Fig. 4 Deduced amino acid sequences of the VP1 proteins (aa 1-100) of the type O FMDVs in this study. Only sequences different from the consensus are shown. Strains with similar profiles of the antigenic site are grouped together. Similar amino acid sequences of the two reference trains, the O/MYA/2/2000 strain (for the SEA group) and the O/A/CHA/58 strain (for the ME-SA group), are shadowed. The VP1 antigenic sites are at amino acid position 43, 44, 45, 144, 148, 149, 154, and 208. The "n" represents the strains with similar profiles to the antigenic site

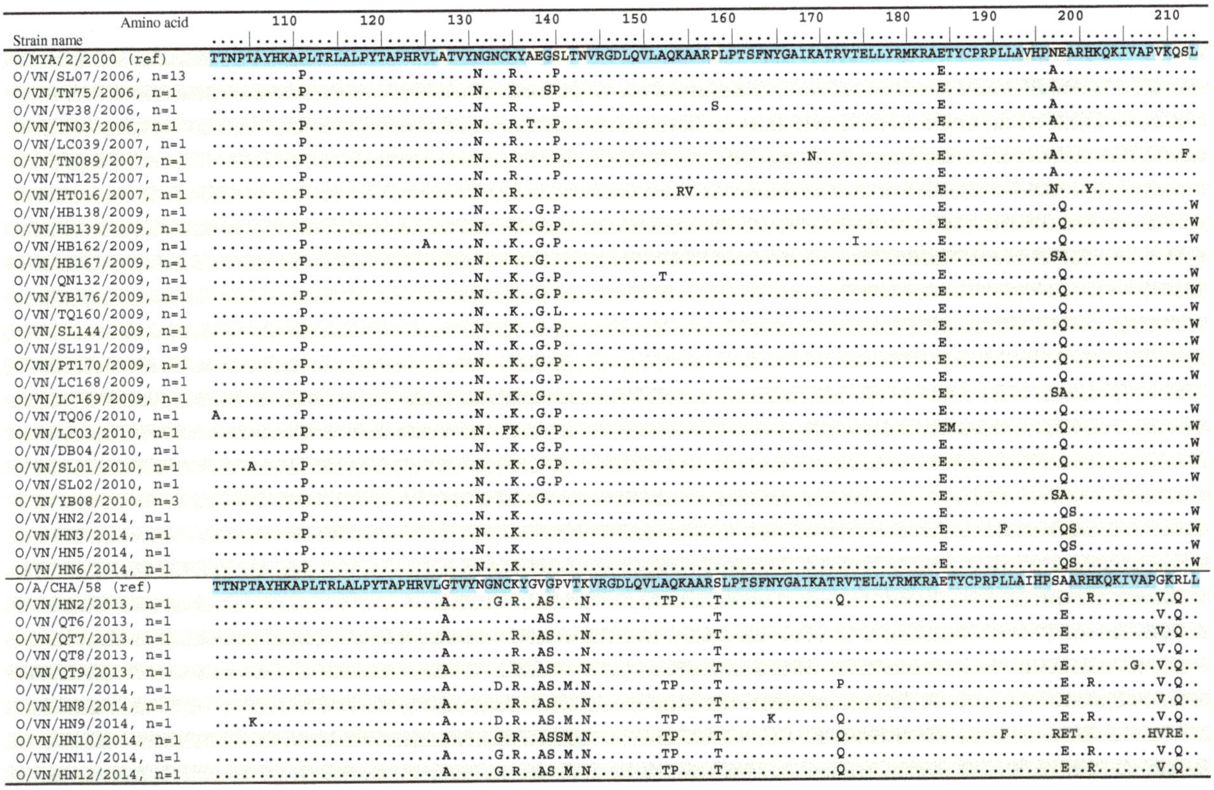

Fig. 5 Deduced amino acid sequences of the VP1 proteins (aa 101-213) of the type O FMDVs in this study. Only sequences different from the consensus are shown. Strains with similar profiles of the antigenic site are grouped together. Similar amino acid sequences of the two reference trains, the O/MYA/2/2000 strain (for the SEA group) and the O/A/CHA/58 strain (for the ME-SA group), are shadowed. The VP1 antigenic sites are at amino acid position 43, 44, 45, 144, 148, 149, 154, and 208. The "n" represents the strains with similar profiles to the antigenic site

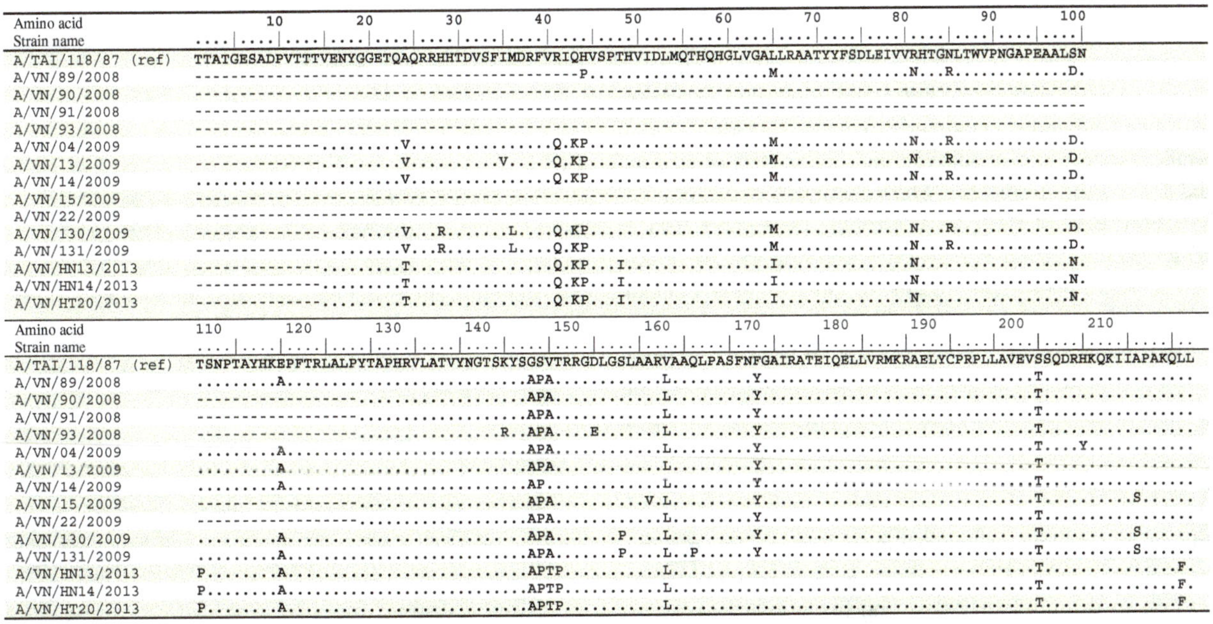

Fig. 6 Deduced amino acid sequences of the VP1 proteins of the type A FMDVs in this study. Only sequences different from the consensus are shown. Strains with similar profiles of the antigenic site are grouped together. The VP1 antigenic sites are at amino acid position 43, 44, 45, 144, 148, 149, 154, and 208.

ME-SA and SEA highly affect China, the Indian subcontinent (India, Pakistan, Bangladesh, Sri Lanka, Nepal, and Bhutan), and Southeast Asian countries (Myanmar, Thailand, Cambodia, Malaysia, Laos, and Vietnam). The Vietnamese FMDV serotype O fell within the ME-SA (lineage Mya-98), SEA (lineage PanAsia), and Cathay topotypes in which the isolates form distinct genetic sublineages and are distant from the prototype isolates. These findings highlight that these topotypes might have adapted in recent years to circulate in Vietnam.

The FMDV serotype A has been reported in all FMDV infected areas around the world. Based on the phylogeny analysis of the VP1 capsid genes, the global FMDV serotype A is divided into 10 major genotypes (designated as I to X) with over 15% nucleotide divergence [5, 6]. In Vietnam, these viruses were reported to be predominant during the outbreaks between 2008 and 2009 in the northern central regions of the country [16]. Genetic characterization of six serotype A strains isolated from the northern central regions of Vietnam revealed that the Vietnamese FMDV serotype A strains were all clustered into the genotype IX (topotype ASIA) and shared close relation to the recent FMDV serotype A strains isolated in Laos, Thailand, and Malaysia [12]. The FMDV serotype A strains in this study were also clustered into the genotype IX, together with the strains isolated between 2004 and 2009. Interestingly, the strains isolated in 2013 clustered into a single sub-cluster and showed distance from previous isolates. These results indicated the genetic variation of the FMDV serotype A strains and their persistent circulation in Vietnam, particularly in the north and northern central regions.

Available since the early 1900s, the FMDV vaccine continues to plays a significant role in the protection of animals against FMDV-related morbidity and mortality; however, the vaccinated individuals could only be protected for a specific serotype and/or subtype, and this protection is only valid for a short term [25]. The concurrent circulation of FMD outbreaks of the FMDV serotype O, A, and Asia 1 and its genetic variation may apply pressure for the selection of effective vaccine strains to prevent and control FMDV outbreaks in Vietnam [15]. These points also suggest that more FMDV surveillance studies will be necessary in order to evaluate the genetic relationship and efficacy of the current used vaccines against different FMDV serotypes circulating in Vietnam.

Conclusions

This study provides valuable information on the genetic variation among the FMDVs serotype O and A circulating in Vietnam during 2006–2014, and likely indicates transmission between neighboring countries in Southeast Asia, such as Myanmar, Thailand, Cambodia, Malaysia, and Laos.

Abbreviations

CHY: Cathay; EA-1: East Africa 1; EA-2: East Africa 2; EA-3: East Africa 3; EA-4: East Africa 4; ELISA: Enzyme-linked immunosorbent assay; Euro-SA: Europe-South America; FMDV: Foot-and-mouth disease virus; ISA-1: Indonesia-1; ISA-2: Indonesia-2; ME-SA: Middle East-South Asia; RT-PCR: Reverse transcription polymerase chain reaction; SAT: South African Territories; SEA: Southeast Asia; WA: West Africa

Funding

This work was supported by the Vietnam National Project under the Project Code No: SPQG.05b.01.

Authors' contributions

VPL and VTT conceived and designed the proposal. VPL and TTHV performed the experiments. VPL, VTT, TTHV, HQD, and DS participated in analyzing the data. VTT, HQD, and VPL wrote the paper. All authors have read and approved the final manuscript.

Competing interest

The authors declare that they have no competing interests.

Consent for publication

Not applicable.

Author details

[1]Department of Microbiology and Infectious Disease, Faculty of Veterinary Medicine, Vietnam National University of Agriculture, Hanoi, Vietnam. [2]Research and Development Laboratory, Rural Technology Development JSC, Hung Yen, Vietnam. [3]Institute of Research and Development, Duy Tan University, Danang, Vietnam. [4]Department of Microbiology, Chung-Ang University College of Medicine, Seoul, South Korea. [5]College of Pharmacy, Korea University, Sejong, South Korea.

References

1. OIE-FAO. The Global Foot and Mouth Disease Control Strategy. Strengthening Animal Health System through Improved Control of Major Diseases. pp43 FAO, Rome, Italy. 2012.
2. Fry EE, Lea SM, Jackson T, Newman JW, Ellard FM, Blakemore WE, et al. The structure and function of a foot-and-mouth disease virus-oligosaccharide receptor complex. EMBO J. 1999;18(3):543–54.
3. Carrillo C, Tulman ER, Delhon G, Lu Z, Carreno A, Vagnozzi A, et al. Comparative genomics of foot-and-mouth disease virus. J Virol. 2005;79(10):6487–504.
4. Kitching RP, Knowles NJ, Samuel AR, Donaldson AI. Development of foot-and-mouth disease virus strain characterisation–a review. Trop Anim Health Prod. 1989;21(3):153–66.
5. Kitching RP. Global epidemiology and prospects for control of foot-and-mouth disease. Curr Top Microbiol Immunol. 2005;288:133–48.
6. Tosh C, Sanyal A, Hemadri D, Venkataramanan R. Phylogenetic analysis of serotype A foot-and-mouth disease virus isolated in India between 1977 and 2000. Arch Virol. 2002;147(3):493–513.
7. Knowles NJ, Samuel AR, Davies PR, Midgley RJ, Valarcher JF. Pandemic strain of foot-and-mouth disease virus serotype O. Emerg Infect Dis. 2005;11(12):1887–93.
8. Ayelet G, Mahapatra M, Gelaye E, Egziabher BG, Rufeal T, Sahle M, et al. Genetic characterization of foot-and-mouth disease viruses, Ethiopia, 1981–2007. Emerg Infect Dis. 2009;15(9):1409–17.

9. Jamal SM, Ferrari G, Ahmed S, Normann P, Belsham GJ. Molecular characterization of serotype Asia-1 foot-and-mouth disease viruses in Pakistan and Afghanistan; emergence of a new genetic Group and evidence for a novel recombinant virus. Infect Genet Evol. 2011;11(8):2049–62.
10. Kitching P, Hammond J, Jeggo M, Charleston B, Paton D, Rodriguez L, et al. Global FMD control–is it an option? Vaccine. 2007;25(30):5660–4.
11. Jamal SM, Belsham GJ. Foot-and-mouth disease: past, present and future. Vet Res. 2013;44:116.
12. Le VP, Nguyen T, Lee KN, Ko YJ, Lee HS, Nguyen VC, et al. Molecular characterization of serotype A foot-and-mouth disease viruses circulating in Vietnam in 2009. Vet Microbiol. 2010;144(1–2):58–66.
13. Lee KN, Nguyen T, Kim SM, Park JH, Do HT, Ngo HT, et al. Direct typing and molecular evolutionary analysis of field samples of foot-and-mouth disease virus collected in Viet Nam between 2006 and 2007. Vet Microbiol. 2011;147(3–4):244–52.
14. Gleeson LJ. A review of the status of foot and mouth disease in South-East Asia and approaches to control and eradication. Rev Sci Tech. 2002;21(3):465–75.
15. Le VP, Nguyen T, Park JH, Kim SM, Ko YJ, Lee HS, et al. Heterogeneity and genetic variations of serotypes O and Asia 1 foot-and-mouth disease viruses isolated in Vietnam. Vet Microbiol. 2010;145(3–4):220–9.
16. Hoang VN. In: Proceedings of the 15th Meeting of the OIE Sub-Commission for Foot and Mouth Disease in South-East Asia, Kota Kinabalu, Sabah, Malaysia, 9–13 March. 2009.
17. Jamal SM, Ferrari G, Hussain M, Nawroz AH, Aslami AA, Khan E, et al. Detection and genetic characterization of foot-and-mouth disease viruses in samples from clinically healthy animals in endemic settings. Transbound Emerg Dis. 2012;59(5):429–40.
18. Jackson T, King AM, Stuart DI, Fry E. Structure and receptor binding. Virus Res. 2003;91(1):33–46.
19. Burman A, Clark S, Abrescia NG, Fry EE, Stuart DI, Jackson T. Specificity of the VP1 GH loop of Foot-and-Mouth Disease virus for alphav integrins. J Virol. 2006;80(19):9798–810.
20. Larkin MA, Blackshields G, Brown NP, Chenna R, McGettigan PA, McWilliam H, et al. Clustal W and Clustal X version 2.0. Bioinformatics. 2007;23(21):2947–8.
21. Tamura K, Stecher G, Peterson D, Filipski A, Kumar S. MEGA6: Molecular Evolutionary Genetics Analysis version 6.0. Mol Biol Evol. 2013;30(12):2725–9.
22. Xin A, Li H, Li L, Liao D, Yang Y, Zhang N, et al. Genome analysis and development of infectious cDNA clone of a virulence-attenuated strain of foot-and-mouth disease virus type Asia 1 from China. Vet Microbiol. 2009;138(3–4):273–80.
23. Abdul-Hamid NF, Hussein NM, Wadsworth J, Radford AD, Knowles NJ, King DP. Phylogeography of foot-and-mouth disease virus types O and A in Malaysia and surrounding countries. Infect Genet Evol. 2011;11(2):320–8.
24. Samuel AR, Knowles NJ. Foot-and-mouth disease type O viruses exhibit genetically and geographically distinct evolutionary lineages (topotypes). J Gen Virol. 2001;82(Pt 3):609–21.
25. Rodriguez LL, Grubman MJ. Foot and mouth disease virus vaccines. Vaccine. 2009;27 Suppl 4:D90–4.

17

Visual and equipment-free reverse transcription recombinase polymerase amplification method for rapid detection of foot-and-mouth disease virus

Libing Liu[1†], Jinfeng Wang[1†], Ruoxi Zhang[3], Mi Lin[5], Ruihan Shi[1,4], Qingan Han[3], Jianchang Wang[1,4*] and Wanzhe Yuan[2*] (iD)

Abstract

Background: Foot-and-mouth disease (FMD), which is caused by foot-and-mouth disease virus (FMDV), is a highly contagious tansboundary disease of cloven-hoofed animals and causes devastating economic damages. Accurate, rapid and simple detection of FMDV is critical to containing an FMD outbreak. Recombinase polymerase amplification (RPA) has been explored for detection of diverse pathogens because of its accuracy, rapidness and simplicity. A visible and equipment-free reverse-transcription recombinase polymerase amplification assay combined with lateral flow strip (LFS RT-RPA) was developed to detect the FMDV using primers and LF probe specific for the 3D gene.

Results: The FMDV LFS RT-RPA assay was performed successfully in a closed fist using body heat for 15 min, and the products were visible on the LFS inspected by the naked eyes within 2 min. The assay could detect FMDV serotypes O, A and Asia1, and there were no cross-reactions with vesicular stomatitis virus (VSV), encephalomyocarditis virus (EMCV), classical swine fever virus (CSFV), porcine reproductive and respiratory syndrome virus (PRRSV), porcine circovirus 2 (PCV2) and pseudorabies virus (PRV). The analytical sensitivity was 1.0×10^2 copies in vitro transcribed FMDV RNA per reaction, which was the same as a real-time RT-PCR. For the 55 samples, FMDV RNA positive rate was 45.5% (25/55) by LFS RT-RPA and 52.7% (29/55) by real-time RT-PCR. For the LFS RT-RPA assay, the positive and negative predicative values were 100% and 80%, respectively.

Conclusions: The performance of the LFS RT-RPA assay was comparable to real-time RT-PCR, while the LFS RT-RPA assay was much faster and easier to be performed. The developed FMDV LFS RT-RPA assay provides an attractive and promising tool for rapid and reliable detection of FMDV in under-equipped laboratory and at point-of-need facility, which is of great significance in FMD control in low resource settings.

Keywords: FMDV, 3D gene, RPA, LF probe, Lateral flow strip

* Correspondence: jianchangwang1225@126.com;
yuanwanzhe2015@126.com
†Libing Liu and Jinfeng Wang contributed equally to this work.
[1]Center of Inspection and Quarantine, Hebei Entry-Exit Inspection and Quarantine Bureau, Shijiazhuang 050051, People's Republic of China
[2]College of Veterinary Medicine, Agricultural University of Hebei, No.38 Lingyusi Street, Baoding, Hebei 071001, People's Republic of China
Full list of author information is available at the end of the article

Background

Foot-and-mouth disease (FMD) is a highly contagious viral disease of wild and domesticated cloven-hoofed animals. The causative agent, foot-and-mouth disease virus (FMDV), a non-enveloped, single-stranded positive-sense RNA virus, belongs to the genus *Aphthovirus* within the family *Picornaviridae* [1]. FMDV exists in seven distinct serotypes comprising O, A, C, Asia1 and South African Territories (SAT) serotypes SAT1, SAT2 and SAT3 and multiple subtypes due to the high mutational rate of the virus [1]. Although the mortality rate of FMD is generally low, the disease can be economically devastating due to production losses in endemic countries and trade restrictions in FMD-free countries. It is estimated that annual global impact of FMD in endemic regions alone is between US$ 6.5 and 21 billion [2].

The above facts clearly indicate that the early, rapid and robust diagnosis of FMD is imperative in the prevention and control of the disease. FMD is characterized by vesicular lesions and ulcerations on the tongue, mouth, nasal region and coronary bands of infected animals [3]. Nevertheless, reliable diagnosis based on clinical signs alone can sometimes be difficult because the clinical signs are often mild in adult sheep and goats [4] and a number of viral diseases clinically mimic FMD, including vesicular stomatitis (VS), swine vesicular disease (SVD), vesicular exanthema of swine (VES), and Senecavirus A (SVA) infection. Therefore, laboratory diagnostic tools for detection of FMDV are imperative for the effective control and elimination of the disease. Currently, several conventional methods are available for the detection of FMDV, including virus isolation (VI), antigen-capture ELISA (Ag-ELISA), and immunochromatographic lateral flow device (Ag-LFD) [5, 6]. VI is a relatively laborious and time-consuming method that must be performed in a high-containment biosafety laboratory. Ag-ELISA has a limited sensitivity and also requires skilled technicians to perform and interpret the assays. Ag-LFD has only been validated for use with epithelial samples [5]. Molecular diagnostic assays are now recognized as reliable detection methods for FMDV. A number of reverse transcription polymerase chain reaction (RT-PCR) assays have been reported and accepted widely for the detection of FMDV RNA, such as RT-PCR and real-time RT-PCR [7–9]. The RT-PCR assays are designed for use in well-equipped laboratories with reliable electrical supply and highly trained technicians, and unsuitable for being used in under-equipped laboratories and in field. Although several real-time RT-PCR assays have been transferred onto a portable platform and trialled successfully in field settings [7, 10, 11], expensive high precision instrumentation and consistent electrical power are still needed. When compared to current RT-PCR assays, the use of isothermal technologies reduces the need for high precision instrumentation,

consistent electrical power and complex sample preparation [12]. Recently, several field-deployable isothermal DNA amplification assays including the reverse transcription insulated isothermal PCR (RT-iiPCR), reverse transcription loop-mediated isothermal amplification (RT-LAMP), nucleic acid sequence based amplification (NASBA) and reverse transcription helicase dependent amplification (RT-HDA) have been developed for FMDV detection [13–16]. However, RT-LAMP assay requires six primers and has unsatisfactory reliability in detection of highly variable viruses [15, 17], which makes the assay difficult to design for FMDV. Furthermore, the results are usually produced within 60 min–120 min for the above methods, and depend on water baths and specialized instruments [13–16]. Therefore, a simple, rapid, and sensitive method is still needed for the point-of-need (PON) detection of FMDV.

As an isothermal DNA amplification technique, recombinase polymerase amplification (RPA) has been demonstrated to be rapid, specific, sensitive, and cost-effective, and has been applied widely in the detection of different pathogens [18, 19]. Recombinase, single-stranded DNA-binding protein (SSB) and strand-displacing polymerase are three core enzymes employed in RPA. Recombinases form complexes with primers and pair the primers with homologous sequences in the template DNA. SSB binds to the displaced strand and stabilizes the resulting loop, then DNA amplification is initiated by DNA polymerase [18, 19]. Abd El Wahed et al. had developed a real-time RT-RPA assay based on exo probe for rapid detection of FMDV, while the assay still depended on the specialized instrument, ESEQuant tubescanner [20]. With the Endonuclease IV, LF probe and the reverse primer labelled at the 5′ end with a biotin in the RPA reaction system, the products could be detected by the naked eye. The LF probe oligonucleotide backbone includes a 5′- FAM group, an internal tetrahydrofuran residue (THF) and a 3′- C3-spacer (Fig. 1) [18, 19]. The generated amplicons dual labelled with FAM and Biotin are then detected by the naked eye in 'sandwich' assay formats, such as the lateral flow strip (LFS) that contains anti-FAM gold conjugates and biotin-ligand molecules. A series of LFS RPA assays have been developed for the detection of porcine parvovirus (PPV), peste des petits ruminants virus (PPRV), infectious bovine rhinotracheitis virus (IBRV) and bovine ephemeral fever virus (BEFV) [21–24].

In this study, we developed an equipment-free RPA assay for rapid, specific and sensitive detection of FMDV, which was combined with a LFS (USTAR, Hangzhou, China) and performed by incubating the reactions tubes in a closed fist using body heat.

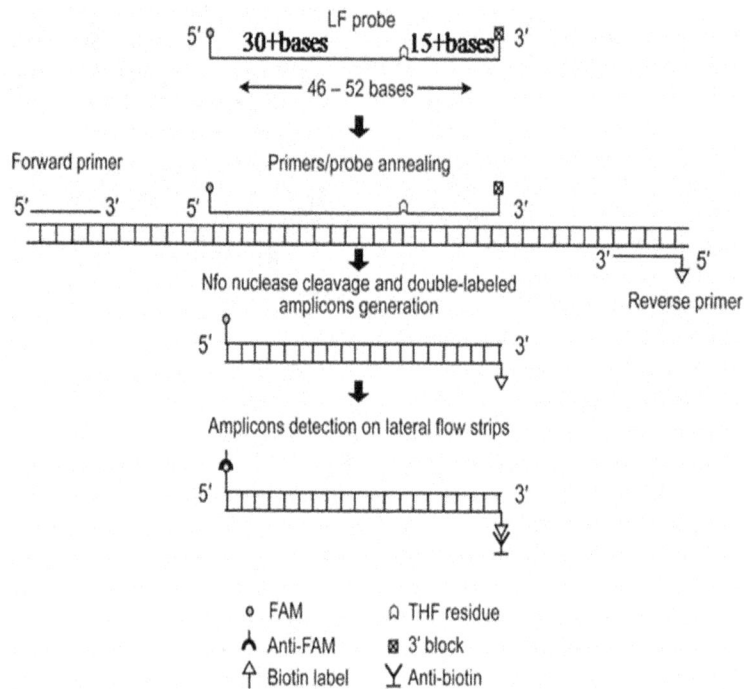

Fig. 1 Diagram of LF probe and post-RPA detection with lateral flow strip. LF probe is typically 46–52 nucleotides long, at least 30 of which are placed 5' to the THF site, and at least a further 15 nucleotides are located 3' to the site. Detection of amplicons is accomplished by capture of tags with anti-FAM antibodies and biotin-ligand molecules generating a visual colored line on LFS

Methods

Virus strains

Different serotypes of foot-and-mouth disease virus (FMDV) and a panel of other pathogens considered dangerous to pigs were used in the study. Denatured cell-free extracts of FMDV (serotype O, A and Asia1) were obtained from the commercial Liquid-phase Blocking ELISA Kit (Lanzhou Veterinary Research Institute, Lanzhou, China). Encephalomyocarditis virus (EMCV, strain BD2), porcine circovirus 2 (PCV2, strain HB-MC1), and the viral RNA of vesicular stomatitis virus (VSV) were maintained in our laboratory. Porcine reproductive and respiratory syndrome virus (PRRSV, strain JXA1-R, Pulike Biological Engineering), classical swine fever virus (CSFV, strain AV1412, Ringpu) and pseudorabies virus (PRV, strain Barth-K61, Ringpu) were from commercial attenuated live vaccines.

Clinical and spiked samples

Twelve RNA extracted from the vesicular fluid and epithelium tissue collected from pigs experimentally infected with FMDV serotype O were provided by the State Key Laboratory of Veterinary Etiological Biology (Animal Ethics Committee of the Lanzhou Veterinary Research Institute, approval number: LVRIAEC 2012–018) and used for the clinical validation of the FMDV LFS RT-RPA assay. Twenty serum samples were

collected from clinically healthy pigs and twenty serum samples were collected from clinically healthy cattle, which were tested to be FMDV RNA negative by a real-time RT-PCR [7]. Eight swine and eight bovine sera were spiked with the denatured cell-free extracts of FMDV serotype O at the ratio of 1:1, 1:10, 1:20, 1:40, 1:80, 1:100, 1:200 and 1:400, respectively, and the other sera were used as control samples. The above samples and denatured cell-free extracts of FMDV serotype O, A and Asia1 were used as the clinical, spiked and control samples in this study.

DNA/RNA extraction

FMDV, EMCV, PRRSV and CSFV viral RNA was extracted using Trizol Reagent (Invitrogen, Waltham, USA), PRV and PCV2 viral DNA was extracted using the TIANamp Virus DNA kit (Tiangen, Beijing, China), which were performed according to manufacturer's instructions, respectively. Two hundred μL of the sera and FMDV type O were used for viral RNA extraction using the Trizol Reagent, and viral RNA was finally eluted in 20 μL of nuclease-free water. Viral RNA and DNA were quantified using a ND-2000c spectrophotometer (NanoDrop, Wilmington, USA). All RNA and DNA templates were stored at − 80 °C until use.

Generation of FMDV standard RNA

The 1104 bp RT-PCR product covering the 3D gene of FMDV was generated from viral genomic RNA of FMDV serotype O using 3D-F and 3D-R primers (Table 1). In vitro transcribed FMDV standard RNA was generated using the 3D gene RT-PCR products as described previously [25], and diluted in ten-fold series to achieve RNA concentrations ranging from 1.0×10^7 to 1.0×10^0 copies/μL.

RPA primers and LF probe

The nucleotide sequences of 3D gene are highly conserved among the different serotypes of FMDV, and the 3D gene was chosen as the target of the RPA. According to the reference sequences of different FMDV strains (accession numbers: serotype O: KF985189; KX712091; NC_004004; HQ412603; JX947859; serotype A: HQ832592; KJ968663; KU127247; serotype Asia I: KC412634; DQ533483; HQ8322592; serotype C:FJ824812;DQ409191; serotype SAT1: KU821590; JF749860; serotype SAT2: JF749862; JX014256; KU821592; serotype SAT3: KM268901; KJ820999), the primers and LF probe were designed based on 3D gene. Primers and LF probe's specificity was also tested in silico with the nucleotide sequence of other picornaviruses, such as VSV (accession numbers: NC_001560, MF196237), SVDV (accession numbers: AF268065, EU151461), VESV (accession numbers: NC_002551, KM26948, U76874), and SVA (accession numbers: NC_011349, DQ641257, KC667560, KR063107). The primers and LF probe were listed in Table 1 and synthesized by a commercial company (Sangon Biotech Co., Shanghai, China).

LFS RT-RPA

LFS RT-RPA reactions were performed in a 50 μL volume containing 29.5 μL rehydration buffer and 2.5 μL magnesium acetate (280 mM) from the TwistAmp™ nfo kit (TwistDX, Cambridge, UK). Other components included 420 nM each RPA primer (FMDV-LFS-F and FMDV-LFS-R), 120 nM LF probe (FMDV-LFS-P), 200 U

MMLV reverse transcriptase (Takara, Dalian, China), 40 U Recombinant RNase Inhibitor (Takara, Dalian, China) and 1 μL of viral RNA or 5 μL of sample RNA. Except for the viral template and magnesium acetate, the other reagents were prepared in a master mix and distributed into a 0.2 mL freeze-dried reaction tube containing a dried enzyme pellet. One μL of viral RNA and 2.5 μL of magnesium acetate were pipetted into the tubes. The RPA was performed in the technician's closed fist at room temperature for 5, 10, 15 and 20 min as described previously [26, 27]. The RPA products, which were dual labelled with FAM and Biotin, were detected using LFS as described previously [26, 27]. A testing sample was considered positive when both the test line and the control line were visible, negative when only the control line was visible, and invalid when the control line was invisible.

Analytical specificity and sensitivity analysis

Ten ng of RNA or DNA was used as template for the analytical specificity analysis of the LFS RT-RPA assay. The assay was evaluated against a panel of pathogens considered dangerous to pigs, FMDV serotype O, A, Asia1, VSV, PRRSV, CSFV, EMCV, PRV and PCV2. Three independent reactions were performed by three different technicians in the laboratory, office or in the field with an ambient temperature of 23.8 °C, 23.0 °C and 19.3 °C, respectively.

The ten-fold serial diluted in vitro transcribed RNA with concentrations ranging from 1.0×10^7 to 1.0×10^0 copies/μL were used as the standard RNA for FMDV LFS RT-RPA assay. One μL of each dilution was then amplified by the LFS RT-RPA to determine the limit of detection (LOD) of the assay. Three independent reactions were performed by three different technicians.

Validation with the clinical, spiked and control samples

The LFS-RPA method was assessed on clinical samples from experimentally infected pigs, bovine and porcine serum samples spiked with FMDV serotype O, and control samples from clinically healthy cattle and pigs.

Table 1 Sequences of primers and probes for FMDV RT-PCR, real-time RT-PCR and LFS RT-RPA assays

Assay	Primers and probes	Sequence 5'-3'	Amplicon size (bp)	References
RT-PCR	3D-F	CCCATTGAGTATCTACGAGG	1104	This study
	3D-R	CAACGCAGGTAAAGTGATC		
real-time RT-PCR	FMDV-F	ACTGGGTTTTACAAACCTGTGA	86	[7]
	FMDV-R	GCGAGTCCTGCCACGGA		
	FMDV-P	FAM-TCCTTTGCACGCCGTGGGAC-BHQ1		
LFS RT-RPA	FMDV-LFS-F	TTGGTCACTCCATTACCGATGTCACTTTCCTC	258	This study
	FMDV-LFS-R	5'-Biotin-AACGCAGGTAAAGTGATCTGTAGCTTGGAAT		
	FMDV-LFS-P	5'-FAM-GCACGCCGTGGGACCATACAGGAGAAGTT GAT(THF)TCCGTGGCAGGACTCG-C3-spacer-3'		

All samples tested by LFS RT-RPA were also tested by a real-time RT-PCR [7]. Positive predictive (the probability that the disease is present when the test is positive) and negative predictive (the probability that the disease is absent when the test is negative) values were calculated for the LFS RT-RPA and real-time RT-PCR. Since the status of the clinical, spiked and control samples were known, FMDV LFS RT-RPA and real-time RT-PCR results were classified as true positive (TP) or true negative (TN) if in agreement with the known status of tested samples. If results differed from the known status of tested samples, they were classified as false positive (FP) or false negative (FN). Positive predictive value was calculated as $TP/(TP + FP)$ and negative predictive value as $TN/(TN + FN)$ and expressed as a percentage.

Results

Optimization of the reaction time

The results from performing the LFS RT-RPA test with different reaction times are shown in Fig. 2. No amplified products were observed in reactions incubated for 5 min and slightly weak amplified product observed at 10 min. When the incubation time increased over 15 min, the assay performance was improved, and there were no clear differences between 15 and 20 min. Similar results were observed in three independent reactions,

and the temperature in the closed fists was 35.8 °C, 36.7 °C and 35.7 °C, respectively. Therefore, 15 min was set as the optimal incubation time for FMDV LFS RT-RPA assay.

Analytical specificity and sensitivity

Using 10 ng of viral RNA and DNA as template, the results showed only the FMDV serotype O, A and Asia1 were detected by LFS RT-RPA while the other viruses were not detected (Fig. 3). No cross-detections were observed, which showed the high analytical specificity of the assay. Three independent reactions were performed with similar results, demonstrating the good repeatability of the assay.

The level of detection was 1.0×10^2 copies as shown in Fig. 3. The results were similar from all three technicians. The LOD was the same as that of the real-time RT-PCR.

Evaluation of LFS RT-RPA with the clinical and spiked samples

Ten out of 12 clinical samples were FMDV RNA positive in the LFS RT-RPA (Table 2). For the 16 spiked samples, 12 samples (the spiked swine and bovine sera from 1:1 to 1:100) were FMDV RNA positive in LFS RT-RPA, while 14 samples (the spiked swine and bovine sera from 1:1 to 1:200) were FMDV RNA positive in real-time RT-PCR (Table 2). At the dilution of 1:200, all spiked samples were negative in LFS RT-RPA, while they were positive in real-time RT-PCR with the Ct values of 37.15 and 37.64, respectively (Table 2). The spiked sera at the dilution 1:400, and the 24 negative control sera were all negative in both assays (Table 2). The denatured cell-free extracts of FMDV serotype O, A and Asia1 were positive in both LFS RT-RPA and real-time RT-PCR,

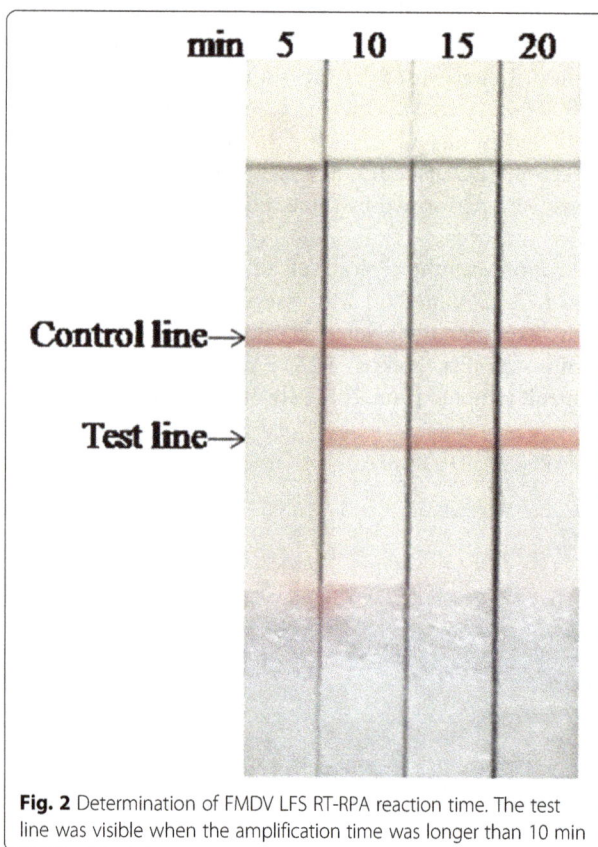

Fig. 2 Determination of FMDV LFS RT-RPA reaction time. The test line was visible when the amplification time was longer than 10 min

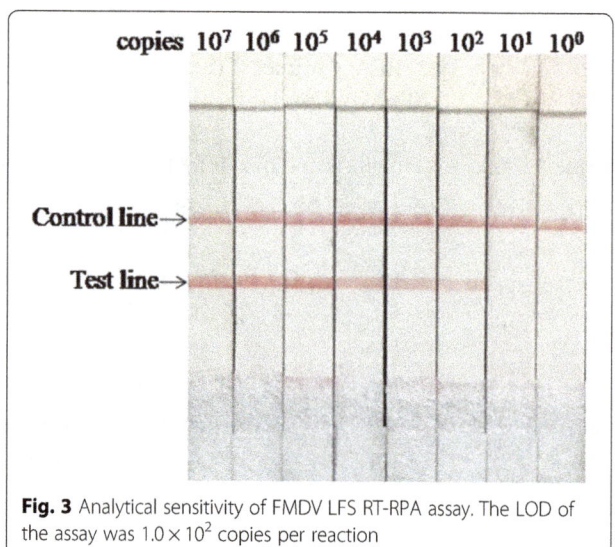

Fig. 3 Analytical sensitivity of FMDV LFS RT-RPA assay. The LOD of the assay was 1.0×10^2 copies per reaction

Table 2 Comparison of FMDV LFS RT-RPA with real-time RT-PCR assays performed on RNA extracts from the virus strains, clinical samples, spiked serum samples and samples from healthy controls

Table 2 Comparison of FMDV LFS RT-RPA with real-time RT-PCR assays performed on RNA extracts from the virus strains, clinical samples, spiked serum samples and samples from healthy controls (Continued)

Sample type	Sample name	LFS RT-RPA	real-time RT-PCR(Ct)
Virus strains	FMDV type O	+	18.95
	FMDV type A	+	20.63
	FMDV type Asia1	+	20.37
Clinical samples	4	+	23.45
	28	+	32.58
	124	+	26.78
	125	+	30.42
	126	+	30.24
	131	+	24.26
	133	+	30.76
	140	+	33.75
	208	+	25.11
	209	-	35.85
	213	-	36.96
	217	+	32.10
Spiked serum samples	Swine serum 1 (1:1)	+	23.47
	Swine serum 2 (1:10)	+	25.75
	Swine serum 3 (1:20)	+	26.41
	Swine serum 4 (1:40)	+	28.45
	Swine serum 5 (1:80)	+	33.24
	Swine serum 6 (1:100)	+	34.45
	Swine serum 7 (1:200)	-	37.15
	Swine serum 8 (1:400)	-	>40.00
	Bovine serum 1 (1:1)	+	23.08
	Bovine serum 2 (1:10)	+	25.51
	Bovine serum 3 (1:20)	+	26.64
	Bovine serum 4 (1:40)	+	29.56
	Bovine serum 5 (1:80)	+	34.40
	Bovine serum 6 (1:100)	+	35.14
	Bovine serum 7 (1:200)	-	37.64
	Bovine serum 8 (1:400)	-	>40.00
Control samples	Swine serum (9-20)	-	>40.00
	Bovine serum (9-20)	-	>40.00

+ : positive; - : negative

with the Ct values of 18.95, 20.63 and 20.37, respectively (Table 2). The positive predicative values for the LFS RT-RPA and real-time RT-PCR were 100%, and the negative predicative values for the LFS RT-RPA and real-time RT-PCR were 80% and 92.3%, respectively. It took less than 20 min in the LFS RT-RPA assay to obtain the positive results, while it took 30–51 min in the real-time RT-PCR with the Ct values ranging from 18.95 to 37.64. These results indicated that the performance of the LFS RT-RPA assay was comparable to real-time RT-PCR, while the LFS RT-RPA assay was faster.

Discussion

Outbreaks of FMD have caused great economic losses to the livestock farming worldwide, therefore, accurate and rapid diagnosis is imperative for the prevention and control of the disease. Although RT-PCR assays have played an important role in the control of FMD and have been accepted widely for the detection of FMDV in the laboratories, it still needs a lengthy process for the clinical samples being transported to laboratories in suitable cold-chain conditions, which could impose delays on diagnosis and consequently on critical decision making. The PON molecular diagnostic assays for FMDV would be of significant importance for the disease control.

This study describes a visible, equipment-free LFS RT-RPA assay with high sensitivity and specificity for rapid detection of FMDV. The FMDV LFS RT-RPA reaction tubes were held in a closed fist for 15 min, and the results were inspected directly by the naked eyes within 2 min. FMDV serotype C, SAT1-3, SVDV, VESV and SVA were not included in the analytical specificity analysis, which is a shortcoming of this study. RPA is tolerant to 5–9 mismatches in primer and probe showing no influence on the performance of the assay [20, 28], and the maximum number of mismatches found within one sequence was four in some FMDV serotypes available in GenBank (e.g. accession numbers DQ533483, HQ412603, KU821590, JF749862, KJ820999, and KT968663). It is assumed the assay would detect all the seven serotypes of FMDV, based on the facts that the LFS RT-RPA assay targets the conserved 3D gene of FMDV and that the in silico analysis of the primers and probe shows their high specificity for FMDV.

RPA operates at a wide range of temperatures, and does not require the reaction temperature to be precisely controlled [19]. TwistDx recommends an incubation temperature of 37 °C (the temperature of the human body), others studies have shown that RPA retains reliable functionality between 31 °C and 43 °C [22, 29, 30], even between 30 °C and 45 °C [22, 30]. Our previous study also showed that RPA could work well for detection of PCV2 between 34 °C and 42 °C [31]. Normal human body temperature (36.1–37 °C) is within the above temperature range, and several RPA assays had been developed to perform the reaction using body heat either holding in the axilla or in closed fists [32, 33]. In this study, FMDV LFS RT-RPA assay was performed by holding the reaction tubes in the closed fists, which is one feature of the assay.

Most of the published LFS RT-RPA assays are either developed for DNA or performed using water baths [21–23, 29, 30, 33]. In the PPRV and BEFVLFS RT-RPA assays, the viral RNA was reverse transcribed to cDNA firstly, then the viral cDNA but not viral RNA was used as the template [21, 24]. In this assay, the MMLV (4 U/μL) and RNase inhibitor (0.8 U/μL) were added into the RPA reaction system and the LFS RT-RPA worked well with FMDV RNA as the template directly, which is the other feature of the assay.

The LFS RT-RPA assay demonstrated the same positive predicative value as the real-time RT-PCR, while the negative predicative of the LFS RT-RPA (80%) was lower than real-time RT-PCR (92.3%). For two clinical samples and two spiked samples, the testing results were FMDV RNA positive in the real-time RT-PCR, while negative in the LFS RT-RPA. The sensitivity of the LFS RT-RPA was lower than the real-time PCR, nevertheless, the assay showed distinct advantages in other respects, especially the detection time and equipment requirement. Although the above results are inspiring, the assay should still be validated by analysis of more FMDV RNA positive clinical samples.

As in the real-time RT-PCR, RNA extraction is necessary in the LFS RT-RPA in this study. One of the main reasons for developing such assay is its potential use in the field or, at least, in the absence of a reliable power supply. Presently, the cost per reaction performed in FMDV LFS RT-RPA and real-time RT-PCR are approximately $9.8 and $4.8, respectively. While considering no

requirement of any incubation instruments, the rapidness of the reaction, the LFS RT-RPA is still a very promising tool in the FMD control. With the offering of TwistAmp® Liquid RPA kits and the wide application of RPA technology, the cost would be further reduced and the RPA would be closer to become a true PON isothermal molecular assay.

Conclusions
A rapid, visible and equipment-free method using body heat is developed successfully for PON diagnosis of FMD. The good specificity, sensitivity, and easy sample-to-answer protocol make the developed LFS RT-RPA assay ideal for the accurate and rapid detection of FMDV RNA in under-equipped laboratory and at PON facility, especially in low resource settings.

Abbreviations
CSFV: Classical swine fever virus; CT: Cycle threshold; EMCV: Encephalomyocarditis virus; FMDV: Foot-and-mouth disease virus; LFS: Lateral flow strip; MMLV: Moloney Murine Leukemia virus; PCV2: Porcine circovirus 2; PON: Point-of-need; PRRSV: Porcine reproductive and respiratory syndrome virus; PRV: Pseudorabies virus; RPA: Recombinase polymerase amplification; SVA: Senecavirus A; TT: Threshold time; VSV: Vesicular stomatitis virus

Acknowledgements
The authors thank the laboratory staff in the animal hospital of Agricultural University of Hebei and the staff in the Hebei Animal Disease Control Center for samples collection.

Funding
This work was supported by Natural Science Foundation Youth Project of Hebei Province (C2017325001), Science and Technology Project Foundation of Hebei Province (16226604D), Earmarked Fund foe Hebei Sheep&Goat Innovation team of Modern Agro-industry technology Research System (HBCT2018140204) and partially funded by the Fund for One-hundred Outstanding Innovative Talents from Hebei Institution of Higher Learning (SLRC2017039). The funding agencies had no role in study design; in the collection, analysis and interpretation of data; in the writing of the report; or in the decision to submit the article for publication.

Authors' contributions
JCW and WZY conceived and designed the study. LBL, JFW and RXZ developed the LFS RT-RPA assay and analyzed the data. ML, RSH and QAH made the spiked samples, performed the clinical samples testing. ML, RSH and QAH helped in the data analysis and manuscript revision. JCW and WZY wrote the manuscript. All authors read and approved the final manuscript.

Consent for publication
Not applicable.

Competing interests
The authors declare that they have no competing interests.

Author details
¹Center of Inspection and Quarantine, Hebei Entry-Exit Inspection and Quarantine Bureau, Shijiazhuang 050051, People's Republic of China. ²College of Veterinary Medicine, Agricultural University of Hebei, No.38 Lingyusi Street, Baoding, Hebei 071001, People's Republic of China. ³Hebei Animal Disease Control Center, Shijiazhuang 050050, People's Republic of China. ⁴Hebei Academy of Science and Technology for Inspection and Quarantine, Shijiazhuang 050051, People's Republic of China. ⁵State Key Laboratory of Veterinary Etiological Biology, Lanzhou Veterinary Research Institute, Chinese Academy of Agricultural Sciences, Lanzhou 730046, People's Republic of China.

References

1. Alexandersen S, Zhang Z, Donaldson AI, Garland AJ. The pathogenesis and diagnosis of foot-and-mouth disease. J Comp Pathol. 2003;129(1):1–36.
2. Knight-Jones TJ, Rushton J. The economic impacts of foot and mouth disease - what are they, how big are they and where do they occur? Prev Vet Med. 2013;112(3–4):161–73.
3. Alexandersen S, Mowat N. Foot-and-mouth disease: host range and pathogenesis. Curr Top Microbiol Immunol. 2005;288:9–42.
4. Donaldson AI, Sellers RF: Foot-and-mouth disease. In: Martin WB, Aitken ID, editors. Diseases of Sheep, 3 edn. Oxford: Blackwell Science; 2000. p. 254–8.
5. Ferris NP, Nordengrahn A, Hutchings GH, Reid SM, King DP, Ebert K, Paton DJ, Kristersson T, Brocchi E, Grazioli S, et al. Development and laboratory validation of a lateral flow device for the detection of foot-and-mouth disease virus in clinical samples. J Virol Methods. 2009;155(1):10–7.
6. Jamal SM, Belsham GJ. Foot-and-mouth disease: past, present and future. Vet Res. 2013;44:116.
7. Callahan JD, Brown F, Osorio FA, Sur JH, Kramer E, Long GW, Lubroth J, Ellis SJ, Shoulars KS, Gaffney KL, et al. Use of a portable real-time reverse transcriptase-polymerase chain reaction assay for rapid detection of foot-and-mouth disease virus. J Am Vet Med Assoc. 2002;220(11):1636–42.
8. Reid SM, Ferris NP, Hutchings GH, Zhang Z, Belsham GJ, Alexandersen S. Diagnosis of foot-and-mouth disease by real-time fluorogenic PCR assay. Vet Rec. 2001;149(20):621–3.
9. Reid SM, Ferris NP, Hutchings GH, Zhang Z, Belsham GJ, Alexandersen S. Detection of all seven serotypes of foot-and-mouth disease virus by real-time, fluorogenic reverse transcription polymerase chain reaction assay. J Virol Methods. 2002;105(1):67–80.
10. Howson ELA, Armson B, Madi M, Kasanga CJ, Kandusi S, Sallu R, Chepkwony E, Siddle A, Martin P, Wood J, et al. Evaluation of two lyophilized molecular assays to rapidly detect foot-and-mouth disease virus directly from clinical samples in field settings. Transbound Emerg Dis. 2017;64(3):861–71.
11. Madi M, Hamilton A, Squirrell D, Mioulet V, Evans P, Lee M, King DP. Rapid detection of foot-and-mouth disease virus using a field-portable nucleic acid extraction and real-time PCR amplification platform. Vet J. 2012;193(1):67–72.
12. LaBarre P, Hawkins KR, Gerlach J, Wilmoth J, Beddoe A, Singleton J, Boyle D, Weigl B. A simple, inexpensive device for nucleic acid amplification without electricity-toward instrument-free molecular diagnostics in low-resource settings. PLoS One. 2011;6(5):e19738.
13. Ambagala A, Fisher M, Goolia M, Nfon C, Furukawa-Stoffer T, Ortega Polo R, Lung O. Field-deployable reverse transcription-insulated isothermal PCR (RT-iiPCR) assay for rapid and sensitive detection of foot-and-mouth disease virus. Transbound Emerg Dis. 2017;64(5):1610–23.
14. Collins RA, Ko LS, Fung KY, Lau LT, Xing J, Yu AC. A method to detect major serotypes of foot-and-mouth disease virus. Biochem Biophys Res Commun. 2002;297(2):267–74.
15. Dukes JP, King DP, Alexandersen S. Novel reverse transcription loop-mediated isothermal amplification for rapid detection of foot-and-mouth disease virus. Arch Virol. 2006;151(6):1093–106.
16. Jingwei J, Baohua M, Suoping Q, Binbing L, He L, Xiaobing H, Yongchang C, Chunyi X. Establishment of reverse transcription helicase-dependent isothermal amplification for rapid detection of foot-and-mouth disease virus. Guizhou Journal of Animal Husbandry & Veterinary Medicine. 2014;38(5):1–5.
17. Aebischer A, Wernike K, Hoffmann B, Beer M. Rapid genome detection of Schmallenberg virus and bovine viral diarrhea virus by use of isothermal amplification methods and high-speed real-time reverse transcriptase PCR. J Clin Microbiol. 2014;52(6):1883–92.
18. Daher RK, Stewart G, Boissinot M, Bergeron MG. Recombinase polymerase amplification for diagnostic applications. Clin Chem. 2016;62(7):947–58.
19. Piepenburg O, Williams CH, Stemple DL, Armes NA. DNA detection using recombination proteins. PLoS Biol. 2006;4(7):e204.
20. Abd El Wahed A, El-Deeb A, El-Tholoth M, Abd El Kader H, Ahmed A, Hassan S, Hoffmann B, Haas B, Shalaby MA, Hufert FT, et al. A portable reverse transcription recombinase polymerase amplification assay for rapid detection of foot-and-mouth disease virus. PLoS One. 2013;8(8):e71642.
21. Yang Y, Qin X, Song Y, Zhang W, Hu G, Dou Y, Li Y, Zhang Z. Development of real-time and lateral flow strip reverse transcription recombinase polymerase amplification assays for rapid detection of peste des petits ruminants virus. Virol J. 2017;14(1):24.
22. Yang Y, Qin X, Zhang W, Li Y, Zhang Z. Rapid and specific detection of porcine parvovirus by isothermal recombinase polymerase amplification assays. Mol Cell Probes. 2016;30(5):300–5.
23. Hou P, Wang H, Zhao G, He C, He H. Rapid detection of infectious bovine Rhinotracheitis virus using recombinase polymerase amplification assays. BMC Vet Res. 2017;13(1):386.
24. Hou P, Zhao G, Wang H, He C, Huan Y, He H. Development of a recombinase polymerase amplification combined with lateral-flow dipstick assay for detection of bovine ephemeral fever virus. Mol Cell Probes. 2017.
25. Wang J, Wang J, Li R, Liu L, Yuan W. Rapid and sensitive detection of canine distemper virus by real-time reverse transcription recombinase polymerase amplification. BMC Vet Res. 2017;13(1):241.
26. Liu L, Wang J, Geng Y, Wang J, Li R, Shi R, Yuan W. Equipment-free recombinase polymerase amplification assay using body heat for visual and rapid point-of-need detection of canine parvovirus 2. Mol Cell Probes. 2018;39:41–6.
27. Wang J, Wang J, Li R, Shi R, Liu L, Yuan W. Evaluation of an incubation instrument-free reverse transcription recombinase polymerase amplification assay for rapid and point-of-need detection of canine distemper virus. J Virol Methods. 2018;260:56–61.
28. Daher RK, Stewart G, Boissinot M, Boudreau DK, Bergeron MG. Influence of sequence mismatches on the specificity of recombinase polymerase amplification technology. Mol Cell Probes. 2015;29(2):116–21.
29. Lillis L, Lehman D, Singhal MC, Cantera J, Singleton J, Labarre P, Toyama A, Piepenburg O, Parker M, Wood R, et al. Non-instrumented incubation of a recombinase polymerase amplification assay for the rapid and sensitive detection of proviral HIV-1 DNA. PLoS One. 2014;9(9):e108189.
30. Wu YD, Xu MJ, Wang QQ, Zhou CX, Wang M, Zhu XQ, Zhou DH. Recombinase polymerase amplification (RPA) combined with lateral flow (LF) strip for detection of toxoplasma gondii in the environment. Vet Parasitol. 2017;243:199–203.
31. Wang J, Wang J, Liu L, Li R, Yuan W. Rapid detection of porcine circovirus 2 by recombinase polymerase amplification. J Vet Diagn Invest. 2016;28(5):574–8.
32. Crannell ZA, Rohrman B, Richards-Kortum R. Equipment-free incubation of recombinase polymerase amplification reactions using body heat. PLoS One. 2014;9(11):e112146.
33. Wang R, Zhang F, Wang L, Qian W, Qian C, Wu J, Ying Y. Instant, visual, and instrument-free method for on-site screening of GTS 40-3-2 soybean based on body-heat triggered recombinase polymerase amplification. Anal Chem. 2017;89(8):4413–8.

18

Identification of canine norovirus in dogs in South Korea

Kwang-Soo Lyoo[1†], Min-Chul Jung[2,3†], Sun-Woo Yoon[2], Hye Kwon Kim[2] and Dae Gwin Jeong[2,3*] (iD)

Abstract

Background: Canine noroviruses (CaNoVs) are classified into genogroups GIV, GVI, and GVII and have been detected in fecal samples from dogs since their first appearance in a dog with enteritis in Italy in 2007. CaNoVs may be a public health concern because pet animals are an integral part of the family and could be a potential reservoir of zoonotic agents. Nonetheless, there was no previous information concerning the epidemiology of CaNoV in South Korea. In the present study, we aimed to detect CaNoV antigens and to investigate serological response against CaNoV in dogs.

Results: In total, 459 fecal samples and 427 sera were collected from small animal clinics and animal shelters housing free-roaming dogs in geographically distinct areas in South Korea. For the detection of CaNoV, RT-PCR was performed using target specific primers, and nucleotide sequences of CaNoV isolates were phylogenetically analyzed. Seroprevalence was performed by ELISA based on P domain protein. CaNoVs were detected in dog fecal samples (14/459, 3.1%) and were phylogenetically classified into the same cluster as previously reported genogroup GIV CaNoVs. Seroprevalence was performed, and 68 (15.9%) of 427 total dog serum samples tested positive for CaNoV IgG antibodies.

Conclusion: This is the first study identifying CaNoV in the South Korean dog population.

Keywords: Canine norovirus, Dog, Korea

Background

Noroviruses (NoVs) belong to the family *Caliciviridae*, which are non-enveloped viruses approximately 35 nm in diameter with a positive single-strand RNA genome of about 7.7 kb [1]. Human NoVs are responsible for gastroenteritis in humans across all age groups, and typical symptoms of nausea, abdominal pain, fever, vomiting, and diarrhea can be demonstrated in an infected person [2]. The fecal-oral route through contaminated food or water is the major pathway of the viral transmission [3]. Several characteristics of NoVs such as low infectious doses or high resistance to harsh environmental conditions facilitate the spread and infection of the virus [1, 4].

NoVs has been genetically classified into seven major genogroups designated GI through GVII based on sequence analysis of the VP1 protein (ORF2; major capsid protein) [5]. Recently, genotypes are classified by the sequence

diversity of the ORF1-ORF2 junction region and the RNA-dependent RNA polymerase (RdRp) region within the ORF1 and VP1 region [5–8]. In humans, three genogroups GI, GII, and GIV are found, and GII strains are the most frequently detected. Additionally, bovine, porcine, and murine NoVs are classified in GIII, GII, and GV, respectively [9].

Canine norovirus (CaNoV) was first identified in a single dog with enteritis in Italy in 2007 [10]. The detection of CaNoVs in fecal samples from dogs was followed in Portugal circa 2007, and then these viruses have been described in dogs from several countries in Europe and Asia [11–14]. A recent prevalence study demonstrated that seropositivity for CaNoV was estimated to be 39% in dog serum samples from 14 different European countries [15]. In addition, it was suggested that CaNoV may infect humans, particularly veterinary workers with potential risk factors for virus exposure. Mesquita et al. reported that they tested serum samples collected from 373 small animal veterinarians and 120 controls in the general human population for the presence for CaNoV antibodies, and the seropositive rate of the veterinarian group was 22.3% compared to 5.8% in control [16].

---

* Correspondence: dgjeong@kribb.re.kr
†Kwang-Soo Lyoo and Min-Chul Jung contributed equally to this work.
[2]Infectious Disease Research Center, Korea Research Institute of Bioscience and Biotechnology, Daejeon 305-806, South Korea
[3]University of Science and Technology (UST), Daejeon, South Korea
Full list of author information is available at the end of the article

The emergence of CaNoVs may be a public health concern because pet animals are an integral part of family life in most industrialized countries, and their close relationship with humans needs to have special attention with potential reservoir of zoonotic agents [17, 18]. Thus, in the present study, we aimed to detect CaNoVs in fecal samples from dogs and to investigate its seroprevalence in the dog population of South Korea.

Results

Canine norovirus detection and phylogenetic analysis
Fecal samples from 14 (3.1%) of the 459 dog stool samples tested positive for CaNoV. Unfortunately, the completeness of basic data associated with stool samples, including age, sex, breed, and any clinical history of enteric disease varied among the small animal clinics and animal shelters. This precludes further statistically valid analysis of the positive samples. The sequences of partial RdRp region of CaNoVs were analyzed with other CaNoVs sequences from the GenBank database (Fig. 1). The CaNoV isolates in South Korea were classified into the same clade with CaNoV strains detected in Italy and Costa Rica and showed a high nucleotide identity (range 98–100%).

P domain proteins produced in *E. coli*
The recombinant P domain protein of CaNoV which was successfully expressed as a soluble protein with a

C-terminus His6-tag in BL21 Star *E.coli* cells. Expressed P domain proteins were purified by a commercial metal affinity chromatography kit. The purified P domain protein was detected by SDS-PAGE staining with Coomassie brilliant blue which gave a band corresponding to the expected molecular mass (Fig. 2a). To determine the specificity of the expressed P domain protein, western blotting was demonstrated using negative control dog serum and CaNoV positive dog serum which was tested by the ELISA developed in this study (Fig. 2b).

Seroprevalence
To determine the seropositivity of serum samples, positive serum and negative control serum were serially diluted two-fold from 1:50 of starting dilution to 1:12,800 and tested by ELISA. A range of titers were identified, and the cut-off threshold was calculated as mean plus three times the standard deviation of the OD450 reading of negative control serum (Fig. 3). Dog serum samples were obtained from small animal clinics in seven provinces; Seoul, Gyeonggi, Gangwon, Chungbuk, Chungnam, Jeonbuk, and Busan, in South Korea and tested by ELISA for antibodies to CaNoV. From the 427 serum samples, 68 (15.9%) tested positive for IgG antibodies against CaNoV. The prevalence varied with respect to the provinces; Seoul (18.0%), Gyeonggi (14.1%), Gangwon (11.6%), Chungbuk (18.9%), Chungnam (11.5%), Jeonbuk (15.4%), and Busan (16.1%).

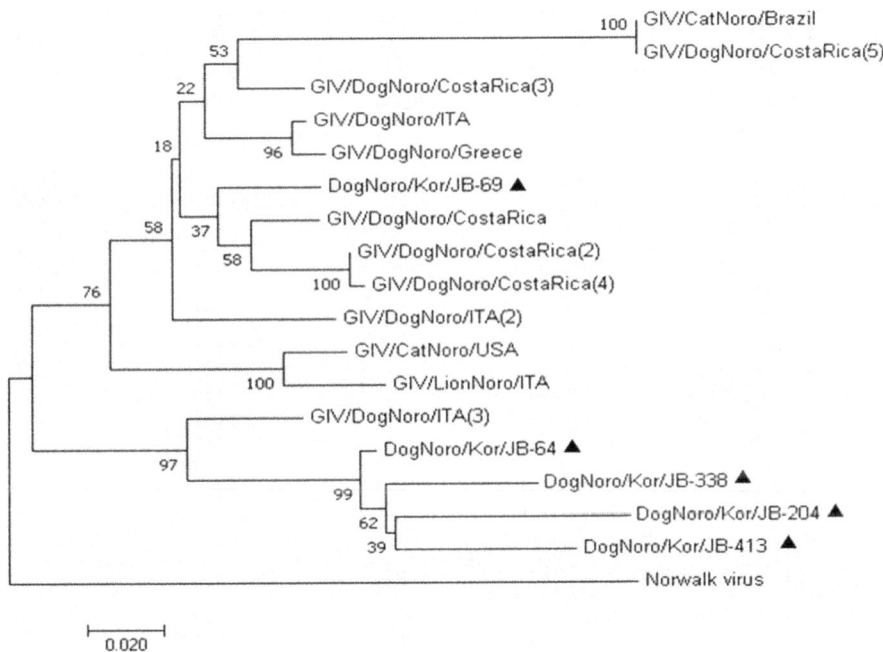

Fig. 1 Phylogenetic analysis of CaNoV strains via comparison of nucleotide sequences of partial RdRp region. The analysis was conducted using the MEGA program, and the branches indicate bootstrap values calculated from 1000 bootstrapping replicates. ▲: CaNoV isolates identified in the present study

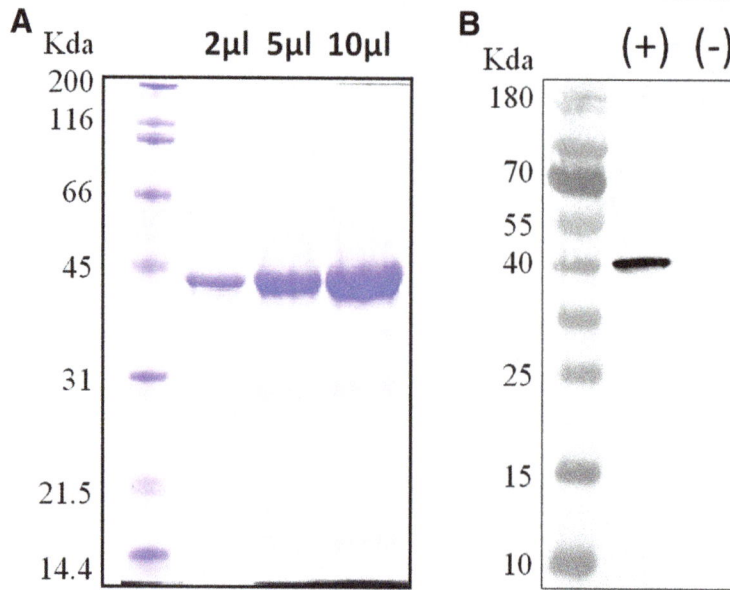

Fig. 2 Expression of P domain protein (**a**). Concentration of the purified P domain protein was determined by Bradford assay, and SDS-PAGE gel loaded with 2 µl, 5 µl, and 10 µl of the protein (1 mg/ml) were stained with Coomassie blue. Immunoblotting with CaNoV positive antibody (**b**). The recombinant N proteins were transferred to nitrocellulose membrane, and the membrane was incubated with CaNoV-positive dog serum (lane +) or negative control dog serum (lane -)

Discussion

Recently, dogs were suggested to be potential zoonotic vectors of HuNoV, since HuNoV was detected in stool samples from pet dogs that had been in direct contact with the owner infected with identical HuNoV strain, and the presence of antibodies to HuNoV in dogs was demonstrated in across Europe [15, 17]. In South Korea, the National Agricultural Cooperative Federation reported that companion animal market in South Korea is continuously expanding and is expected to grow at a significant pace in coming years. They forecast that the market will increase from 1 billion U.S. dollars in 2012 to 5.3 billion

U.S. dollars in 2020. Regardless of the increased exposure risk to humans with companion animals, there was no previous information concerning the epidemiology of CaNoV in the country. Although a markedly high detection rate of CaNoV in dogs was not demonstrated in the present study, this is the first study describing the detection and prevalence of CaNoV in dogs in South Korea and provides good baseline information for the design of further studies.

CaNoVs detected in the present study were classified into the same cluster with previously reported genogroup GIV CaNoVs in the phylogenetic tree. It was expected that

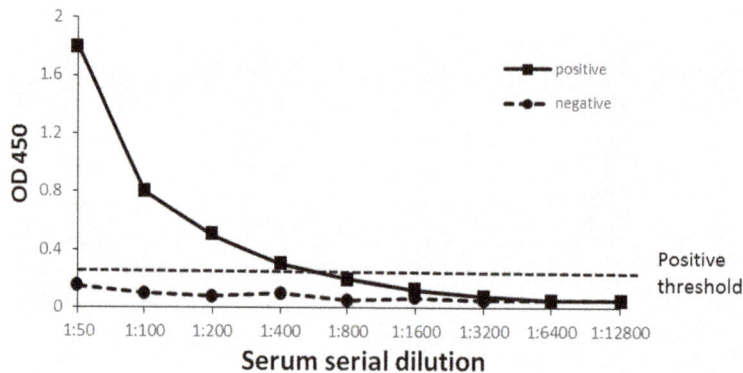

Fig. 3 Antibody titers in dog serum samples. Positive serum and negative control serum were diluted from 1:50 to 1:13200 used in the ELISA assay. The corrected OD450 was obtained by subtracting the background signal from the VLP coated well OD450 value. The positive threshold was determined by calculating the mean OD450 of buffer coated wells with the highest serum dilution, plus 3 standard deviations

the CaNoV prevalence in fecal samples from the animal shelters would be relatively higher when compared to the samples from small animal clinics as environmental factors such as hygiene reasons or crowded space would be expected to increase transmission between dogs. However, there was no difference in the prevalence between these both dog populations. Seroprevalence of CaNoV was performed by ELISA based on P domain protein, and 15.9% of 427 dog serum samples collected from small animal clinics tested positive for IgG against genogroup GIV CaNoV. These results demonstrate that CaNoV has notably circulated in the dog population in South Korea.

It was previously suggested that CaNoV can easily infect and spread within dog population and its homologous host, and viral characteristics such as incubation and shedding period of this virus resemble to human norovirus [19]. Indeed, dogs contacted with imported dogs from Russia showed diarrhea and diagnosed with CaNoV infection in Portugal, even though the dogs had been vaccinated against major gastroenteritis pathogens [20]. Although the origin of CaNoV in South Korea could not be established in the present study, it is clear that epidemiologic data should be collected from other geographic regions including South Korea.

The P domain of noroviruses comprises a protruded structure of capsid protein which is believed to play the receptor-binding site for virus attachment and viral determinant for immune-response in the host [21, 22]. In the present, the P domain proteins of a genogroup GIV CaNoV were produced in E. coli and purified using a Ni-NTA column to develop ELISA, although multiple studies have previously demonstrated to identify antibodies against CaNoV based on virus-like-particle (VLP) antigens produced in a recombinant baculovirus system [15]. The antigenic activities between P domain proteins and VLPs were previously compared, and both antigens showed a similar binding effect and detectable affinity for norovirus specific antibodies [22]. Moreover, the production procedure of the P domain protein in an E. coli system is simpler and more cost-effective than in baculovirus expression systems [22, 23].

Conclusion

This is the first study identifying CaNoV in the South Korean dog population. We successfully developed ELISA based on the P domain protein of CaNoV produced in E. coli to detect IgG antibodies and demonstrated the seroprevalence and co-circulation of CaNoV within the dog population using the ELISA.

Methods
Sample collection

We collected fecal samples from small animal clinics and animal shelters housing free-roaming dogs in geographically

distinct areas in South Korea. The sampling was followed the General Animal Care Guideline as required and approved by the Institutional Animal Care and Use Committee of Chonbuk National University (# CBNU-2018-063). Clinic veterinarians collected only the stools of dogs with consenting owners in their clinics, whereas managers of the animal shelters collected their dog stools. A total of 459 fecal samples were collected from healthy or unhealthy dogs, and then stored at – 20 °C until and during transportation to the laboratory. A total of 427 dog serum samples were collected from small animal clinics in geographically distinct areas of South Korea. Veterinarians with owner's consent retained the serum remaining after running necessary diagnostic tests on hospitalized dogs. All sera were stored at – 20 °C until transportation to the laboratory.

RT-PCR and sequence analysis

Fecal samples were diluted 10% (wt/vol) in phosphate-buffered saline (PBS) (pH 7.2). After centrifugation at 8000 g for 5 min, viral nucleic acid was extracted from 140 µl of each supernatant using Viral RNA Isolation Kit (QIAamp Viral RNA Mini Kit, QIAGEN, Hilden, Germany) according to the manufacturer's instructions. For the detection of CaNoV, reverse transcription PCR (RT-PCR) was performed with the One Step RT-PCR kit (QIAGEN) with previously reported primers: JV102 (5′-TGG GAT TCA ACA CAG CAG AG-3′) and JV103 (5′-TGC GCA ATA GAG TTG ACC TG-3′) [12]. The amplified PCR fragments were purified using the QIAquick Gel Extraction kit (QIAGEN) and sequenced (Bionear, Korea). Analysis of the nucleotide sequences of the CaNoV isolates was performed using BLAST/Align (bl2seq) on the NCBI website and the MEGA (ver.4.1) program. Nucleotide sequences of the isolates were aligned using the ClustalX (ver. 1.81) software, and a phylogenetic tree was constructed by the neighbor-joining method with other CaNoV isolates from GenBank. The phylogenetic distances were determined using bootstrapping with 1000 repeats, and molecular evolutionary analysis was performed using the MEGA (ver.5.0) program.

Viral protein synthesis

To express the P domain of CaNoV capsid protein, the VP1 gene of a CaNoV (GenBank accession No. GQ443611, NCBI) originated from a dog with diarrhea was synthesized. The P domain (222~570 amino acids) was amplified by PCR using primers: CNVPD-NdeI forward (5′-GTTG CCCATATGGAGTCTCGTGTCACCCCTTTTTC-3′) and CNVPD-BamHI reverse (5′-CGCGGATCCTTAGGAGCC AGTTCCCACAGGGCTG-3′). The PCR products were cloned into the NdeI/BamHI restriction site of pET21a (Novagen), which attached a His6-tagged at the C-terminus of the protein. Expression of the P domain protein was performed in BL21 Star *E.coli* cells (Invitrogen). The pET21a-P domain vector-containing *E.coli* cells were incubated with 0.4 mM IPTG (final concentration) to induce protein expression. After 16~18 h of incubation, the cells were harvested via centrifugation. The supernatant was removed, and the cell pellets were suspended in 30 ml of lysis/binding buffer (50 mM Tris-HCl (pH 7.5), 500 mM NaCl, 1 mM phenylmethylsulfonyl fluoride (PMSF), 4 mM 2-mercaptoethanol, 10% glycerol per liter culture volume and sonicated on ice 3x30sec at 60% power. The protein was purified via Ni-NTA Superflow (Qiagen) metal affinity chromatography. The eluate was concentrated and further purified by Superdex 200 size exclusion chromatography (GE Healthcare). The protein was isolated from the final purification step in the twofold PBS buffer and the concentration and purity were determined by Bradford assay. The purified protein and standard BSA protein were separated on 10% SDS-polyacrylamide gels, and then the gel was stained with Coomassie Blue for 1 h at room temperature.

ELISA procedure

For serology, 96-well polystyrene microtiter plates (Nunc Maxisorp; Fisher Scientific, Waltham, MA, USA) were coated with 50 ng of the purified P domain protein in 0.05 M carbonate-bicarbonate buffer (pH 9.6) overnight at 4°. The plates were blocked with 5% skimmed milk in 0.05% Tween20-PBS (PBS-T) for 1 h at 37 °C and washed three times with PBS-T. Dog serum samples diluted in 1:50 in blocking buffer were incubated for 1 h at 37 °C. Pooled sera from SPF dogs (provided by Green Cross Veterinary Product, Suwon, Korea) was used as a negative control. After three washes with PBS-T, the wells were incubated with goat anti-dog IgG conjugated with horseradish peroxidase (HRP) (Abcam, Cambridge, MA, USA) diluted 1:10,000 in blocking buffer for 1 h at 37 °C. The plates were washed three times with PBS-T, and the antibody reaction was developed with tetramethylbenzidine (TMB) (Sigma-Aldrich), followed by incubation at room temperature for 10 min. The reaction was stopped with 1 N H_2SO_4, and the optical density (OD) at 450 nm was assessed in a microplate reader (Model 680; Bio-Rad, Hercules, CA). All sera including negative control serum were tested in duplicate, and they were considered as positive when the corrected OD value was above the mean of the ODs from the negative control wells plus three standard deviations.

Western blotting

To confirm the specificity of the expressed CaNoV P domain protein, an immunoblotting assay was performed with dog sera tested by the ELISA which was developed in the present study. The protein loaded by SDS-PAGE was transferred onto a nitrocellulose membrane for western blotting. The membrane was blocked in 5% skimmed milk

buffer, and then incubated with CaNoV-positive or -negative dog serum at a dilution of 1:500 in TBS-T containing 1% skimmed milk at 4 °C overnight. After washing three times with PBS-T, the membrane was incubated for 1 h with goat anti-dog IgG HRP-conjugated secondary antibodies, and a specific band was detected using ECL reagents (GE Healthcare Life Sciences, Buckinghamshire, UK).

Abbreviations

CaNoV: Canine noroviruses; HRP: horseradish peroxidase; PBS: phosphate-buffered saline; RdRp: RNA dependent RNA polymerase; TMB: tetramethylbenzidine

Acknowledgements

We would like to thank Prof. Daesub Song (College of pharmacy, Korea University) for helping with sampling.

Consent to publication

Not applicable.

Funding

This research was supported by National Research Foundation of Korea (NRF) funded by the Ministry of Science and ICT (NRF-2014R1A1A1005112). BioNano Health-Guard Research Center, funded by the Ministry of Science and ICT (MSIT) of Korea as a Global Frontier Project (Grant number HGUARD_2013M3A6B2078954(1711073748)).

Authors' contributions

KS wrote the manuscript. MC and SY collected samples, MC and HK analyzed and interpreted experiment data. DG and KS designed this study. All authors critically revised the manuscript, approved the final version and agreed to be accountable for all aspects of the work.

Competing interests

The authors declare that they have no competing interests.

Author details

[1]Korea Zoonosis Research Institute, Chonbuk National University, Iksan, South Korea. [2]Infectious Disease Research Center, Korea Research Institute of Bioscience and Biotechnology, Daejeon 305-806, South Korea. [3]University of Science and Technology (UST), Daejeon, South Korea.

References

1. Glass RI, Parashar UD, Estes MK. Norovirus gastroenteritis. N Engl J Med. 2009;361:1776–85.
2. Thornton AC, Jennings-Conklin KS, McCormick MI. Noroviruses: agents in outbreaks of acute gastroenteritis. Disaster Manag Response. 2004;2:4–9.
3. Teunis PF, Moe CL, Liu P, Miller SE, Lindesmith L, Baric RS, Le Pendu J, et al. Norwalk virus: how infectious is it? J Med Virol. 2008;80:1468–76.
4. Koopmans M, Duizer E. Foodborne viruses: an emerging problem. Int J Food Microbiol. 2004;90:23–41.
5. Vinjé J. Advances in laboratory methods for detection and typing of norovirus. J Clin Microbiol. 2015;53:373–81.

6. Green J, Vinje J, Gallimore CI, Koopmans M, Hale A, Brown DW, et al. Capsid protein diversity among Norwalk-like viruses. Virus Genes. 2000;20:227–36.

7. Wang J, Jiang X, Madore HP, Gray J, Desselberger U, Ando T, et al. Sequence diversity of small, round-structured viruses in the Norwalk virus group. J Virol. 1994;68:5982–90.

8. Zheng DP, Ando T, Fankhauser RL, Beard RS, Glass RI, Monroe SS. Norovirus classification and proposed strain nomenclature. Virology. 2006;346:312–23.

9. Martella V, Decaro N, Lorusso E, Radogna A, Moschidou P, Amorisco F, et al. Genetic heterogeneity and recombination in canine noroviruses. J Virol. 2009;83:11391–6.

10. Martella V, Lorusso E, Decaro N, Elia G, Radogna A, D'Abramo M, et al. Detection and molecular characterization of a canine norovirus. Emerg Infect Dis. 2008;14:1306–8.

11. Mesquita JR, Nascimento MS. Molecular epidemiology of canine norovirus in dogs from Portugal, 2007-2011. BMC Vet Res. 2012;8:107.

12. Mesquita JR, Barclay L, Nascimento MS, Vinje J. Novel norovirus in dogs with diarrhea. Emerg Infect Dis. 2010;16:980–2.

13. Caddy S, Emmott E, El-Attar L, Mitchell J, de Rougemont A, Brownlie J, et al. Serological evidence for multiple strains of canine norovirus in the UK dog population. PLoS One. 2013;8:e81596.

14. Soma T, Nakagomi O, Nakagomi T, Mochizuki M. Detection of norovirus and Sapovirus from diarrheic dogs and cats in Japan. Microbiol Immunol. 2015;59:123–8.

15. Mesquita JR, Delgado I, Costantini V, Heenemann K, Vahlenkamp TW, Vinje J, et al. Seroprevalence of canine norovirus in 14 European countries. Clin Vaccine Immunol. 2014;21:898–900.

16. Mesquita JR, Costantini VP, Cannon JL, Lin SC, Nascimento MS, Vinje J. Presence of antibodies against genogroup VI norovirus in humans. Virol J. 2013;10:176.

17. Summa M, von Bonsdorff CH, Maunula L. Pet dogs--a transmission route for human noroviruses? J Clin Virol. 2012;53:244–7.

18. Day MJ. One health: the importance of companion animal vector-borne diseases. Parasit Vectors. 2011;4:49.

19. Ntafis V, Xylouri E, Radogna A, Buonavoglia C, Martella V. Outbreak of canine norovirus infection in young dogs. J Clin Microbiol. 2010;48:2605–8.

20. Mesquita JR, Nascimento MS. Gastroenteritis outbreak associated with faecal shedding of canine norovirus in a Portuguese kennel following introduction of imported dogs from Russia. Transbound Emerg Dis. 2012;59:456–9.

21. Choi JM, Hutson AM, Estes MK, Prasad BV. Atomic resolution structural characterization of recognition of histo-blood group antigens by Norwalk virus. Proc Natl Acad Sci U S A. 2008;105:9175–80.

22. Koho T, Huhti L, Blazevic V, Nurminen K, Butcher SJ, Laurinmaki P, et al. Production and characterization of virus-like particles and the P domain protein of GII.4 norovirus. J Virol Methods. 2012;179:1–7.

23. Lu Y, Welsh JP, Swartz JR. Production and stabilization of the trimeric influenza hemagglutinin stem domain for potentially broadly protective influenza vaccines. Proc Natl Acad Sci U S A. 2014;111:125–30.

19

Seroprevalence of border disease virus and other pestiviruses in sheep in Algeria and associated risk factors

Naouel Feknous[1*†] (iD), Jean-Baptiste Hanon[2†], Marylène Tignon[2], Hamza Khaled[1], Abdallah Bouyoucef[3] and Brigitte Cay[2]

Abstract

Background: Border disease virus (BDV) is a pestivirus responsible for significant economic losses in sheep industry. The present study was conducted between 2015 and 2016 to determine the flock seroprevalence of the disease in Algeria and to identify associated risk factors. 56 flocks from nine departments were visited and 689 blood samples were collected from adult sheep between 6 and 24 months of age ($n = 576$) and from lambs younger than 6 months ($n = 113$). All samples were tested by RT-PCR as well as by Ag-ELISA, to detect Persistently Infected (PI) animals. Serum samples from adults were tested by Ab-ELISA (Enzyme Linked Immuno-Sorbent Assay), to detect specific antibodies against pestivirus and 197 of them were further characterized by VNT (virus neutralization test) for the detection of neutralizing antibodies specific for BDV and for Bovine virus diarrhea virus (BVDV-1 and BVDV-2).

Results: No PI animals were found among the 689 sheep tested. 144/197 sera were positive in VNT for BDV, and 2 sera were strongly positive BVDV-2. Fifty-five flocks (98%) had at least one seropositive animal and the apparent within-flock seroprevalence was estimated to be 60.17% (95% C.I.: 52.96–66.96). The true seroprevalence based on estimated sensitivity and specificity of the Ab-ELISA was 68.20% (95% C.I.; 60.2–76.3). Several risk factors were identified as linked to BDV such as climate, landscape, flock management and presence of other ruminant species in the farm.

Conclusion: These high seroprevalence rates suggest that BDV is widespread and is probably endemic all over the country. Further studies are needed to detect and isolate the virus strains circulating in the country and understand the distribution and impact of pestiviruses in the Algerian livestock.

Keywords: Border disease virus, Pestivirus, Seroprevalence, Sheep, Algeria, Persistently infected

Background

In accordance with the ICTV (International Committee on Taxonomy of Viruses), BDV (Border Disease Virus) belongs to the *Flaviviridae* family which includes four genera: *Flavivirus, Hepacivirus, Pegivirus,* and *Pestivirus*; the latter previously included four species: bovine viral diarrhea virus 1, (BVDV1), bovine viral diarrhea virus 2 (BVDV2), classical swine fever virus (CSFV), D and border disease virus (BDV). Since 2017, these four species have been renamed *Pestivirus A, B, C, and D,* respectively. Seven other species have been added in the

genus, namely *Pestivirus E* to *K*, including giraffe pestivirus (*Pestivirus G*), Hobi-like pestivirus (*Pestivirus H*) and other atypical species isolated in wild and domestic mammals. The 11 currently recognized pestivirus species are now named in relation to molecular and antigenic relatedness in a host-independent scheme [1]. BVDV can infect cattle, sheep, goats, pigs and other ungulate species [2] and infection of sheep by BVDV-1 and BVDV-2 in natural and experimental conditions was demonstrated [3]; in some regions BVDV prevalence in sheep can be higher than BDV [4]. CSFV seems to be restricted to pigs and wild boars [5]. Although BDV is generally considered as an agent for a sheep disease, it is not strictly host specific and can cross infect cattle, sheep, goats, pigs and non-domesticated species [6]; transmission of BDV between small ruminants and

* Correspondence: feknousnaouel33@gmail.com
†Feknous Naouel and Hanon Jean Baptiste contributed equally to this work.
[1]LBRA, Institute of Veterinary Sciences, Saad Dahlab University, Soumaa Road, BP 270, 09000 Blida, Algeria

cattle has been described by several authors [7–9]. BDV infection can cause significant economic losses to sheep industry due to its impact on reproduction and health. Clinical signs in sheep are dominated by infertility, abortions, stillbirths, or even the birth of lambs with hairy fleeces called "hairy-shaker" or "blurred" or an abnormal body conformation. BDV can also cause a condition similar to mucosal disease [10]. The main source of infection in a flock are the PI sheep, which are born infected and spread the virus during their whole life. PI lambs result from transplacental infection of the fetus before the 60th day up to the 80th day of gestation, when the immunological system is still immature [6]. Border disease is present in several continents and seroprevalence rates in sheep range from 5 to 50% depending on the country or the regions within a same country [11]. However, prevalence of BDV in Algeria, where vaccination is not practiced remains completely unknown and there has been no scientific publication on the topic so far to our knowledge. The purpose of this study conducted between 2015 and 2016 was to estimate the BD seroprevalence and shedding in Algerian sheep flocks and to identify associated risk factors. Such epidemiological data should contribute to improve the visibility of this neglected disease and to develop a monitoring plan for the country.

Results

Flock and within-flock seroprevalence

A flock was considered positive for ruminant pestivirus when at least one animal was positive in Ab-ELISA. All flocks except one were seropositive, therefore the flock seroprevalence was estimated to be 55/56 = 98.2% (95% C.I. 90.5–99.6). The proportion of positive sheep in each flock ranged between 1 and 100%. Out the 576 sera tested, 344 samples were considered as seropositive (304 positive

+ 40 doubtful in Ab-ELISA). The apparent overall within-flock seroprevalence, based on the GEE model was estimated to be 60.17% (95% CI: 52.96–66.96). The true overall seroprevalence, taking into account our estimation of Se (84.0%) and Sp (92.4%) of the Ab-ELISA (see below) was calculated to be 68.20% (95% C.I. 60.2–76.3).

The within-flock seroprevalence by departments is described in Table 1. There were marked regional differences in the flock prevalence, ranging from 18% in Chlef (95% C.I. 5.1–30.9) to 100% in Setif. However, our sampling design was built to estimate with a reasonable precision the overall within-flock seroprevalence but cannot provide an accurate estimation at department level (this would have required a larger number of flocks in each department). This is the reason why the 95% CI at regional level were large and the differences of within-flock prevalence between regions were not statistically significant except for the prevalence in Chlef which was significantly lower ($p < 0.0001$) than the prevalences in Djelfa, Al Bayadh, Msila, Saida and Laghouat.

Comparison between ab-ELISA and VNT results

The number of positive, doubtful and negative samples among the 576 sera tested by ELISA-Ab was 304, 40, and 232 respectively. A list of all samples with their respective Ab-ELISA and VNT results is provided in Additional file 1: Table S1.

To estimate the performances (Se and Sp) of the pestivirus Ab-ELISA compared to VNT, 197 sera were tested in parallel by Ab-ELISA and by three different VNT (BDV, BVDV-1 and BVDV-2). Table 2 gives the number of positive, doubtful and negative samples when tested by Ab-ELISA compared to the BDV-VNT titer. Table 3 gives the number of positive, doubtful and negative samples in

Table 1 Ovine population (number of flocks), sampling performed and estimated (apparent) seroprevalence (with 95% CI) of Border disease, by department, according to GEE model

Department	Ovine flocks	N Fl	N Lb	N Ad	N Pos	Prev %	95% CI
El Bordj	9000	3	0	30	15	50.00	[27.33; 72.67]
Setif	4230	1	0	10	10	100.00	n.d.
Msila	23,000	7	31	71	51	71.90	[60.36; 81.13]
Djelfa	21,000	6	5	60	35	58.33	[43.12; 72.11]
Laghouat	23,000	8	24	80	39	51.07	[36.55; 65.41]
El Bayadh	25,000	7	0	70	57	81.43	[64.40; 91.40]
Tizi Ouzou	18,000	5	1	50	23	46.00	[20.87; 73.34]
Chlef	17,230	5	30	50	9	18.00	[09.90; 30.50]
Saida	50,000	15	22	155	105	67.68	[55.81; 77.64]
Total	190,460	57	113	576	344	60.17	[52.96; 66.96]

N Fl: Number of sampled flocks
N Lb: Number of sampled lambs (animals < 6 months, plasma)
N Ad: number of sampled adults (serum + plasma)
N Pos: number of positive sera (adults)
Prev %: within-herd seroprevalence

Ab-ELISA compared to the ratio [VNT titer for BDV / VNT titer for BVDV], the latter being calculated compared to the highest titer in BVDV, whether it was BVDV-1 or BVDV-2. This ratio was split into three categories: (a) BDV titer = four folds the BVDV-1 or BVDV-2 titer, (b) BDV titer = two to three folds the BVDV-1 or BVDV-2 titer, (c) BDV titer = less than two folds the BVDV-1 or BVDV-2 titer. Samples from category (a) were considered as specifically positive for BDV.

Among the 197 sera tested in parallel by Ab-ELISA and VNT, 144 were positive in VNT for BDV (titer ≥1/8) (Table 2) including 103 sera with a titer four folds higher for BDV than for BVDV-1 or BVDV-2 (Table 3). Of these BDV VNT positive samples, 89 were positive, 32 doubtful and 23 negative in Ab-ELISA (Table 3). Out of the 36 sera with a doubtful result in Ab-ELISA, 32 were positive in VNT for BDV (Table 2) and most of these (31/32) had a high VNT titer (1/32 up to 1/192) (Table 2) including 19 samples with a BDV titer four folds higher than for BVDV-1 or BVDV-2 (Table 3). Based on this observation we considered the sera with a doubtful result in Ab-ELISA as seropositive samples for the rest of our analysis. Therefore, the relative sensitivity of the Ab-ELISA compared to the BDV-VNT was estimated to be [(89 positive ELISA + 32 doubtful ELISA)/ 144 positive VNT] = 84.0% and the relative specificity to be [49 negative ELISA/ 53 negative VNT] = 92.4% (Table 2). If only samples with a BDV titer four folds higher than the BDV titer, then the sensitivity of the Ab-ELISA compared to BD VNT = [(63 positive ELISA + 19 doubtful ELISA) / 118 positive BDV-VNT] = 69.5.3% (Table 3). Some positive sera with the BDV-VNT cross-reacted with BVDV-1 and BVDV-2 VNT but the titers observed for BVDV were generally low except for two samples from two different flocks which had very high titers for BVDV-2 (titer = 1/480 and 1/640 respectively).

The agreement (Cohen's kappa coefficient) between Ab-ELISA and BDV-VNT test was 0.68 (95% C.I. 0.58–0.79).

Seroprevalence and risk factors
The differences in the proportion of seropositive animals were not statistically significant for the following studied variables: flock size, sheep breed, presence of cattle in the farm, purchase of breeding females, purchase of sheep for fattening, abortion history, sharing breeding rams, and vaccination for sheep pox and brucellosis and other diseases. On the contrary, a significant difference in seroprevalence was found for the following variables: climate (arid versus Mediterranean; OR = 4.04), landscape (mountain versus plateau; OR = 0.49), flock management (sedentary versus transhumant; OR = 0.59), presence of goat versus no goat (OR = 0.58), other clinical diseases (OR = 0.66). The detailed results including odds ratios for these risk factors are presented in Table 4.

Ag-ELISA and RT-PCR
All 689 individual samples were tested negative by Ag-ELISA and these negative results were confirmed by the fact that all pools of plasma samples were tested negative by RT-PCR.

Discussion
Seroprevalence study
After infection with a ruminant pestivirus, the detection of antibodies against the highly conserved pestivirus-NS2–3 (p80) protein by competitive ELISA provides reliable results to confirm seroconversion. Such assays have been used in

Table 3 Number of positive, doubtful and negative samples in Ab ELISA compared to the ratio [VNT titer for BDV / VNT titer for BVDV-1 or BVDV-2]

	VNT Titre BDV/ Titre BVDV				Total
	> 4 x	2–3 x	< 2 x	Neg	
ELISA Ab results					
Doubt	19	7	6[a]	4	36
Pos	63	8	18	–	89
Neg	21	–	2[a]	49	72
Total	103	15	26	53	197

> 4 x / 2–3 x / < 2 x: VNT titer for BDV compared to VNT titer for BVDV-1 or BVDV-2 more than fourfold higher / between two and threefold higher / less than twofold higher
Pos: positive
Neg: negative
Doubt: doubtful
[a]including 1 sample with high BVDV-2 Titer

Table 2 Number of positive, doubtful and negative samples in Ab ELISA, according to the manufacturer's recommended cut-off, compared to VNT titer for BDV, considering 1/8 titer as cut-off for VNT

ELISA result	Neg VNT < 1/8	Pos VNT (titer ≥1/8)											Total Pos (VNT)	Total samples
		1/8	1/12	1/16	1/24	1/32	1/48	1/64	1/96	1/128	1/192	1/256		
Doubt	4		1			7	1	5	3	10	5		32	36
Pos	0	1		1	1	9	5	20	3	15	6	28	89	89
Neg	49	2		2		7	2	5	1	2		2	23	72
Total	53	3	1	3	1	23	8	30	7	27	11	30	144	197

Neg: negative
Pos: positive
Doubt: doubtful

Table 4 Risk factors for being seropositive for Border Disease with corresponding number and proportion of positive samples (apparent seroprevalence), number of negative samples and associated Odds-ratio (with 95% CI)

	Positive samples (%)	Negative samples	p value	OR	95% CI
Climate					
arid	312 (65.5%)	164	0.0001	4.04	2.55–6.39
mediterranean	32 (32.0%)	68		1	
Landscape					
mountain	31 (44.3%)	39	0.005	0.49	0.29–0.80
plateau	313 (61.9%)	193		1	
Flock management					
sedentary	126 (52.5%)	114	0.0001	0.59	0.42–0.83
transhumant	218 (64.9%)	118		1	
Herd Composition					
mixed (goat or cattle)[a]	243 (59.1%)	168	0.644[a]	0.92[a]	0.63-1.33
sheep only	101 (61.2%)	64		1	
sheep with cattle[a]	206 (60.6%)	134	0.611[a]	1.09[a]	0.78-1.53
no cattle	138 (58.5%)	98		1	
sheep with goat	143 (52.8%)	128	0.001	0.58	0.41–0.81
no goat	201 (65.9%)	104		1	
Clinical diseases					
yes	126 (59.0%)	108	0.017	0.66	0.47–0.93
no	218 (63.7%)	124		1	

[a]Non significant

different countries to conduct seroprevalence surveys for pestivirus in small ruminants [4, 12, 13]. Our results indicate an estimated flock prevalence of 98.20% and an apparent within-flock prevalence of 60.17%. The true overall prevalence was estimated to be 68.20%. In Tunisia, similar results are reported, with 95% or 52/55 of positive flocks and an animal seroprevalence of 54% ± 4% [14]. Such high levels of prevalence were also found in France, where a recent study revealed that 38 sheep flocks tested in Ab-ELISA were positive in Border disease and individual seroprevalence reached 76.5% (95% CI = 74.2–78.8%) [15]. Other serological surveys carried out in Spain, Ireland, Austria, and India revealed a high seroprevalence of ruminant pestivirus at flock level with rates varying between 58 and 70% and at the individual level between 49.3 and 83% [13, 16–18].

Several factors may be in favor of the high rates observed in Algerian flocks and thus may participate in the dissemination of the virus: keeping animals in poor housing conditions, insufficient knowledge of livestock breeders about biosecurity rules, common use of transhumance and mixing flocks of different origins, lack of periodic laboratory investigations and illegal exchanges of animals from the neighboring countries. Our results were potentially biased by the fact that random selection was performed at the municipality level and not at the flock level, as there was no available sheep flock database. The selection of the sampled flock in each municipality was done by private vets and could be considered as a convenient sampling. To minimize this bias the vets were asked to select flocks as much representative as possible of the local context.

Regional differences in the seroprevalence

In this survey, we observed marked regional differences in the within-flock seroprevalence of BD in sheep with estimated rates between 18 and 100% depending on the department. However, our sampling design was calibrated to estimate with a reasonable precision the overall within-flock seroprevalence but cannot provide an accurate estimation at department level (this would have required a larger number of flocks in each department). Indeed, the differences observed between departments were not statistically significant except for Chlef (P = 18%; 95% CI 9.90–30.50) which has a lower prevalence compared to five other departments. The lower prevalence in this department can be explained by the fact that most flocks over there are sedentary, as grazing is available throughout the year thanks to favorable environment and Mediterranean climate. Such conditions can limit the contacts with potentially infected flocks. It has been reported that the transmission of the virus depends also on the degree of contacts between animals and may be more important in animals kept in buildings with nose-to-nose contact [19] than in animals that remain in the open air. A study in

Northern Ireland also reported significant regional variations in flock prevalence and attributed such differences to the levels of movement, differences between the regions in management practices, and the density of the populations of sheep in the flocks [13].

Virus circulation

The high seroprevalence rates observed in our study can be considered as indicative of a recent infection in some flocks given that only animals aged between 6 and 24 months were sampled. By this age, maternal antibodies have waned and the presence of antibodies is due to a recent exposure to a pestivirus [20]. Animals older than 2 years were not sampled, to exclude a bias due to the seropositivity in older animals which remain lifelong seropositive after seroconversion.

Vaccination

Vaccination of ruminants against pestivirus can also induce seroconversion but it is not practiced in Algeria. There is no standard vaccine for BDV, but a commercial killed whole-virus vaccine has been produced [21]. However, in Algeria there is extensive vaccination of small ruminants against sheep pox virus using a locally made vaccine prepared with cellular lineage resulting from a strain of sheep embryo. Isolation of pestivirus strains from several batches of anti- sheep pox vaccines has been reported in Tunisia [22], which could be at the origin of a wide spread of the virus in this country. According to OIE (World Organisation for Animal Health) [21], contamination of modified live virus vaccines by pestivirus have been found to be a cause of serious disease following their use in sheep (including sheep pox vaccine) and other livestock. However, in our study we did not observe a significant increase of BD seroprevalence in vaccinated flocks compared to unvaccinated ones.

Comparison between VNT and ab-ELISA

We used a BVDV-1 and BVDV-2 cattle isolate and a BDV sheep isolate for cross neutralization study. Previous studies using a commercially available indirect ELISA (SVANOVIR BD-Ab-ELISA; Svanova Biotech), comparing Ab-ELISAs to VNT have reported a sensitivity of 94.3% and 100% and a specificity of 93.7 and 100% for sheep and goats respectively from BD virus [23, 24]. In our study, 197 samples from 20 different flocks were tested in parallel with Ab-ELISA and VNT for BVDV-1, BVDV-2 and BDV. We observed low performances of the Ab-ELISA especially for the sensitivity estimated to be 84%. A possible explanation is that the commercial ELISA we used is more adapted to European strains of pestivirus and may not detect well BDV strains circulating in Algeria, as it is well known that there is a large antigenic variability in BDV strains generally [11].

The majority of the positive samples (103/144) tested in parallel by VNT for BDV, BVD-1 and BVD-2 had a VNT titer for BDV four folds higher than the titer for BVDV-1 or BVDV-2. We can therefore conclude that the prevailing pestivirus circulating in the sheep population in Algeria is Border disease virus rather than BVDV. Surprisingly, two of the sera tested in parallel had high titer for BVDV-2. These samples came from two different farms in two separate departments (Chlef and Saida); these flocks were sedentary but share grazing with cattle. BVDV-2 is a pestivirus usually specific to cattle and rarely identified in sheep. It was initially detected in cattle of North America [25] and later in other countries. In India, a cross neutralization study on sheep and goat samples exhibited a titer more than fourfold higher to BVDV-2 in one sheep and one goat [20]. Recently, a study in Spain revealed that six of eight fetuses / lambs were positive from BVDV-2 [26]. This virus may cause abortions, and probably be highly virulent, in naturally infected sheep. However, in most cases, the primary source of BVDV in non-bovid species is unknown, although direct contact with cattle appears to be the source of initial contamination [27].

Detection of PI animals by ag-ELISA and RT-PCR

Ag-ELISA and RT-PCR were performed in our study in order to detect PI animals but no viral antigen could be detected among the 689 samples tested, despite serological findings that showed the presence of recent infection in the flocks. There are several hypotheses to explain why we were not able to detect PI BDV among the tested sheep. First, only a limited number of animals younger than 6 months were tested in each flock ($n = 113$), which is the age category in which there is a greater chance to detect a PI. Given the low prevalence of PI sheep commonly observed and reported in previous studies, the probability to detect PI animals in a small sample size is low: in Austria, the PI prevalence was only 0.32% in sheep, and in Spain, it has been described a prevalence of 0.3; 0.6 and 0.24% [15, 28]. In addition, many lambs are slaughtered at a young age for economic purposes, decreasing the chance to detect young PI animals at the time of sampling. Finally, we did not observe typical clinical signs of BD such as nervous signs, paralysis and muscle tremor on the lambs sampled so we cannot exclude that most PI animals were dead or culled at the time of the sampling. With blood samples taken from young animals younger than six months, one could expect false negative Ag-ELISA results due to the presence of maternal antibodies derived from colostrum intake. However, these samples were tested in parallel by RT-PCR, a method which is not influenced by maternal antibodies when performed on full blood.

Risks factors

Several significant risk factors for high BDV seroprevalence were identified in our study. Due to the limited number of flocks and animals tested and the limited study zone, these risk factors should be considered as specific to the Algerian context and generalized cautiously to other endemic countries.

The seroprevalence (P) was significantly higher in inland areas with cold and arid climate characteristics (P = 65.5%; OR = 4.04) than in the coastal zone with Mediterranean climate (P = 32.0%). A similar observation was made in a seroprevalence study in Turkey [29]. Although pestiviruses are endemic in many countries with very different climatic conditions, one cannot exclude that climatic factors such as outside temperature or hygrometry could influence the survival and dissemination of the virus in the environment (feces, fomites) and have in impact on virus transmission. In our study, a significant lower seroprevalence was observed for sheep flocks raised in the mountainous regions (P = 44.3%; OR = 0.49) compared to flocks raised in the plateaus (P = 61.9%). Our results could be explained by the fact that flocks from the coastal zone are predominantly sedentary and therefore are rarely in contact with other flocks while in inland regions, flock movements are more intense, resulting in a higher infection rate.

Indeed, a significant lower seroprevalence was observed in flocks managed in a sedentary system (P = 52.5%; OR = 0.59) compared to transhumant flocks (P = 64.9%). Transhumance is a system widely practiced in Algeria, it concerns flocks located in the steppe region, where shepherds carry their flocks in the north of the country during summer season (May to September) for more pasture and return in autumn (October) to their farms. Another movement of transhumance is observed at the beginning of winter (second half of December) a little towards the south because of the enormous temperature decreases in the steppe region.

Transhumance has already been identified as a risk factor by previous studies. The seasonal migration of flocks and the use of communal pastures for grazing make the direct or indirect exposure to other species possible, including free-living ruminants [4, 15]. In another study carried out in Syria, it has been reported that transhumant flocks, especially those travelling long distances, have a significantly higher seroprevalence with an increase of 14% compared to sedentary flocks or those moving on short distances [30].

We examined in our study the possible impact of mixing sheep with other ruminant species (cattle and/or goat). Natural cases of pestiviruses transmission from cattle to sheep and vice versa have been reported [31] and the presence of sheep is a recognized as a risk factor for the introduction of BVDV into cattle herds [32, 33] and vice-versa [4]. However, the higher seroprevalence of seropositive beef and dairy herds in Northern Ireland (> 85%) suggests that the infection pressure is more important from cattle to sheep than from sheep to cattle [34]. A Swiss serological study confirms the fact that housing sheep and cattle separately significantly reduces the seroprevalence of BVDV infection in sheep but not of BDV [35]. In our study, there was no statistically significant association between serological status at the individual level and the presence or absence of cattle on farms. On the other hand, our study revealed a significantly lower seroprevalence in flocks where sheep were mixed with goats (P = 52.8%; OR = 0.58) compared to flocks without goats (P = 65.9%). Although PI goats infected with BDV have rarely been reported [36], this does not explain the apparent protecting factor of the presence of goat observed in our study. Other confusing factors, linked to the presence of goats, could explain this observation such as flock management or environmental conditions. Surprisingly, a significant lower seroprevalence was observed in flocks where other clinical diseases were reported by the owner (diarrhea, respiratory problems, weak lambs) (P = 63.7%; OR = 0.66) compared to flocks without clinical signs. However, the presence of other disease in our survey was based on the declarations of the breeders and not on clinical observations leading to possible bias. Moreover, in flocks with high BDV seroprevalence the disease can be considered as endemic and not in acute phase so that the immune status of animals will make the disease circulate at low noise compared to flocks that have more naïve animals. We did not find any statistical association between the occurrence of abortion cases and the BDV seroprevalence, but again this was based on the declarations of the breeders which could be biased. In Northern Ireland it is reported that among 186 fetuses serologically tested; only one was positive for BD, concluding that pestiviruses are not an important cause of sheep abortion in in this country [13]. A similar result was drawn from a study in Tunisia [14] which concludes that despite the high prevalence of BD, it is only implicated in the abortion syndrome of a single flock out of 20 tested.

Conclusion

This work is the first epidemiological study estimating the BDV seroprevalence in sheep flocks in Algeria. Our results provide serological evidence of widespread BDV infection in Algerian sheep population but also the presence of BVDV-2 infection. As a consequence, we can propose the following recommendations for the Algerian context: (i) a control program of pestivirus infections should be considered in sheep and other ruminant species; (ii) diagnostic tests to differentiate between BVDV-1, BVDV-2 and BDV need to be available as second line, since this can have an impact on disease control measures; (iii) locally produced live vaccine batches against other diseases should be controlled for the risk of pestivirus contamination even though vaccination did not appear as a risk factor in our study. This survey also shows the impact of some risk factors on the spread and maintenance of BDV infection, such as climate, landscape, flock management, flock composition and other concurrent diseases. On the other hand, this survey did not reveal any significant effect of the flock size, the presence of cattle, the introduction of new animals into flocks and the occurrence of abortions. Further studies aiming at the isolation of the circulating strains of Border Disease virus in Algerian sheep will help to better understand the origin and dynamics of pestivirus infections in the country.

Methods
Study site
The study was carried out in nine departments (regions) of Algeria, covering the geographical and climatic diversity of the country (Al Bayedh, Saida, Laghouat, Djelfa, Msila, El Bordj, Setif, Chlef and Tizi Ouzou). These departments were selected because of their relatively high density of small ruminants (Fig. 1). The sheep population in Algeria accounts for 80% of the total number of ruminants; it increased by more than 25%, from 21 million heads to 28 million between 2010 and 2014 [37], of

which 57% is present in the area investigated. This type of sheep production is more concentrated in the steppe zone (in the north-central part of the country). During warm season, the transhumance and nomadic activities are necessary especially from May to September when pastures can no longer satisfy the food requirements of the flocks.

Sampling
Sampling design
A two-stage cluster sampling was performed with a total of 576 animals (6–24 months) sampled from 56 sheep flocks originating from 56 different municipalities spread over 9 departments (Fig. 1). The sample size was calculated based on an estimated within-flock prevalence of 54% taking into account the prevalence found in a similar study in Tunisia [14] and a desired relative precision of 5%. The initial sample size ($n = 382$) was increased to 554 animals to take into account the design effect due to the cluster sampling with the following parameters: number of clusters (flocks) = 56; number of sampling units (animals) in each cluster = 10; intra-class correlation (ICC or ρ) = 0.05. We chose an ICC of 0.05 which is an intermediate value for a range of values (0.01–0.36) mentioned for BVDV in [38]. The number of flocks sampled in each department was proportionate to the number of ovine flocks estimated by department (based on

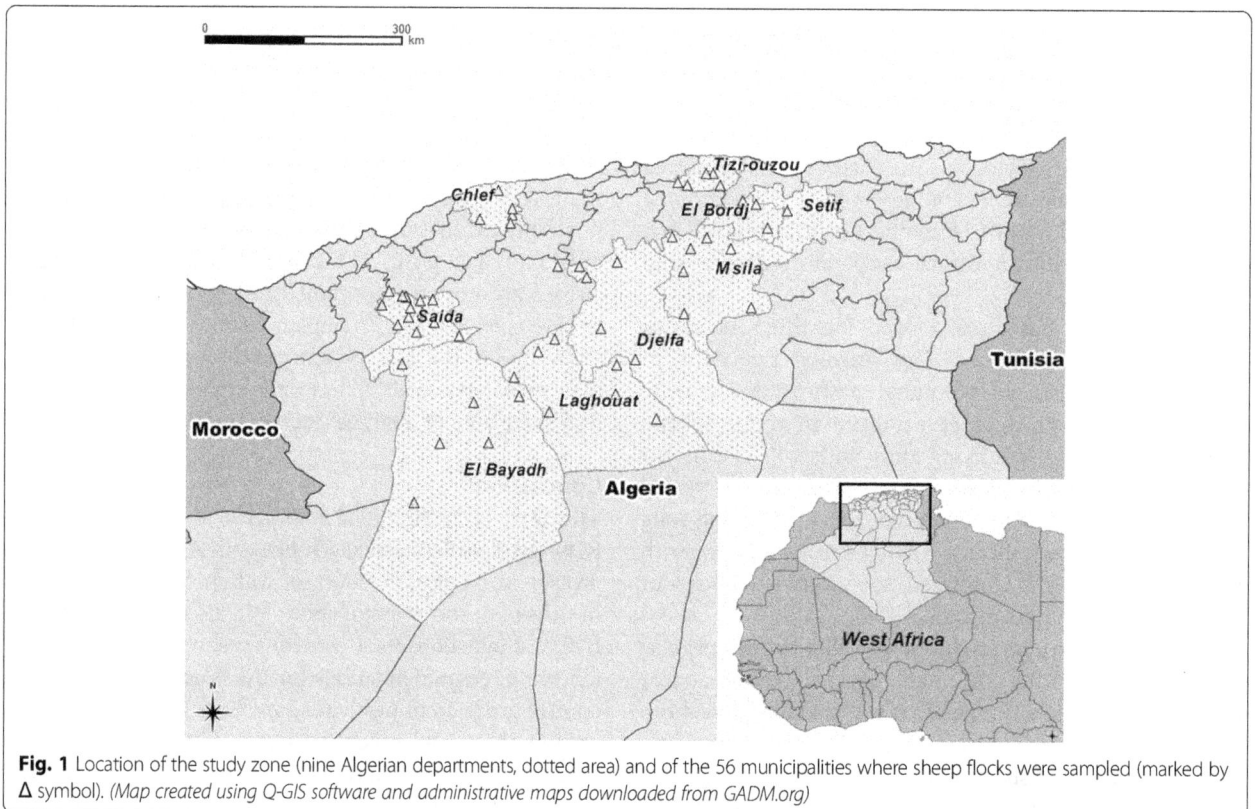

Fig. 1 Location of the study zone (nine Algerian departments, dotted area) and of the 56 municipalities where sheep flocks were sampled (marked by Δ symbol). *(Map created using Q-GIS software and administrative maps downloaded from GADM.org)*

data from regional agricultural services). As there was no detailed list of sheep flocks in Algeria we used instead the list of all municipalities from the studied regions ($n = 342$) as a sampling frame and performed a proportionate random sampling of 56 municipalities with SAS 9.2 (procsurveyselect, strata = region); one flock, supposed to be representative of the municipality, was then selected in each of the 56 municipalities by the private veterinarians working there.

Sampling collection

In each flock, at least 10 apparently healthy adult animals, aged between 6 and 24 months, were randomly selected by the veterinarians for blood collection to detect BD antibodies and virus, in addition to some young animals (less than 6 months), depending on the owner's cooperation, which were sampled for virus detection only. In total, 689 animals were sampled, 576 adults (serum and plasma), and 113 lambs (plasma only). The distribution and number of samples taken in each department are given in Table 1. Each sample (5 ml) was taken from the jugular vein using a vacutainer EDTA tube to collect plasma and a simple vacutainer tube to collect serum. Sera and plasmas were separated from the clotted blood by centrifugation at 1500 g for 15 min, aliquoted into sterile eppendorf tubes of 1 ml. The storage was realized in − 80 °C freezer. Each specimen was marked with a code comprising an individual sample number with the flock identity.

Survey

A questionnaire was conducted and discussed directly with the sheep owners, during the same visit as the blood sampling, to provide information concerning flocks' characteristics and epidemiological data: flock size and composition, animal movements and contacts, reproductive management, sanitary situation (abortion, other diseases, and vaccination). All answers were recorded on paper and later on registered in an Excel data base. The number and proportion of flocks according to the main investigated characteristics and management practices are summarized in Table 5.

Laboratory testing

ELISA for the detection of antibodies to ruminant pestivirus (ab enzyme linked Immuno sorbent assay)

The specific anti-pestivirus antibodies were measured in 576 sera from adult sheep using a commercially available ELISA kit (SERELISA® BVD NS2–3 (p80) Ab Mono Blocking, Synbiotics (Zoetis) according to the manufacturer's instructions. This kit allows the detection of anti-BVDV and anti-BDV antibodies in ruminants. Optical density (OD) was measured in bichromatism at 450 and 630 nm. Results are expressed as competition percentage resulting from the difference of OD between the negative control and the sample reported to the difference of

Table 5 Distribution of the investigated flocks ($n = 56$) according to the main investigated characteristics (number and percentage)

Parameter	Category	N Flocks (%)
Climate	Arid	46 (82%)
	Mediterranean	10 (18%)
Landscape	Mountain	7 (12.5%)
	Plateau	49 (87.5%)
Breed	El Hamra + other	27 (48%)
	OuledDjellal + other	22 (39%)
	Local breed	2 (4%)
	Rimbi	5 (9%)
Herd size	< 50	8 (14%)
	51–100	19 (34%)
	101–200	20 (36%)
	> 200	9 (16%)
Management	Sedentary	24 (43%)
	Transhumant	32 (57%)
Herd species	Mixed (sheep/goat/cattle)	40 (71%)
	Ovine only	16 (29%)

OD between the negative control and the positive control. According to the manufacturer's instructions, a sample was considered positive if the competition percentage was superior or equal to 40%, negative below 20% and doubtful between 20 and 40%.

Virus neutralization test

Among the 576 sera tested in Ab ELISA, 197 samples (including positive, doubtful and negative samples in Ab ELISA) were tested in parallel for the presence of neutralizing antibodies against BVDV-1, BVDV-2 and BDV using two strains of BVDV (BVDV-1 strain NADL [39], BVDV-2 strain 3534 [40] and the BDV strain AV [41]. The samples were inactivated at 56 °C for 30 min before testing. The inactivated sera were then diluted in minimum essential medium (MEM) in a two two-fold dilution series starting from 1:5 dilution for BVDV-1 and BVDV-2 and from 1:2 for BDV. A fixed virus dose containing 100 TCID50/50 µl (between 30 and 300 TCID50) was incubated for 2 h at 37 °C with each dilution in an antibiotic enriched growth medium (i.e. penicillin, gentamicin and amphotericin B). MDBK cells (ATCC Number CCL-22) (3x10E7 cclls/100 µl) were added and the cultures were grown for 72H at 37 °C in a CO2 incubator. All sera were tested in duplicate. Viruses were titrated in all assays. After incubation the cell cultures were evaluated directly for cytopathogenic effects by optical microscopy (BVDV-1) or after immunolabelling with an anti-pestivirus polyclonal serum (BVDV-2 and BDV). The virus neutralizing titers were calculated according to the Reed-Muench method [42]. Titers were expressed as the reciprocal of the highest serum dilution yielding virus growth neutralization and considered as positive for BVDV and for BDV when greater than or equal to 1/10 or 1/8 respectively.

ELISA for the detection of ruminant pestivirus antigens (ag enzyme linked Immuno sorbent assay)

Plasma collected from the 689 blood samples were tested for the presence of pestivirus antigen using the SERELISA® Kit BVD NS2–3 (p80) Ag Indirect Mono, Synbiotics (Zoetis). This kit allows the detection of BVDV and BDV antigens in individual samples from PI animals, using a monocupule indirect immuno-enzymatic technique for antigen detection (non-structural protein NS2–3 (p80)/ 125 common to all strains of BVD and BD viruses). OD was measured in bichromatism at 450 and 630 nm. Results are expressed as an index = 0.5 x OD sample – OD Positive control (P). Any plasma sample having an index ≥ (0.15 x OD P) was considered positive. Any plasma sample with an index < (0.3 x OD P) was considered negative. Any plasma sample with an index between (0.15 x OD P) and (0.3 x OD P) was considered doubtful according to the manufacturer's instructions.

RT-PCR (real time-polymerase chain reaction)

RT-PCR were performed on pools of plasma samples from 10 different animals which were constituted by mixing together 100 µl of each individual sample. RNA (Ribonucleic acid) was extracted from each pool using a volume of 100 µl. Extraction was performed with the QIAamp Viral RNA Mini Kit (Qiagen GmbH, Hilden, Germany) according to the manufacturer's instructions. Four µl of the total extracted RNA were used for the reverse transcription in the presence of hexanucleotides [43]. For real-time PCR amplification, 5 µl of the resulting cDNAs were included in the reaction mix. The primers (F2: CTCGAGATGCCATG TGGAC and PESTR: CTCCATGTGCCATGTACAGCA) and TaqMan probes used in this study targeted the 5′UTR conserved regions of BD and BVDV genotype 1 (probe BVDV-1: [5′] FAM-CAGCCTGATAGGGTGCTGCAGAGG C-TAMRA [3′]) and of BVDV genotype 2 (probe BVDV-2: [5′] VIC-CACAGCCTGATAGGGTGTAGCAGAGACCT G-TAMRA [3′]) [44]. PCR reaction was run in 25 µl containing 2X FastStart DNA Taqman probe Master Mix (LifeScience), 450 nM of both primers and 50 nM of both fluorescent probes. The PCR conditions were as followed: 10 min at 95 °C and 45 cycles with 15 s at 95 °C and 45 s at 60 °C. Fluorescent measurements were carried out during the elongation step.

Statistical analysis
Seroprevalence estimation

In order to take into account the clustering effect (~ 10 animals were sampled in each flock), the within-flock seroprevalence was estimated at the overall level and at department level with a generalized estimating equation model (GEE) using SAS 9.2 software ("proc genmod"). In this model, the flock was taken as repeated subject, the department as an independent variable and the prevalence was estimated as the predictive probability to be seropositive; an exchangeable correlation matrix was assumed. Doubtful Ab-ELISA results, based on the cut-off recommended by the kit manufacturer, were considered in our analysis as positive, given that we found that most samples that were doubtful in Ab-ELISA were positive with the VNT (see results section). These prevalence rates were apparent prevalence (Pa), not taking into account the sensitivity (Se) and specificity (Sp) of the Ab-ELISA. The overall true seroprevalence (Pt) was then calculated taking into account the Se and Sp that we estimated relatively to the VNT (see results section). The true overall prevalence and 95% CI (Rogan and Gladen method) were calculated using the on-line epidemiological calculator EpiTools (Estimated true prevalence and predictive values from survey testing, [45]).

Comparison between ab-ELISA and VNT results

A comparative study was performed on part of the serum samples (196/576) to estimate the performances of the Ab-ELISA compared to the VNT for BDV, BVDV-1 and BVDV-2. These samples included negative, doubtful and positive sera in Ab ELISA originating from 20 different flocks and 8 different departments. The relative sensitivity and specificity of the ELISA were calculated as the number of positive or negative samples in ELISA divided by the number of positive or negative samples in VNT, respectively. The Cohen's kappa coefficient test was used to measure the agreement between Ab-ELISA and the VNT and was calculated using the on-line epidemiological calculator EpiTools [45]. Doubtful Ab-ELISA results were considered as positive in the calculations mentioned above.

Descriptive statistics and risk factors

Descriptive statistics were performed to establish the proportion of flocks according to the different characteristics studied through the survey and the corresponding proportion of seropositive animals. Doubtful Ab-ELISA results were considered as positive as explained above. The following parameters, considered as potential risk factors, were compared in terms of seroprevalence: climate (arid vs/ Mediterranean), landscape (mountain vs/ plateau) flock management (sedentary vs/transhumant), flock size (< 100 vs > 100), flock composition (sheep/cattle/ goats), sheep breed, purchase of breeding females (yes/no), purchase of fattening lambs (yes/no), origin of breeding rams (external vs/own ram), contacts with other flocks at pasture, contacts with wild animals, number of abortions, clinical diseases, vaccination. A chi-squared test was used to detect significant differences in seroprevalence for the studied characteristics; a probability of less than 5% was considered as statistically significant. The odds ratio (OR) and chi-square were calculated with the software XLSTAT version 2014 to quantify the association between positive Ab-ELISA and the identified risk factors. The 95% CI were calculated using the Miettinen method.

Abbreviations
Ab-ELISA: Antibody enzyme-linked immunosorbent assay; Ag-ELISA: Antigen enzyme-linked immunosorbent assay; BD: Border Disease; BDV: Border Disease Virus; BVDV: Bovine Viral Diarrhea virus; C.I.: Confidence interval; CSFV: Classical Swine Fever virus; GEE: Generalized estimating equation model; ICC: Intra-class correlation; MDBK cells: Madin-Darby Bovine Kidney cells; MEM: Minimum essential medium; OD: Optical Density; OIE: World Organisation for Animal Health; OR: Odds ratio; P: Seroprevalence; Pa: Apparent seroprevalence; PI: Persistently Infected; Pt: True seroprevalence; RNA: Ribonucleic acid; RT-PCR: Real Time-polymerase chain reaction; Se: Sensitivity; Sp: specificity; TCID: Tissue culture infectious dose 50%; VNT: Virus neutralization test

Acknowledgements
We are most grateful to the staff of the laboratory of virology at Sciensano for their precious help and support, in particular: Celia Thoraval, Muriel Verhoeven, Laurent Rosar, and Annebel De Vleeschauwer. We would like to thank Dr. Bekara Amine for her advice in the sampling plan.
Special thanks are due to farmers and veterinary practitioners for their cooperation and their valuable help during samples collection. Feknous Naouel was supported by Institute of Veterinary Sciences of Blida, Algeria and Sciensano in Brussels, Belgium.

Funding
Sciensano, Infectious animal diseases directorate, Service of enzootic, vector-borne and bee diseases,
Groeselenberg 99, 1180 Brussels, Belgium.
LBRA, Institute of Veterinary Sciences, University Blida 1, Algeria.

Authors' contributions
FN and HJB designed the study and contributed equally under the supervision of CAB an BA. KH facilitated the use of specimens collected. TM and FN performed the laboratory analyses. HJB performed the data analyses. FN and HJB wrote the manuscript. CAB, TM, KH and BA were involved in some sections of the draft manuscript and revised it critically. All co-authors read and approved the final manuscript.

Ethics approval and consent to participate
The collection of blood samples and the recording of herd data was carried out with the verbal consent of the owners of the animals, who were assisting to hold their animals during sampling. Samples were taken by qualified private and state veterinarians, in a professional manner and respecting animal welfare. No animals were euthanized and the study was approved by the ethical review board of the Institute of veterinary medicine of the University of Blida in Algeria.

Consent for publication
Not Applicable.

Competing interests
The authors declare that they have no competing interests.

Author details
[1]LBRA, Institute of Veterinary Sciences, Saad Dahlab University, Soumaa Road, BP 270, 09000 Blida, Algeria. [2]Sciensano, Infectious animal diseases directorate, Service of enzootic, vector-borne and bee diseases, Groeselenberg 99, 1180 Brussels, Belgium. [3]ENSV, National superior veterinary school, Bab ezzouar, El allia, Algeria.

References
1. Simmonds P, Becher P, Bukh J, Gould EA, Meyers G, Monath T, et al. ICTV report consortium. J Gen Virol. 2017;98:2–3.
2. Becher P, Orlich M, Kosmidou A, Konig M, Baroth M, Thiel MJ. Genetic diversity of pestiviruses: identification of novel groups and implications for classification. Virology. 1999;262:64–71.
3. Sullivan DG, Akkina RK. A nested polymerase chain reaction assay to differentiate pestiviruses. Virus Res. 1995;38:231–9.
4. Krametter-Froetscher R, Loitsch A, Kohler H, Schleiner A, Schiefer P, Möstl K, et al. Serological survey for antibodies against pestiviruses in sheep in Austria. Vet Rec. 2007;160:726–30.
5. Vilcek S, Nettleton PF. Pestiviruses in wild animals. Vet Microbiol. 2006;116: 1–12.
6. Nettelton PF, Entrican G. Ruminant pestiviruses. Br Vet J. 1995;151:615–42.
7. Paton DJ, Carlsson U, Lowings JP, Sands JJ, Vilcek S, Alenius S. Identification of herd-specific bovine viral diarrhoea virus isolates from infected cattle and sheep. Vet Microbiol. 1995;43:283–94.
8. Braun U, Reichle SF, Reichert C, Hässig M, Stalder HP, Bachofen C, et al. Sheep persistently infected with border disease readily transmit virus to calves seronegative to BVD virus. Vet Microbiol. 2014;168(1):98–104.
9. Braun U, Hilbe M, Janett F, Hässig M, Zanoni R, Frei S, et al. Transmission of border disease virus from a persistently infected calf to seronegative heifers in early pregnancy. BMC Vet Res. 2015;11:43.
10. Monies RJ, Paton DJ, Vilcek S. Mucosal disease-like lesions in sheep infected with border disease virus. Vet Rec. 2004;155:765–9.
11. Nettleton PF, Gilary AJ, Russo P. Delissi E; border disease of sheep and goats. Vet Res. 1998;29:327–240.
12. Mishra N, Rajukumar K, Tiwari A, Nema RK, Behera SP, Satav JS, et al. Prevalence of bovine viral diarrhoea virus (BVDV) antibodies among sheep and goats in India. Trop Anim Health Prod. 2009;41:1231–9. https://doi.org/10.1007/s11250-009-9305-z.
13. Graham DA, Calvert V, German A, McCullough SJ. Pestiviral infections in sheep and pigs in Northern Ireland. Vet Rec. 2001;148:69–72.
14. Rekiki A, Thabti F, Dlissi I, Russo P, Sanchis R, Pepin M, et al. Enquête sérologique sur les principales causes D'avortements infectieux chez les petits ruminants en Tunisie. Rev Méd Vét. 2005;156:7,395–401.
15. Martin C, Duquesne V, Adam G, Belleau E, Gauthier D, Champion JL, et al. Pestiviruses infections at the wild and domestic ruminants interface in the French southern Alps. Vet Microbiol. 2015;175:341–8.
16. Valdazo-González B, Alvarez-Martinez M, Greiser-Wilke I. Genetic typing and prevalence of border disease virus (BDV) in small ruminant flocks in Spain. Vet Microbiol. 2006;117:141–53.
17. Krameter-Froetscher R, Duenser M, Preyler B, Theiner A, Benetka V, Moestl K, et al. Pestivirus infection in sheep and goats in West Austria. Vet J. 2010;186:342–6.
18. Mishra N, Rajukumara K, Vilcek S, Kalaiyarasua S, Beheraa SP, Dubeya P, et al. Identification and molecular characterization of border disease virus (BDV) from sheep In India. Comp Immunol Microbiol Infect Dis. 2016;44:1–7.
19. Nettleton PF, Gilmour JS, Herring AJ, Sinclair AJ. The production and survival of lambs persistently infected with border disease virus. Comp Immunol Microbiol Infect Dis. 1992;15(3):179–88.
20. Mishra N, Rajukumar K, Vilcek S, Tiwari A, Satav JS, Dubey SC. Molecular characterization of bovine viral diarrhea virus type 2 isolate originating from a native Indian sheep (Oviesaries). Vet Microbiol. 2008;130:88–98.
21. World organisation for animal health (OIE). Ovidae and caprinae. Border disease. OIE terrestrial manual. 2017. Section 2.7. Chapter 2.7.1. 1–13. Available at: http://www.oie.int.

22. Thabti F, Fronzaroli L, Dlissi E, Guibert JM, Hammami S, Pepin M, et al. Experimental model of border disease virus infection in lambs: comparative pathogenicity of pestiviruses isolated in France and Tunisia. Vet Res. 2002;33:35–45.

23. Krametter-Froetscher R, Loitsch A, Moestl K, Sommerfeld-Stur I, Baumgartner W. Seroprävalenz von Border Disease und Boviner Virus diarrhöbeiSchafen und Ziegen in ausgewählten Region en Österreich. Wien Tierärztl Monatsschr. 2005;92:238–44.

24. Mohammadi A, Ghane M, Kadivar E, Ansari-Lari M. Seroepidemiology of border disease and risk factors in small ruminants of shiraz suburb, Fars Province. South of Iran Global Veterinaria. 2011;6:383–8.

25. Pellerin C, Van den Hurk J, Lecomte J, Tijssen P. Identification of a new group of bovine viral diarrhea virus strains associated with severe outbreaks and high mortalities. Virology. 1994;203:260–8.

26. Elvira Partida L, Fernandez M, Gutiérrez J, Esnal A, Benavides J, Pérez V, de la Torre A, Alvarez M, Esperon F. Detection of bovine viral Diarrhoea virus 2as the cause of Abortion outbreaks on commercial sheep flocks. Transbound Emerg Dis. 2017;64:19–26.

27. Nelson DD, Duprau JL, Wolff PL, Evermann JF. Persistent bovine viral diarrhea virus infection in domestic and wild small ruminants and camelids including the mountain goat (Oreamnosamericanus). Front Microbiol. 2016;6:1415.

28. Valdazo-González B, Alvarez-Martınez M, Sandvik T. Prevalence of border disease virus in Spanish lambs. Vet Microbiol. 2008;128:269–78.

29. Okur-Gumusova S, Yazici Z, Albayarakl H. Pestivirus seroprevalence in sheep populations from inland and coastal zones of Turkey. Revue Méd Vét. 2006; 157:595–8.

30. Tabbaa D, Giangaspero M, Nishikawa H. Seroepidemiological survey of border disease (BD) in Syrian Awassi sheep. Small Rumin Res. 1995;15:273–7.

31. Paton DJ, Sands JJ, Lowings JP, Smith JE, Ibata G, Edwards S. A proposed division of the pestivirus genus using monoclonal antibodies, supported by cross-neutralisation assays and genetic sequencing. Vet Res. 1995;26:92–109.

32. Lindberg ALE, Alenius S. Principles for eradication of bovine viral diarrhea virus (BVDV) infections in cattle populations. Vet Microbiol. 1999;64:197–222.

33. Kaiser V, Nebel L, Schüpbach-Regula G, Zanoni RG, Schweizer M. Influence of border disease virus (BDV) on serological surveillance within the bovine virus diarrhea (BVD) eradication program in Switzerland. BMC Vet Res. 2017;13:21.

34. McCullough SJ, Adair MB, McKillop ER. A survey of serum antibodies to respiratory viruses in cattle in Northern Ireland. Ir Vet J. 1987;41:342–4.

35. Braun U, Bachofen C, Schenk B, Hässig M, Peterhans E. Investigation of border disease and bovine virus diarrhoea in sheep from 76 mixed cattle and sheep farms in eastern Switzerland. Schweiz Arch Tierheilkd. 2013;155:293–8.

36. Rosamilia A, Grattarola C, Caruso C, Peletto S, Gobbi E, Tarello V, et al. Detection of border disease virus (BDV) genotype 3 in Italian goat herds. Vet J. 2014;199:446–50.

37. FAO. FAOSTAT database collections. Food and Agriculture Organization of the United Nations. Rome; 2014. Accessed 03 July 2017 URL: http://faostat.fao.org

38. McDermott JJ, Schukken YH. A review of methods used to adjust for cluster effects in explanatory epidemiological studies of animal populations. Prev Vet Med. 1994;18:155–73.

39. Gutekunst DE, Malmquist WA. Separation of a soluble antigen and infectious particles of bovine viral diarrhea viruses and their relationship to hog cholera. Can J Comp Med Vet Sci. 1963;27(5):121–3.

40. Letellier C, Pardon B, Van der Heyden S, Deprez P. Circulation in Belgium of a bovine viral diarrhoea virus type 2 closely related to north American hypervirulent viruses. Vet Rec. 2010;166(20):625–6.

41. Dubois E, Russo P, Prigent M, Thiéry R. Genetic characterization of ovine pestiviruses isolated in France, between 1985 and 2006. Vet Microbiol. 2008; 130(1–2):69–79.

42. Reed L, Muench H. A simple method of estimating fifty per cent endpoints. Am J Hyg. 1938;27:493–7.

43. Letellier C, Kerkhofs P, Wellemans G, Vanopdenbosch E. Detection and genotyping of bovine diarrhea virus by reverse transcription-polymerase chain amplification of the 5′ untranslated region. Vet Microbiol. 1999;64(2–3):155–67.

44. Letellier C, Kerkhofs P. Real-time PCR for simultaneous detection and genotyping of bovine viral diarrhea virus. J Virol Methods. 2003;114:21–7.

45. Sergeant ESG. Epitools epidemiological calculators: Ausvet Pty Ltd; 2017. Available at: http://epitools.ausvet.com.au

20

In vitro immune responses of porcine alveolar macrophages reflect host immune responses against porcine reproductive and respiratory syndrome viruses

Nadeem Shabir[1,2†], Amina Khatun[1†], Salik Nazki[1], Suna Gu[3], Sang-Myoung Lee[3], Tai-Young Hur[4], Myoun-Sik Yang[1], Bumseok Kim[1] and Won-Il Kim[1*] (iD)

Abstract

Background: Currently, an in vitro immunogenicity screening system for the immunological assessment of potential porcine reproductive and respiratory syndrome virus (PRRSV) vaccine candidates is highly desired. Thus, in the present study, two genetically divergent PRRSVs were characterized in vitro and in vivo to identify an in vitro system and immunological markers that predict the host immune response. Porcine alveolar macrophages (PAMs) and peripheral blood mononuclear cells (PBMCs) collected from PRRSV-negative pigs were used for in vitro immunological evaluation, and the response of these cells to VR2332c or JA142c were compared with those elicited in pigs challenged with the same viruses.

Results: Compared with VR2332c or mock infection, JA142c induced increased levels of type I interferons and pro-inflammatory cytokines (TNF-α, IL-1α/β, IL-6, IL-8, and IL-12) in PAMs, and these elevated levels were comparable to the cytokine induction observed in PRRSV-challenged pigs. Furthermore, significantly greater numbers of activated CD4$^+$ T cells, type I helper T cells, cytotoxic T cells and total IFN-γ$^+$ cells were observed in JA142c-challenged pigs than in VR2332c- or mock-challenged pigs.

Conclusions: Based on these results, the innate immune response patterns (particularly IFN-α, TNF-α and IL-12) to specific PRRSV strains in PAMs might reflect those elicited by the same viruses in pigs.

Keywords: PRRSV, Immune response, Alveolar macrophages, Peripheral blood mononuclear cells, Flow cytometry

Background

Porcine reproductive and respiratory syndrome virus (PRRSV), a positive-sense RNA virus belonging to the family *Arteriviridae* of the order *Nidovirales*, exerts a significant economic impact on the swine industry worldwide [1–3], with an estimated annual loss of at least $664 million in the USA alone [4]. PRRSV possesses a 15-kb polycistronic genome containing two large open reading frames (ORFs 1a and 1b) that encode non-structural proteins (NSPs) and eight structural protein-encoding ORFs (2a, 2b, 3, 4, 5a, 5, 6 and 7) [5, 6]. PRRSV is broadly classified into two divergent genotypes, European (Type I) and North American (Type II), with Type II PRRSVs further sub-divided into at least nine different genetic lineages with inter-lineage genetic distances varying from 11 to 18% [7, 8]. The high genetic variability that exists between different PRRSV strains has resulted in increased antigenic and immunological diversity [9, 10]. This diversity among PRRSVs not only hinders our understanding of PRRSV-induced immunity in pigs but also poses a serious hurdle for the selection of immunogenic strains for vaccine development [11]. Moreover, due to the lack of immunological classification of divergent PRRSV strains and reliable parameters for predicting vaccine protection, the immunogenicity and protective efficacy of PRRSV vaccine

* Correspondence: kwi0621@jbnu.ac.kr
†Nadeem Shabir and Amina Khatun contributed equally to this work.
[1]College of Veterinary Medicine, Chonbuk National University, 79 Gobong-ro, Iksan, Jeonbuk, Korea

candidates can be only evaluated through challenge experiments in pigs [12].

The immune responses elicited by PRRSV in vitro and in vivo have been well-studied, but researchers have not reached a consensus on immune response patterns. PRRSV reportedly up-regulates IL-10 [13–15] and induces insignificant amounts of or inhibits the expression of type I interferons (IFNs) and TNF-α in vitro and/or in vivo [16–20]. However, some in vitro and in vivo studies have also demonstrated that the induction of these cytokines is strain dependent [21–24]. According to the majority of studies, PRRSVs induce IL-8 [25, 26], and the in vitro/ex vivo induction of IL-6 by PRRSV has also been reported [27–29].

Our current understanding of cell-mediated immune responses to PRRSV is unclear. IFN-γ has been used as a marker of Th1 polarization during PRRSV infection [30, 31]. Cytotoxic T cell activity against PRRSV-infected macrophages was only induced at 49 days post-infection [32], and weak PRRSV-specific cytotoxic T cell responses (CD8+IFN-γ+) were reported in another study [33]. Furthermore, PRRSV infection increases the number of TGF-β producing regulatory T cells (T_{Regs}) [34]. In addition, monocyte-derived dendritic cells (MoDCs) infected with PRRSV do not increase the frequency and proliferation of T_{Regs} in an in vitro co-culture system [35]. The negligible innate immune responses induced by PRRSV may severely affect adaptive immunity; along with other irregular adaptive immune processes, this impaired response leads to overall immune inefficiency and co-infections [36]. Therefore, it is important to understand early innate immunity events to gain insight into the immune response to PRRSV infection [37, 38].

The role of neutralizing antibodies (NAbs) during PRRSV infection is not completely understood [39]. PRRSV NAbs usually appear during the third or fourth week post-infection [36], and an NAb titre of 1:8 is sufficient to block viremia in PRRSV-infected pigs. However, NAb induction may be dependent on the strain of PRRSV [40, 41].

Given the strain-dependent elicitation of innate, cell-mediated and humoral immune responses by PRRSV, multiple virus strains should be investigated to gain a better understanding of PRRSV immunobiology [9]. Thus, there is a need to select an in vitro system to immunologically classify diverse PRRSVs and to predict the protection and immune responses induced by vaccine candidates. The majority of previous studies examining PRRSV-induced immune responses have used a single PRRSV strain that was tested either in vitro or in vivo but rarely both. These studies also failed to take into account the effects of PRRSV strain diversity, immune evaluation systems, or both when drawing general conclusions regarding PRRSV immunobiology. To fill in these gaps, the current study focused on two important goals: i) in vitro and in vivo characterization of the immune responses elicited by two divergent Type II PRRSV strains, and ii) identification of a feasible in vitro model or specific immunological markers to predict the in vivo immune responses induced by a PRRSV strain.

Results

PRRSVs replicated at similar levels in PAMs and pigs

A multi-step growth curve for PRRSV replication in PAMs was constructed by measuring viral titres in cell lysates at 6, 12, 24, 36 and 48 h post-infection (hpi). Slightly higher levels of viral replication were observed for JA142c than for VR2332c in PAMs (Fig. 1a) and in pigs (Fig. 1b), but these differences were not statistically significant. Neither strain produced viral titres at 6 hpi in PAMs, but JA142c and VR2332c exhibited titres of $10^{2.5}$ and $10^{1.5}$ $TCID_{50}$/mL at 12 hpi, gradually increasing to $10^{3.75}$ and $10^{3.5}$ $TCID_{50}$/mL, respectively, at 48 hpi. In a pig challenge model, the animals were tested for viremia at 0, 3, 7, 14, 21 and 28 dpc. The mean levels of viremia in

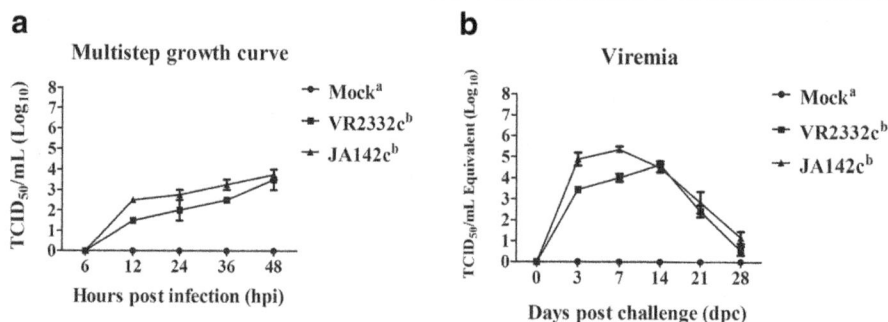

Fig. 1 PRRSV replication in PAMs and pigs. **a** Multi-step growth curve of PRRSVs in PAMs ($n = 6$) at 6, 12, 24, 36 and 48 hpi, as determined by titration using MARC-145 cells. **b** Viral loads in pig sera ($n = 5$) at 0, 3, 7, 14, 21 and 28 dpc, as determined by a quantitative real-time PCR for PRRSV. The bars represent the means, and the error bars represent the standard errors of the mean (SEM). Bars showing different letters represent values that differ significantly from each other ($p < 0.05$)

pigs challenged with VR2332c or JA142c peaked at 14 and 7 dpc, with values of $10^{4.6}$ and $10^{5.3}$ TCID$_{50}$/mL, and then gradually decreased to $10^{0.5}$ and $10^{1.1}$ log$_{10}$TCID$_{50}$/mL, respectively. Overall, JA142c and VR2332c demonstrated similar levels of replication in pigs and PAMs.

JA142c exhibited increased cytokine transcript and protein levels in PAM and PBMC cultures

Cytokine levels in PAMs, PBMCs and cell supernatants or lysates were evaluated by real-time PCR and/or ELISA. Compared to those of mock infection or VR2332c, JA142c and poly I:C induced significantly increased transcript levels of IFN-β and pro-inflammatory cytokines (IL-1α, IL-1β, IL-6, IL-8, IL-12, and TNF-α) at either two or all three time points assayed (12, 24 and 36 hpi) (Fig. 2a). Furthermore, significantly higher TNF-α and IFN-α levels were observed in the cell lysates of JA142c-infected PAMs than in the cell lysates of the mock- or VR2332c-infected groups for at least two out of

three time points (Fig. 2b). However, there was no significant difference in IL-12 protein expression in PAM lysates between groups. In naïve PBMCs, JA142c induced significantly higher IL-1α, IL-1β and TNF-α mRNA transcript levels at 12 and/or 36 hpi (Fig. 3a). Further, TNF-α protein expression in JA142c-stimulated naïve PBMCs was significantly higher at 12 and 36 hpi (Fig. 3b).

JA142c induced higher levels of type I IFNs and pro-inflammatory cytokines in pigs

Regarding cytokine mRNA transcript levels in PBMCs collected from virus-challenged pigs, JA142c-infected pigs had significantly higher transcript levels of IFN-β and IL-6 that pigs subjected to other treatments at 7 or 14 dpc. The transcript levels of IL-8, IL-12 and IFN-γ at 7, 21 or 28 dpc were higher in JA142c-infected pigs than in the other groups, but these differences were not statistically significant (Fig. 4). Serum cytokines in pigs were evaluated by performing ELISA (Fig. 5).

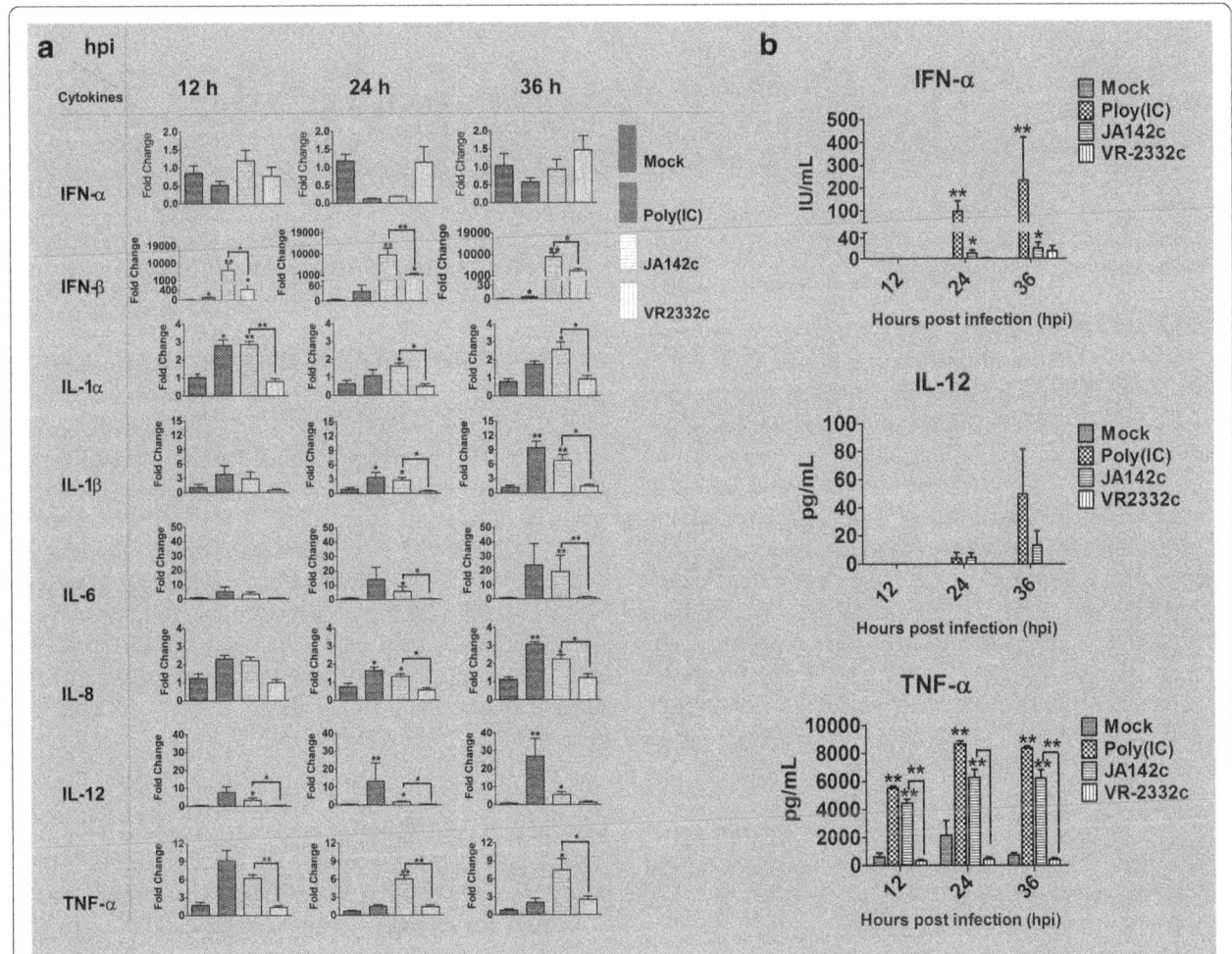

Fig. 2 Cytokine expression in PAM cultures. **a** mRNA and **b** protein expression of cytokines in PAMs and supernatants ($n = 6$), as determined by real-time PCR and ELISA, respectively, at 12, 24 and 36 hpi. Asterisks indicate significant differences in the cytokine expression induced by each virus or stimulant compared with mock treatment or significant differences between JA142c and VR2332c (* indicates $p \leq 0.05$, ** indicates $p \leq 0.01$)

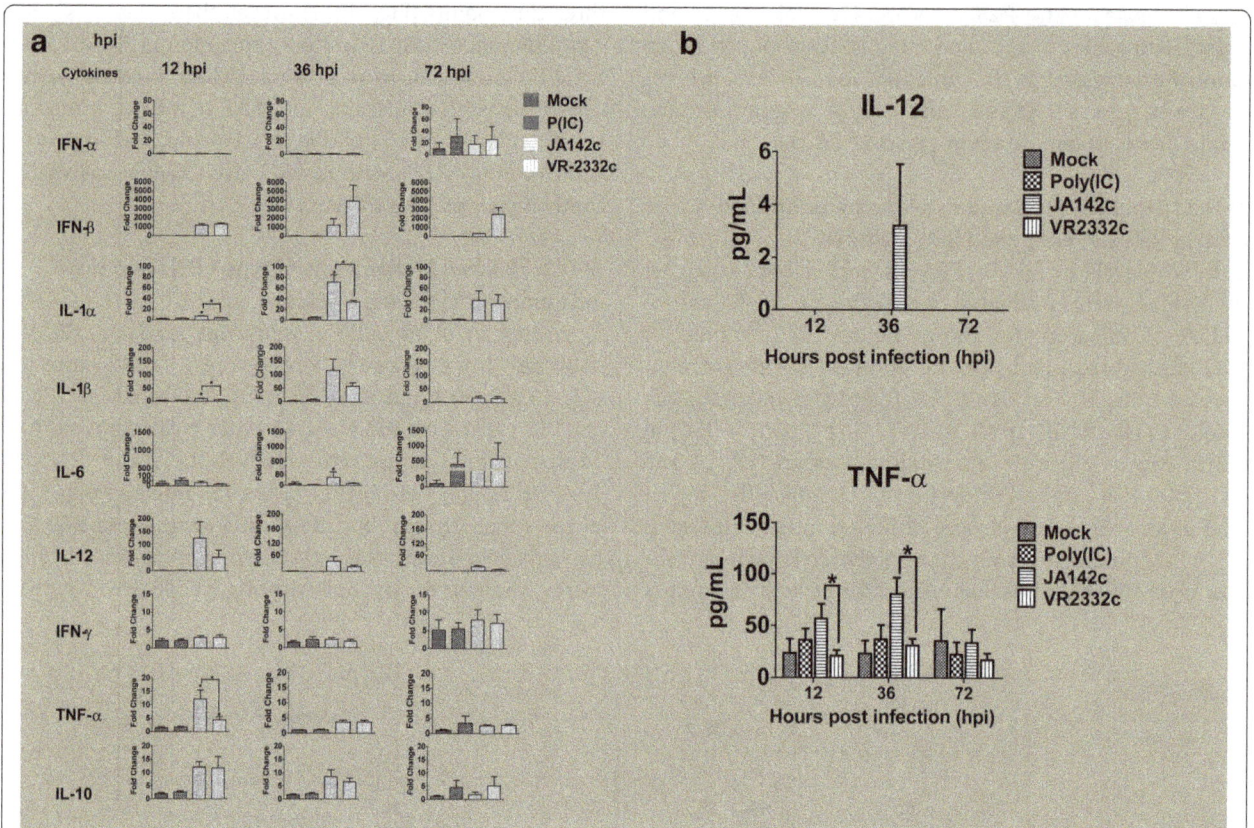

Fig. 3 Cytokine expression in PBMC cultures. **a** mRNA and **b** protein expression of cytokines in PBMCs and supernatants (n = 6), as determined by real-time PCR and ELISA, respectively, at 12, 36 and 72 hpi. Asterisks indicate significant differences in cytokine expression induced by each virus or stimulant compared with mock treatment or significant differences between JA142c and VR2332c (* indicates $p \leq 0.05$, ** indicates $p \leq 0.01$)

Although a significantly stronger induction of serum IFN-α was observed in VR2332c-infected pigs than in mock-infected pigs at 3 dpc, JA142c induced serum IFN-α levels that were significantly and consistently higher than those in either the VR2332c- or mock-infected groups until 14 dpc. Moreover, serum pro-inflammatory cytokine induction was enhanced. Specifically, TNF-α and IL-12 levels were significantly higher in JA142c-infected pigs from 7 to 14 dpc than in VR2332c- and mock-infected pigs; however, comparable levels of these cytokines were found in JA142c- and VR2332c-infected pigs at 21 and 28 dpi. JA142c also induced significantly higher serum levels of IL-4 at 21 dpc than did VR2332c.

A type I IFN bioassay using sera from virus-challenged pigs at 3 dpc revealed significantly higher antiviral activity in JA142c-challenged pigs than in mock- and VR2332c-challenged pigs (Fig. 6).

JA142c induced an enhanced cell-mediated immune response in pigs

As shown in Fig. 7, JA142c induced a significantly higher proportion of activated CD4$^+$ T cells (CD4$^+$CD25$^+$FoxP3$^-$)

at 21 and 28 dpc than did mock or VR2332c challenge in pigs. In addition, significantly higher numbers of type I helper T cells (CD4$^+$IFN-γ^+) and cytotoxic T cells (CD8$^+$IFN-γ^+) among PBMCs were observed in JA142c-infected pigs at 14 and 21 dpc than in mock- or VR2332c-challenged pigs. Total IFN-γ^+ lymphocytes in the JA142c-infected group were significantly higher at 14 and 21 dpc. Furthermore, regulatory T cell (CD4$^+$CD25$^+$FoxP3$^+$) numbers were significantly lower in JA142c-infected pigs at 3 dpc and in JA142c- and VR2332c-infected pigs at 14 dpc than in the mock-infected group, although these levels gradually increased by 28 dpc in both groups.

Induction of PRRSV-specific antibodies after challenge

The PRRSV-specific antibody (IgG) response was measured by performing a nucleocapsid (N) protein-based ELISA for the two virus-challenged groups. The pigs seroconverted after 2 weeks of exposure to PRRSV and remained seropositive until the end of the study (Fig. 8a). The negative control group remained seronegative until the end of the study. No significant differences in the

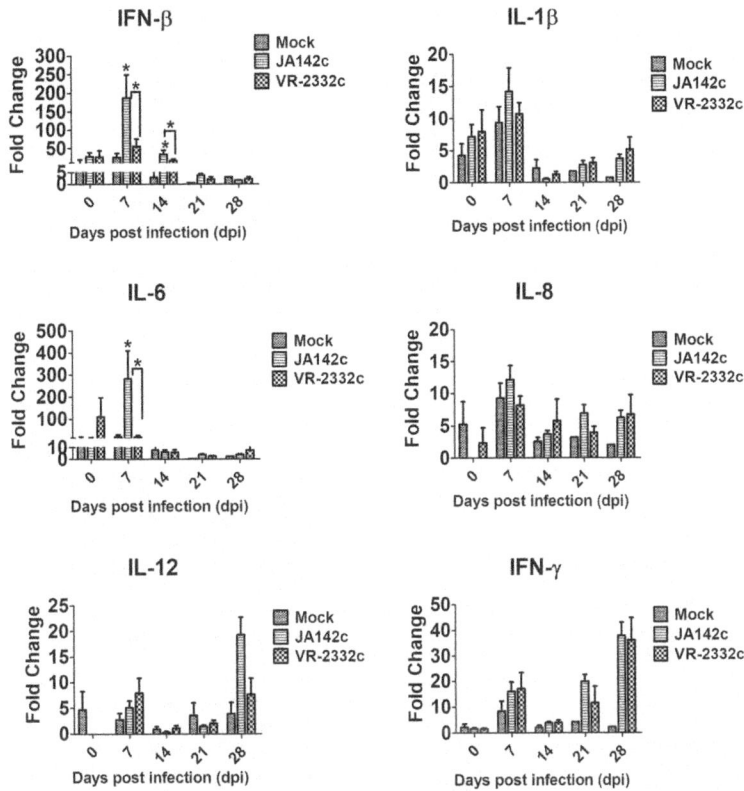

Fig. 4 mRNA transcription of cytokines in PBMCs collected from pigs infected with the VR2332c and JA142c PRRSV strains at 0, 7, 14, 21 and 28 dpc. Analysis of mRNA transcription in PBMCs (n = 5) collected from infected and control pigs using real-time PCR. Asterisks indicate significant differences in cytokine expression induced by each virus compared with mock treatment or significant differences between JA142c and VR2332c (* indicates $p \leq 0.05$)

IgG response were detected between the JA142c- and VR2332c-challenged groups.

NAb levels in the collected serum samples were determined using an FFN assay. At 28 dpc, only homologous NAbs were induced, and no significant cross-NAbs were induced by either VR2332c or JA142c (Fig. 8b and c).

Lung pathology and viral load
JA142c-infected pigs (3.1 ± 0.67 TCID$_{50}$/mL) had higher residual viral loads in the lung than VR2332c-infected pigs (1.662 ± 0.22 TCID$_{50}$/mL); however, this difference was not statistically significant (Fig. 9a). Furthermore, there were no significant differences in gross or microscopic lung lesion scores among the groups challenged with JA142c or VR2332c (Fig. 9b & c).

Discussion
It has been demonstrated that type I IFNs (IFN-α and IFN-β) play an important role in the induction of innate and adaptive immunity during PRRSV infection [42, 43]. In the current study, JA142c induced higher IFN-α production in PAMs than did VR2332c. This differential IFN-α induction in PAMs is consistent with the results of previous study that demonstrated variable IFN-α

production in response to different Type II PRRSV isolates [22]. Moreover, compared with VR2332c, JA142c induced the production of significantly increased serum IFN-α levels in pigs from 3 to 14 dpc, which is in partial agreement with a previous study demonstrating lower levels of serum IFN-α in VR2332c-infected pigs [44]. In addition, JA142c induced significantly higher IFN-β transcription than did VR2332c in naïve PAMs, similar to the results of previous studies describing high IFN-β transcription in PAMs following infection by several PRRSV strains [45–47]. Because PRRSV could modulate and block type I IFN induction [39], these results indicated that VR2332c might exert enhanced inhibitory effects on type I IFN production in PAMs as well as in pigs. Furthermore, in the present study, JA142c induced significantly higher TNF-α production in PAMs and pigs than did VR2332c. This observation is consistent with previous studies that reported various levels of TNF-α induction in PRRSV-infected PAMs, PBMCs, or pigs [9, 19, 20, 25, 44, 48–50]. TNF-α is another major pro-inflammatory cytokine that promotes the antiviral state in uninfected neighbouring cells and recruits lymphocytes to sites of infection [48]. Similarly, significantly higher mRNA transcription of other pro-inflammatory cytokines (IL-1α/β, IL-6, IL-8 and IL-12)

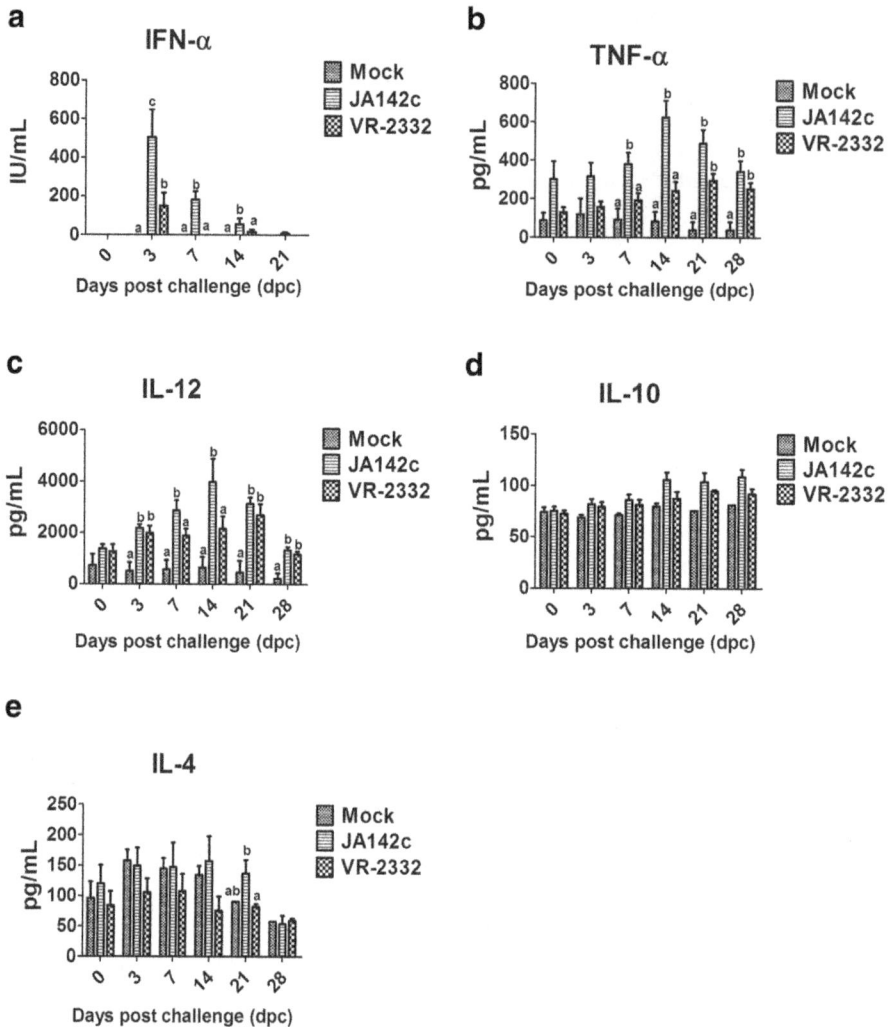

Fig. 5 Serum cytokine levels in PRRSV-infected pigs. Sera collected from pigs (n = 5) at 0, 3, 7, 14, 21 and 28 dpc were analysed for **a** IFN-α; **b** TNF-α; **c** IL-12; **d** IL-10; and **e** IL-4 by ELISA. Each bar represents the average cytokine amount from 5 pigs ± SEM. Bars showing different letters represent values that differ significantly from each other ($p < 0.05$)

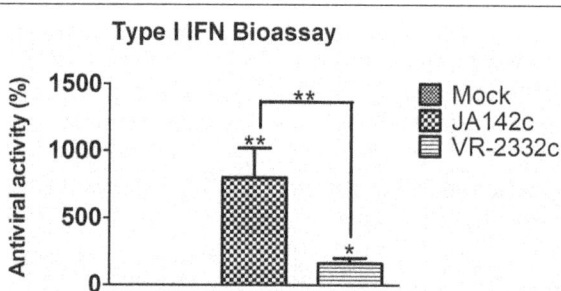

Fig. 6 Type I IFN bioassay using sera from PRRSV-challenged pigs at 3 dpc. MDBK cells (n = 6) in 96-well plates were infected with VSV and incubated until a CPE was evident in mock-treated control cells. Type I IFN bioactivity in the serum was measured by performing a type I IFN bioassay

was also observed in PAMs and JA142c-challenged pigs at 7 dpc than in VR2332c-challenged pigs, which is partially consistent with previous studies that demonstrated lower levels of IL-1β, IL-6 and IL-8 in VR2332c-challenged pigs [44, 51]. Therefore, it was concluded that compared with VR2332c, JA142c consistently induced significantly higher levels of important pro-inflammatory cytokines in PAMs and pigs.

When T cell responses were analysed in pigs challenged with JA142c and VR2332c, compared with those from VR2332c- or mock-challenged pigs, PBMCs from JA142c-challenged pigs contained significantly larger populations of activated CD4+ T cells (CD4+CD25+) and exhibited significant and early increases in type I helper T cells (CD4+IFN-γ+) and total IFN-γ+ cells. Cytotoxic T cells (CD8+IFN-γ+) in both JA142c- and VR2332c-infected pigs increased at 14 dpc, but the JA142c group

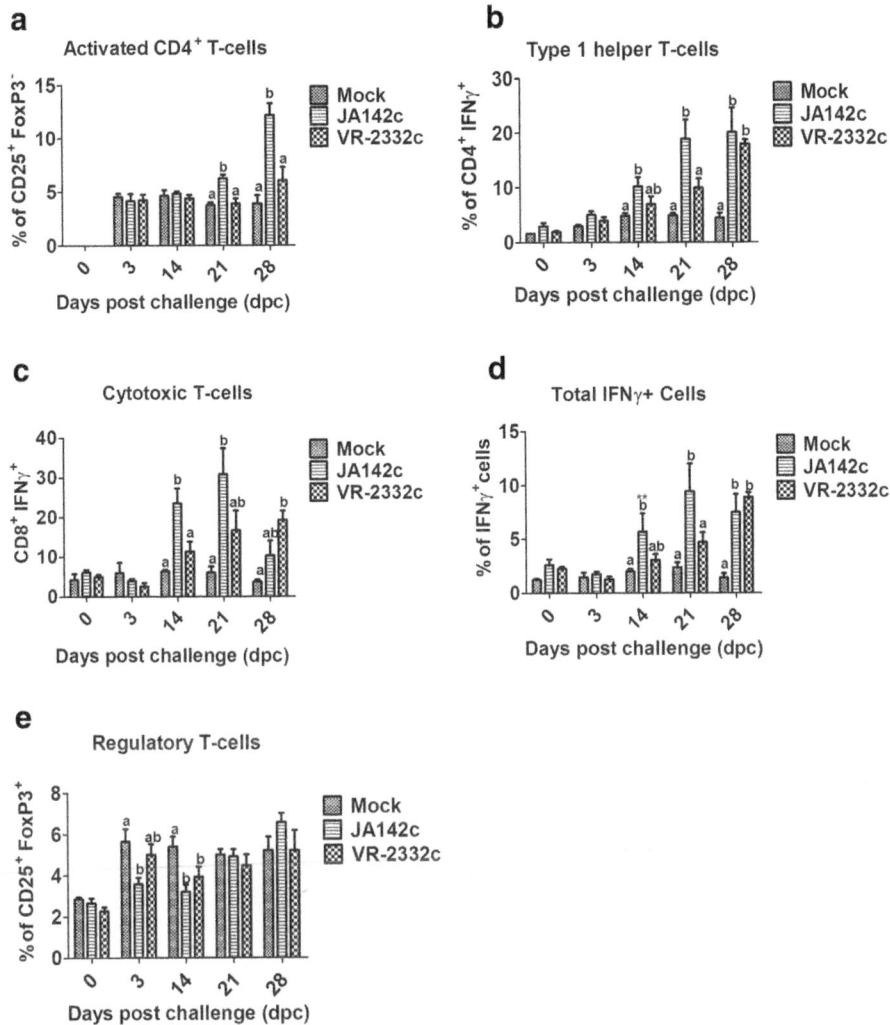

Fig. 7 Frequency of immune cells in PBMCs derived from PRRSV-infected pigs. **a** Activated CD4+ T cells (CD4+CD25+Foxp3−), **b** type I helper T cells (CD4+IFN-γ+), **c** cytotoxic T cells (CD8+IFN-γ+), **d** total PBMCs expressing IFN-γ+ and **e** T_Regs (CD4+CD25+Foxp3+). Each bar represents the average percentage of immune cells from 5 pigs ± SEM. Bars showing different letters represent values that differ significantly from each other ($p < 0.05$)

had a significantly larger population of cytotoxic T lymphocytes (CTLs) than the mock- and VR2332c-infected groups. In a previous study, low numbers of PRRSV-specific IFN-γ+ CD8+ T cells were associated with a delayed CTL response [33].

Conclusions

Collectively, based on the results of the current study, JA142c induced more efficient innate and adaptive immune responses in vitro and in vivo than did VR2332c. Furthermore, of the two in vitro screening systems for immunogenicity (PAMs and PBMCs), PAMs generated immune response patterns to both PRRSVs that were similar to those elicited by these viruses in pigs. Thus, PAMs may be used as an immunogenicity screening tool for the selection of vaccine candidates against PRRSV. Finally, the induction of IFN-α, TNF-α and IL-12 in PAMs might

represent vital immune markers to predict the immunogenicity of PRRSV vaccine candidates in pigs.

Methods

Cells and viruses

MARC-145, an African green monkey kidney cell line that is highly permissive to PRRSV, was used for virus propagation and maintained in Roswell Park Memorial Institute (RPMI)-1640 medium (Gibco® RPMI 1640, Life Technologies, Carlsbad, CA, USA) supplemented with heat-inactivated 10% fetal bovine serum (FBS, Life Technologies), 2 mM L-glutamine, and 100X Antibiotic -Antimycotic (Anti-anti, Life Technologies) containing 100 IU/mL penicillin, 100 μg/mL streptomycin, and 0.25 μg/mL Fungizone® (amphotericin B) in a 5% CO_2 humidified chamber at 37 °C. In this study, this medium is referred to as complete RPMI (cRPMI) medium. Two

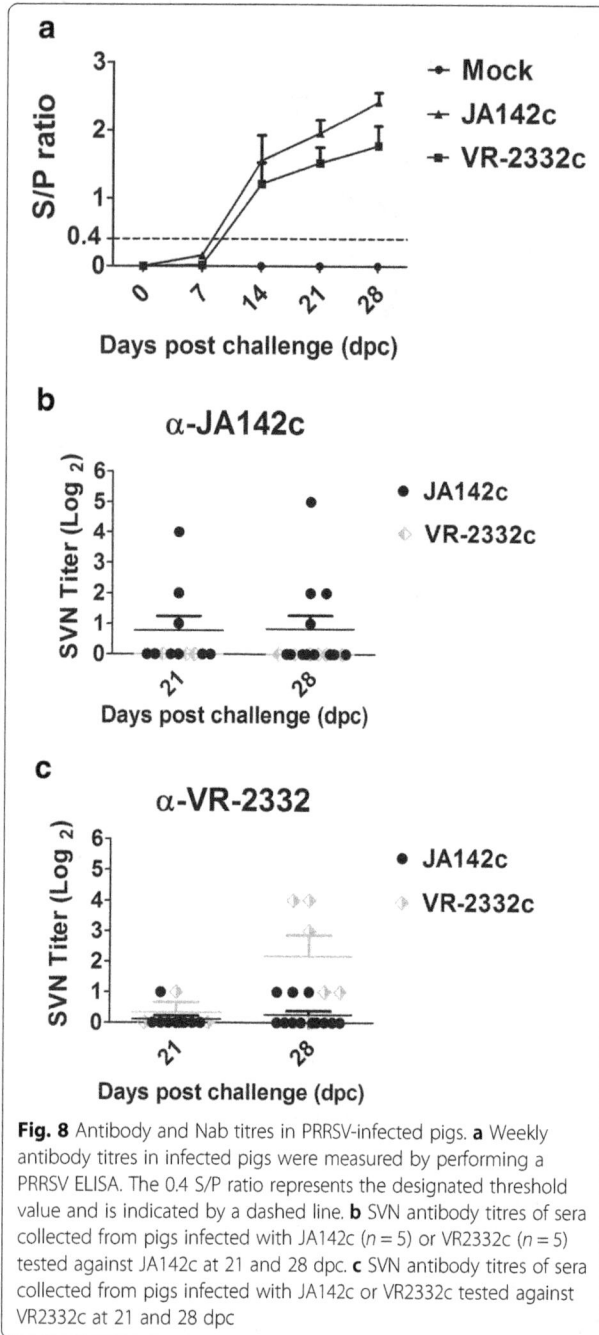

Fig. 8 Antibody and Nab titres in PRRSV-infected pigs. **a** Weekly antibody titres in infected pigs were measured by performing a PRRSV ELISA. The 0.4 S/P ratio represents the designated threshold value and is indicated by a dashed line. **b** SVN antibody titres of sera collected from pigs infected with JA142c (*n* = 5) or VR2332c (*n* = 5) tested against JA142c at 21 and 28 dpc. **c** SVN antibody titres of sera collected from pigs infected with JA142c or VR2332c tested against VR2332c at 21 and 28 dpc

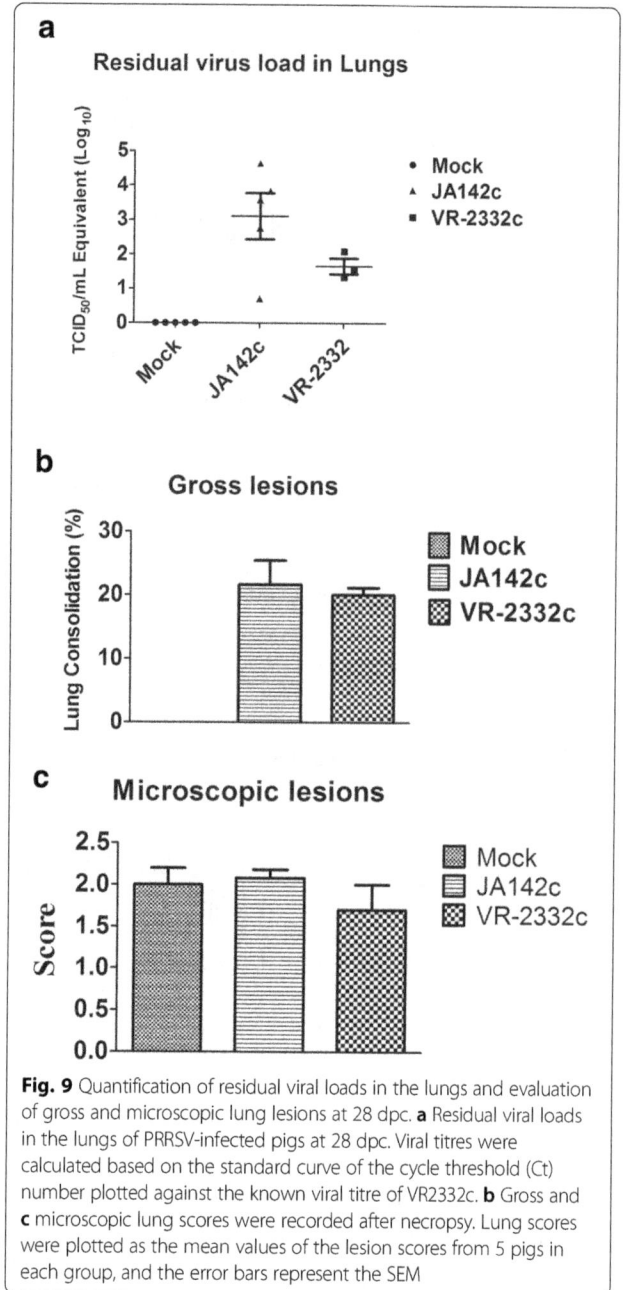

Fig. 9 Quantification of residual viral loads in the lungs and evaluation of gross and microscopic lung lesions at 28 dpc. **a** Residual viral loads in the lungs of PRRSV-infected pigs at 28 dpc. Viral titres were calculated based on the standard curve of the cycle threshold (Ct) number plotted against the known viral titre of VR2332c. **b** Gross and **c** microscopic lung scores were recorded after necropsy. Lung scores were plotted as the mean values of the lesion scores from 5 pigs in each group, and the error bars represent the SEM

infectious clones of Type II PRRSV strains, JA142 and VR2332 [52, 53], were re-cloned into a vector (pOpti-VEC™-TOPO® TA Cloning Kit, Life Technologies) to construct modified infectious clones by inserting viral sequences between the human cytomegalovirus (CMV) promoter and the internal ribosomal entry site (IRES) that are present in the vector [54, 55]. The viruses rescued from each of the modified infectious clones were named JA142c and VR2332c. The infectious clone-driven viruses were used in this study to minimize mutations caused by passaging the virus in cells.

Harvesting porcine alveolar macrophages (PAMs) and collecting peripheral blood mononuclear cells (PBMCs)

PAMs were harvested from six 6-week-old PRRSV-negative pigs, as described previously [56]. The pigs were euthanized, and the lungs, trachea and bronchus were aseptically collected. The lungs were lavaged three times with 50 mL of phosphate-buffered saline (PBS, pH 7.4), and the harvested wash fluid was centrifuged for 10 min at 1000 *g*. The resulting pellet was washed three times with PBS and re-suspended in 5 mL of sterile PBS. To evaluate cell number and viability, the cells were diluted 100-fold in PBS, mixed with 0.4% trypan blue at a 1:1 ratio and counted using a Countess™ Automated Cell Counter (Invitrogen, Carlsbad, CA, USA). After counting, freezing medium (70% Dulbecco's modified Eagle's medium (DMEM) with high glucose, pyruvate, 20% FBS [Gibco, California, USA], and 10% dimethyl sulfoxide [Sigma-Aldrich, St. Louis, Missouri, USA]) was added to the cells to maintain the final cell density at 5×10^7 cells/mL. The cells were collected in cryovials and stored at -80 °C until use.

PBMCs were isolated from 6-mL blood samples collected from six-week-old PRRSV-free pigs in lithium-heparin-containing vacutainers using a density gradient method in Histopaque-1077® solution (Sigma, St. Louis, MO, USA) according to the manufacturer's instructions. The blood samples were briefly stratified in Histopaque-1077® solution at a ratio of 1:1 (blood:Histopaque) and were centrifuged at 400 g for 30 min. Purified PBMCs were collected, washed twice with sterile PBS (pH 7.0) supplemented with 1% FBS (Gibco, Carlsbad, CA, USA) and re-suspended in 0.5 mL of sterile PBS.

In vitro evaluation of the growth kinetics of JA142c and VR2332c

The growth of JA142c and VR2332c was evaluated in MARC-145 cells. Confluent monolayers of MARC-145 cells were prepared in nine 25 cm² flasks, with three flasks inoculated with each virus or mock inoculated at a multiplicity of infection (MOI) of 0.01 and incubated for 1 h in a humidified chamber at 37 °C. After incubation, the cells were replenished with cRPMI and incubated for an additional 48 h. Supernatants were collected from each flask at 6, 12, 24, 36 and 48 h, and the virus in these supernatants was titrated as described previously [57].

Animal study

A total of 15 three-week-old PRRSV-free pigs were purchased from a farm which has been PRRS-negative over last 10 years and randomly divided into 3 groups and housed separately in the animal research facility in Chonbuk National University. Five pigs in each group were challenged with VR2332c or JA142c diluted in RPMI to 10^3 TCID$_{50}$/mL in a volume of 2 mL or with mock inoculum (sterile PBS) injected intramuscularly and housed for 4 weeks. All pigs were bled at 0, 3, 7, 14, 21 and 28 days post-challenge (dpc), and whole blood was collected for PBMC isolation to evaluate cytokine mRNA transcript levels and perform flow cytometry. Sera were separated immediately after bleeding and stored at -80 °C until use. Viremia and serum cytokine levels in all pigs were evaluated weekly, at which point all pigs were weighed. Serum virus neutralizing antibody induction was evaluated at 21 and 28 dpc. All pigs were humanly euthanized at 28 dpc by electrocution after intramuscular injection of 3 ml of Azaperone (40 mg/ml,

StressGuar®, Dong Bang Inc., Seoul, South Korea) for sedation and subjected to pathological evaluation. To evaluate gross and microscopic lung lesions, each lung lobe was scored for the percentage of lung consolidation [58] and interstitial pneumonia resulting from PRRSV infection. Scoring of microscopic lung lesions was recorded as follows: 0 indicates no lesion; 1 indicates mild interstitial pneumonia; 2 indicates moderate multifocal interstitial pneumonia; 3 indicates moderate diffuse interstitial pneumonia; and 4 indicates severe interstitial pneumonia. Lung tissues were collected from each pig and stored at -80 °C until examination.

Transcriptional activation of cellular cytokines

PAMs re-suspended in cRPMI at 5×10^6 cells/mL were seeded in each well of a 24-well plate (BD Falcon, Franklin Lakes, NJ, USA) and cultured overnight in a 5% CO_2 humidified chamber at 37 °C. The cultured PAMs were stimulated with a 0.1 MOI of either JA142c or VR2332c, 10 µg/mL polyinosinic-polycytidylic acid (poly I:C) (Sigma-Aldrich, St. Louis, MO, USA) or mock inoculum (cRPMI) and incubated in a 5% CO_2 humidified chamber at 37 °C. Six PAMs harvested from six PRRSV-free pigs were used in the experiment. The cells were harvested, and the cell lysates were collected at 12, 24 and 36 hpi. Similarly, 1×10^6 naïve PBMCs re-suspended in cRPMI were seeded in each well of a 24-well plate. The cells were stimulated with a 0.1 MOI of either JA142c or VR2332c, 10 µg/mL poly I:C or mock inoculum (cRPMI) and incubated in a 5% CO_2 humidified chamber at 37 °C, after which the cells were harvested, and the cell supernatants were collected at 12, 36 and 72 hpi. Four PBMC cells from 4 PRRSV-free pigs were used for the experiment. Poly I:C, a synthetic analogue of double-stranded RNA, was used as an immune-inducing positive control for PAMs and PBMCs. Cellular RNA was extracted using an RNA isolation kit (GeneAll® Hybrid-RTM kit, GeneAll Biotechnology, Seoul, South Korea) following the manufacturer's instructions. RNA was analysed on agarose gels as initial quality checks and 260/280 and 260/230 ratios were determined to be 1.8–2.2 and 1.7 or higher, respectively. Then, 1 µg of RNA was used for complementary DNA (cDNA) synthesis using a high-capacity cDNA reverse transcription kit (Applied Biosystems, Foster City, CA, USA) following the manufacturer's instructions. Real-time PCR was then performed on a 7500 Fast Real-time PCR system (Applied Biosystems) using various cytokine-specific primers, following the manufacturer's instructions. The primer sequences used in this study are shown in Table 1. Ten microlitres of 2X Power SYBR Green PCR Master Mix (Applied Biosystems, Foster City, CA, USA), 2 µL of cDNA and 1 µL of each forward and reverse primer (each at 10 pm/µL) were used for PCR amplification. All samples were tested in duplicate, and the following

Table 1 Information about the real-time PCR primers used to measure mRNA transcript levels of various cytokines

Genes	Forward Primer (5'- 3')	Reverse Primer (5'- 3')	Accession/Reference
β-Actin	GCGGGACATCAAGGAGAAG	AGGAAGGAGGGCTGGAAGAG	U07786
IFN-α	TCTCATGCACCAGAGCCA	CCTGGACCACAGAAGGGA	[39]
IFN-β	AGTGCATCCTCCAAATCGCT	GCTCATGGAAAGAGCTGTGGT	M86762
TNF-α	TTATTCAGGAGGGCGAGGT	AGCAAAAGGAGGCACAGAGG	NM214022
IL-1α	GTGCTCAAAACGAAGACGAACC	CATATTGCCATGCTTTTCCCAGAA	NM_214029.1
IL-1β	AACGTGCAGTCTATGGAGT	GAACACCACTTCTCTCTTCA	M86725
IL-6	CCACCAGGAACGAAAGAGAG	AGGCAGTAGCCATCACCAGA	NM_214399
IL-8	TAGGACCAGAGCCAGGAAGA	CAGGAAAACTGCCAAGAAGG	M86923.1
IL-10	TGACGATGAAGATGAGGAAGAA	GAACCTTGGAGCAGATTTTGA	NM214041
IL-12	TCAGGGACATCATCAAACCA	GAACACCAAACATCAGGGAAA	NM214013
IFN-γ	GACTTTGTGTTTTTCTGGCTCTTAC	TTTTGTCACTCTCCTCTTTCCA	NM213948
IL-2	TGCACTAACCCTTGCACTCA	CCTGCTTGGGCATGTAAAAT	X56750
IL-4	TTTGCTGCCCCAGAGAAC	TCCTGTCAAGTCCGCTCA	X68330

IL interleukin, *IFN* interferon

cycling conditions were applied: (a) incubation for 10 min at 95 °C; (b) 40 cycles of 15 s at 95 °C and 1 min at 60 °C; and (c) a melting curve stage of 15 s at 95 °C, 1 min at 60 °C, 15 s at 95 °C and 15 s at 60 °C. The relative quantities of cytokine mRNA in infected and non-infected cells were normalized to β-actin mRNA, and the amounts were determined using the $2^{-\Delta\Delta Ct}$ method [59].

Cytokine quantification by ELISA

Commercially available porcine-specific ELISA kits were used to quantify various cytokine protein levels in sera, cell supernatants or lysates, including TNF-α, IL-4, IL-10, and IL-12 (DuoSet® ELISA, R&D Systems, MN, USA), according to the manufacturer's instructions.

To detect IFN-α protein levels in sera and cell lysates, 100 μL (1.8 μg/mL) of a mouse anti-pig IFN-α antibody (Clone F17, PBL Assay Science, NJ, USA) was used as a coating antibody, and a mouse anti-pig IFN-α antibody (Clone K9, PBL Assay Science, NJ, USA) was biotinylated and used as a secondary antibody, with recombinant porcine IFN-α (PBL Assay Science, NJ, USA) as a standard. The procedure was carried out using the provided ELISA reagents (eBioscience, CA, USA) and following the manufacturer's instructions. The results were analysed using SoftMax Pro 5.3 microplate data software (Molecular Devices, CA, USA).

Type I IFN bioassay

A conventional type I IFN bioassay using sera from virus-challenged pigs was performed to assess the amount of virus-induced type I IFNs present by determining anti-vesicular stomatitis virus (VSV) activity. MDBK cells were seeded in 96-well cell culture plates (BD Falcon, Franklin Lakes, NJ, USA) (5×10^5 cells/well) and incubated for 20 h in the presence of two-fold

dilutions of recombinant IFN-α (r-IFN-α) (PBL Assay Science, NJ, USA), as described previously. The cells were infected with propagation-competent VSV (0.01 MOI) and incubated until a cytopathic effect (CPE) was evident in mock-treated control cells. The cells were washed twice with PBS and stained for 1 h with 0.1% crystal violet in 10% formalin. The plates were then washed with tap water to remove excess crystal violet and dried. The dye was dissolved by adding 100 mL of 70% ethanol to each well. Crystal violet absorbance at 595 nm was determined with a microplate reader. Antiviral activity was calculated as described previously [54].

Quantification of PRRSV RNA in sera and lungs

Viral RNA was extracted from 100 μL of serum and 1 g of lung samples using a viral RNA extraction kit (Mag-MAX™ Viral RNA Isolation Kit, Life Technologies) and a total RNA extraction kit (Hybrid-RTM, GeneAll, Seoul, Korea) according to the manufacturer's instructions. The viral loads in sera and lungs were measured by performing real-time RT-PCR employing a one-step reverse transcriptase kit (AgPath-IDTM One-Step RT-PCR Kit, Ambion, Austin, TX, USA) with a 7500 Fast Real-time PCR system (Applied Biosystems, Foster City, CA, USA), although this test cannot differentiate between replication-competent and replication-incompetent virus. Primers and a minor groove binder (MGB) fluorescent probe specific to a conserved region of ORF7 were employed as described previously [60].

Assessment of PRRSV-specific antibodies

A fluorescence focus neutralization (FFN)-based serum virus neutralization (SVN) assay was performed to evaluate NAb titres induced by PRRSVs after challenge. The

SVN assay was performed in MARC-145 cells as described previously [61, 62]. The NAb titres of each anti-serum against each virus were expressed as the reciprocal of the highest dilution for which a 90% or greater reduction in the number of fluorescent focus units (FFU)/mL was observed compared to the wells for each respective virus back-titration.

The presence of PRRSV nucleocapsid-specific antibodies in the sera of infected animals was determined using a direct ELISA kit (HerdCheck® PRRS Antibody Kit 3XR, IDEXX Laboratories, Westbrook, ME, USA) according to the manufacturer's instructions.

Flow cytometry

The frequency and phenotype of the immune cell populations were determined by performing multicolour immunostaining of single-cell suspensions on the day of PBMC isolation. PBMCs were evaluated for the expression of markers of activated $CD4^+$ T cells ($CD4^+CD25^+FoxP3^-$), regulatory T cells ($CD4^+CD25^+FoxP3^+$), type I helper T cells ($CD4^+IFN-\gamma^+$), and cytotoxic T cells ($CD8^+IFN-\gamma^+$). The cells were stained with an appropriate monoclonal antibody (mAb), which was either directly conjugated to a specific fluorochrome or biotinylated, or with a purified antibody targeting porcine-specific immune cell surface markers as described previously [63]. Briefly, 1×10^6 purified PBMCs per sample were re-suspended in fluorescence-activated cell sorting (FACS) buffer (2% FBS in PBS with 0.02% sodium azide), and two replicates of each sample were plated in two separate U-bottom 96-well plates with 1×10^6 cells per well in a 200-μL volume. Subsequently, the cells in one plate were stained with either CD4α-PE (Clone 74–12-4; BD Biosciences, Franklin Lakes, New Jersey, USA) or anti-porcine CD25 (clone K231.3B3; Serotech, Raleigh, NC), followed by washing and staining with an allophycocyanin-conjugated rat anti-mouse IgG1 antibody (Ab) (Clone RMG1–1; Biolegend, San Diego, CA, USA) as a secondary antibody against anti-CD25 according to the manufacturer's instructions. Afterwards, the cells were fixed and stained with a FoxP3-FITC antibody (Clone FJK-16 s; eBiosciences, San Diego, CA, USA), as described. Simultaneously, two other sets of cells were treated with a 1X cell stimulation cocktail (eBiosciences) plus 1X brefeldin A (eBiosciences) in cRPMI media and incubated at 37 °C in a 5% CO_2 humidified chamber for 5 h. Afterwards, one set of cells was stained with CD8α-FITC (Clone 76–2-11; BD Biosciences), and the other set of cells was stained with CD4α-PE and fixed with IC fixation buffer (eBiosciences) according to the manufacturer's instructions. Finally, the cells were stained with IFN-γ-PerCP-Cy™ 5.5 (Clone P2G10; BD Biosciences). A total of 100,000 events (gated by forward and side scatter) were acquired for each sample using an Accuri C6 flow cytometer (BD

Biosciences, Franklin Lakes, New Jersey, USA), and the data were analysed using BD CFlow®Plus software v. 1.0.227.4 (BD Biosciences, Franklin Lakes, New Jersey, USA). Target cell frequencies were expressed as the percentage for a specific cell subset.

Data analysis

Graphs were constructed using GraphPad Prism 5.0.2 (GraphPad, San Diego, CA, USA), and statistical analyses were performed using SPSS Advanced Statistics 17.0 software (SPSS, Inc., Chicago, USA). The Mann-Whitney U test was applied to estimate differences between different groups. Repeated measurements of viremia and PRRSV nucleocapsid-specific antibody levels in challenged pigs were analysed using a repeated ANOVA test (Tukey's post hoc test) to determine the overall difference, and pairwise comparisons were also made between groups.

Abbreviations
CTLs: Cytotoxic T lymphocytes; IFNs: Interferons; Nabs: Neutralizing antibodies; NSPs: Non-structural proteins; PAMs: Porcine alveolar macrophages; PBMCs: Peripheral blood mononuclear cells; PRRS: Porcine reproductive and respiratory syndrome; T_{Regs}: Regulatory T cells

Acknowledgements
The authors wish to acknowledge the assistance of the lab technicians and undergraduate students from the College of Veterinary Medicine, Veterinary Diagnostic Center, Chonbuk National University.

Funding
The authors wish to acknowledge the assistance of the lab technicians and undergraduate students from the College of Veterinary Medicine, Veterinary Diagnostic Center, Chonbuk National University. This study was supported by grants from the Technology Development Program for Bio-industry (315029–3) and the Cooperative Research Program for Agriculture Science & Technology Development (PJ012612) in Rural Development Administration, Ministry of Agriculture, Food and Rural Affairs, Republic of Korea.

Authors' contributions
NS and AK performed the experiments and drafted the manuscript; SN and SG helped with the animal experimentation and the ELISA and type-1 IFN bioassay experiments; SML and TYH provided vital input; MSY and BSK evaluated the lungs for lesions; and WK designed the entire study and assisted in drafting the manuscript. All authors read and approved the final manuscript.

Consent for publication
Not applicable.

Competing interests
The authors declare that they have no competing interests.

Author details
[1]College of Veterinary Medicine, Chonbuk National University, 79 Gobong-ro, Iksan, Jeonbuk, Korea. [2]Division of Animal Biotechnology, Faculty of Veterinary Sciences and Animal Husbandry, Sher-e-Kashmir University of Agricultural Sciences and Technology of Kashmir, Srinagar, India. [3]College of Environmental & Biosource Science, Division of Biotechnology, Chonbuk National University, Iksan, South Korea. [4]Dairy Science Division, National Institute of Animal Science, Rural Development Administration, Cheonan 31000, South Korea.

References
1. Cavanagh D. Nidovirales: a new order comprising Coronaviridae and Arteriviridae. Arch Virol. 1997;142(3):629–33.
2. Collins JE, Benfield DA, Christianson WT, Harris L, Hennings JC, Shaw DP, Goyal SM, McCullough S, Morrison RB, Joo HS, et al. Isolation of swine infertility and respiratory syndrome virus (isolate ATCC VR-2332) in North America and experimental reproduction of the disease in gnotobiotic pigs. J Vet Diagn Investig. 1992;4(2):117–26.
3. Wensvoort G, Terpstra C, Pol JM, ter Laak EA, Bloemraad M, de Kluyver EP, Kragten C, van Buiten L, den Besten A, Wagenaar F, et al. Mystery swine disease in the Netherlands: the isolation of Lelystad virus. Vet Q. 1991;13(3):121–30.
4. Holtkamp DJ, Kliebenstein JB, Neumann E, Zimmerman JJ, Rotto H, Yoder TK, Wang C, Yeske P, Mowrer CL, Haley CA. Assessment of the economic impact of porcine reproductive and respiratory syndrome virus on United States pork producers. J Swine Health Prod. 2013;21(2):72.
5. Firth AE, Zevenhoven-Dobbe JC, Wills NM, Go YY, Balasuriya UB, Atkins JF, Snijder EJ, Posthuma CC. Discovery of a small arterivirus gene that overlaps the GP5 coding sequence and is important for virus production. J Gen Virol. 2011;92(Pt 5):1097–106.
6. Johnson CR, Griggs TF, Gnanandarajah J, Murtaugh MP. Novel structural protein in porcine reproductive and respiratory syndrome virus encoded by an alternative ORF5 present in all arteriviruses. J Gen Virol. 2011;92:1107–16.
7. Shi M, Lam TT, Hon CC, Hui RK, Faaberg KS, Wennblom T, Murtaugh MP, Stadejek T, Leung FC. Molecular epidemiology of PRRSV: a phylogenetic perspective. Virus Res. 2010;154(1–2):7–17.
8. Shi M, Lam TT, Hon CC, Murtaugh MP, Davies PR, Hui RK, Li J, Wong LT, Yip CW, Jiang JW, et al. Phylogeny-based evolutionary, demographical, and geographical dissection of North American type 2 porcine reproductive and respiratory syndrome viruses. J Virol. 2010;84(17):8700–11.
9. Gimeno M, Darwich L, Diaz I, de la Torre E, Pujols J, Martin M, Inumaru S, Cano E, Domingo M, Montoya M, et al. Cytokine profiles and phenotype regulation of antigen presenting cells by genotype-I porcine reproductive and respiratory syndrome virus isolates. Vet Res. 2011;42:9.
10. Silva-Campa E, Cordoba L, Fraile L, Flores-Mendoza L, Montoya M, Hernandez J. European genotype of porcine reproductive and respiratory syndrome (PRRSV) infects monocyte-derived dendritic cells but does not induce Treg cells. Virology. 2010;396(2):264–71.
11. Meng XJ. Heterogeneity of porcine reproductive and respiratory syndrome virus: implications for current vaccine efficacy and future vaccine development. Vet Microbiol. 2000;74(4):309–29.
12. Vu HL, Pattnaik AK, Osorio FA. Strategies to broaden the cross-protective efficacy of vaccines against porcine reproductive and respiratory syndrome virus. Vet Microbiol. 2017;206:29-34.
13. Park JY, Kim HS, Seo SH. Characterization of interaction between porcine reproductive and respiratory syndrome virus and porcine dendritic cells. J Microbiol Biotechnol. 2008;18(10):1709–16.
14. Song S, Bi J, Wang D, Fang L, Zhang L, Li F, Chen H, Xiao S. Porcine reproductive and respiratory syndrome virus infection activates IL-10 production through NF-kappaB and p38 MAPK pathways in porcine alveolar macrophages. Dev Comp Immunol. 2013;39(3):265–72.
15. Wang G, Song T, Yu Y, Liu Y, Shi W, Wang S, Rong F, Dong J, Liu H, Cai X, et al. Immune responses in piglets infected with highly pathogenic porcine reproductive and respiratory syndrome virus. Vet Immunol Immunopathol. 2011;142(3–4):170–8.
16. Albina E, Carrat C, Charley B. Interferon-alpha response to swine arterivirus (PoAV), the porcine reproductive and respiratory syndrome virus. J Interferon Cytokine Res. 1998;18(7):485–90.
17. Buddaert W, Van Reeth K, Pensaert M. In vivo and in vitro interferon (IFN) studies with the porcine reproductive and respiratory syndrome virus (PRRSV). Adv Exp Med Biol. 1998;440:461–7.
18. Han M, Yoo D. Modulation of innate immune signaling by nonstructural protein 1 (nsp1) in the family Arteriviridae. Virus Res. 2014;194:100–9.
19. Gomez-Laguna J, Salguero FJ, Pallares FJ, Carrasco L. Immunopathogenesis of porcine reproductive and respiratory syndrome in the respiratory tract of pigs. Vet J. 2013;195(2):148–55.
20. Lopez-Fuertes L, Campos E, Domenech N, Ezquerra A, Castro JM, Dominguez J, Alonso F. Porcine reproductive and respiratory syndrome (PRRS) virus down-modulates TNF-alpha production in infected macrophages. Virus Res. 2000;69(1):41–6.
21. Baumann A, Mateu E, Murtaugh MP, Summerfield A. Impact of genotype 1 and 2 of porcine reproductive and respiratory syndrome viruses on interferon-alpha responses by plasmacytoid dendritic cells. Vet Res. 2013;44:33.
22. Lee SM, Schommer SK, Kleiboeker SB. Porcine reproductive and respiratory syndrome virus field isolates differ in in vitro interferon phenotypes. Vet Immunol Immunopathol. 2004;102(3):217–31.
23. He Q, Li Y, Zhou L, Ge X, Guo X, Yang H. Both Nsp1beta and Nsp11 are responsible for differential TNF-alpha production induced by porcine reproductive and respiratory syndrome virus strains with different pathogenicity in vitro. Virus Res. 2015;201:32–40.
24. Diaz I, Darwich L, Pappaterra G, Pujols J, Mateu E. Different European-type vaccines against porcine reproductive and respiratory syndrome virus have different immunological properties and confer different protection to pigs. Virology. 2006;351(2):249–59.
25. Aasted B, Bach P, Nielsen J, Lind P. Cytokine profiles in peripheral blood mononuclear cells and lymph node cells from piglets infected in utero with porcine reproductive and respiratory syndrome virus. Clin Diagn Lab Immunol. 2002;9(6):1229–34.
26. Thanawongnuwech R, Young TF, Thacker BJ, Thacker EL. Differential production of proinflammatory cytokines: in vitro PRRSV and mycoplasma hyopneumoniae co-infection model. Vet Immunol Immunopathol. 2001;79(1–2):115–27.
27. Liu CH, Chaung HC, Chang HL, Peng YT, Chung WB. Expression of toll-like receptor mRNA and cytokines in pigs infected with porcine reproductive and respiratory syndrome virus. Vet Microbiol. 2009;136(3–4):266–76.
28. Peng YT, Chaung HC, Chang HL, Chang HC, Chung WB. Modulations of phenotype and cytokine expression of porcine bone marrow-derived dendritic cells by porcine reproductive and respiratory syndrome virus. Vet Microbiol. 2009;136(3–4):359–65.
29. van Gucht S, van Reeth K, Pensaert M. Interaction between porcine reproductive-respiratory syndrome virus and bacterial endotoxin in the lungs of pigs: potentiation of cytokine production and respiratory disease. J Clin Microbiol. 2003;41(3):960–6.
30. Meier WA, Husmann RJ, Schnitzlein WM, Osorio FA, Lunney JK, Zuckermann FA. Cytokines and synthetic double-stranded RNA augment the T helper 1 immune response of swine to porcine reproductive and respiratory syndrome virus. Vet Immunol Immunopathol. 2004;102(3):299–314.
31. Xiao Z, Trincado CA, Murtaugh MP. Beta-glucan enhancement of T cell IFNgamma response in swine. Vet Immunol Immunopathol. 2004;102(3):315–20.
32. Costers S, Lefebvre DJ, Goddeeris B, Delputte PL, Nauwynck HJ. Functional impairment of PRRSV-specific peripheral CD3+CD8high cells. Vet Res. 2009;40(5):46.
33. Ferrari L, Martelli P, Saleri R, De Angelis E, Cavalli V, Bresaola M, Benetti M, Borghetti P. Lymphocyte activation as cytokine gene expression and secretion is related to the porcine reproductive and respiratory syndrome virus (PRRSV) isolate after in vitro homologous and heterologous recall of peripheral blood mononuclear cells (PBMC) from pigs vaccinated and exposed to natural infection. Vet Immunol Immunopathol. 2013;151(3–4):193–206.
34. Silva-Campa E, Mata-Haro V, Mateu E, Hernandez J. Porcine reproductive and respiratory syndrome virus induces CD4+CD8+CD25+Foxp3+ regulatory T cells (Tregs). Virology. 2012;430(1):73–80.
35. Rodriguez-Gomez IM, Kaser T, Gomez-Laguna J, Lamp B, Sinn L, Rumenapf T, Carrasco L, Saalmuller A, Gerner W. PRRSV-infected monocyte-derived dendritic cells express high levels of SLA-DR and CD80/86 but do not stimulate PRRSV-naive regulatory T cells to proliferate. Vet Res. 2015;46:54.
36. Darwich L, Diaz I, Mateu E. Certainties, doubts and hypotheses in porcine reproductive and respiratory syndrome virus immunobiology. Virus Res. 2010;154(1–2):123–32.

37. Kimman TG, Cornelissen LA, Moormann RJ, Rebel JM, Stockhofe-Zurwieden N. Challenges for porcine reproductive and respiratory syndrome virus (PRRSV) vaccinology. Vaccine. 2009;27(28):3704–18.

38. Mateu E, Diaz I. The challenge of PRRS immunology. Vet J. 2008;177(3):345–51.

39. Loving CL, Osorio FA, Murtaugh MP, Zuckermann FA. Innate and adaptive immunity against porcine reproductive and respiratory syndrome virus. Vet Immunol Immunopathol. 2015;167(1–2):1–14.

40. Plagemann PG. Neutralizing antibody formation in swine infected with seven strains of porcine reproductive and respiratory syndrome virus as measured by indirect ELISA with peptides containing the GP5 neutralization epitope. Viral Immunol. 2006;19(2):285–93.

41. Kim WI, Lee DS, Johnson W, Roof M, Cha SH, Yoon KJ. Effect of genotypic and biotypic differences among PRRS viruses on the serologic assessment of pigs for virus infection. Vet Microbiol. 2007;123(1–3):1–14.

42. Brockmeier SL, Loving CL, Nelson EA, Miller LC, Nicholson TL, Register KB, Grubman MJ, Brough DE, Kehrli ME Jr. The presence of alpha interferon at the time of infection alters the innate and adaptive immune responses to porcine reproductive and respiratory syndrome virus. Clin Vaccine Immunol. 2012;19(4):508–14.

43. Royaee AR, Husmann RJ, Dawson HD, Calzada-Nova G, Schnitzlein WM, Zuckermann FA, Lunney JK. Deciphering the involvement of innate immune factors in the development of the host response to PRRSV vaccination. Vet Immunol Immunopathol. 2004;102(3):199–216.

44. Guo B, Lager KM, Henningson JN, Miller LC, Schlink SN, Kappes MA, Kehrli ME Jr, Brockmeier SL, Nicholson TL, Yang HC, et al. Experimental infection of United States swine with a Chinese highly pathogenic strain of porcine reproductive and respiratory syndrome virus. Virology. 2013;435(2):372–84.

45. Gudmundsdottir I, Risatti GR. Infection of porcine alveolar macrophages with recombinant chimeric porcine reproductive and respiratory syndrome virus: effects on cellular gene transcription and virus growth. Virus Res. 2009; 145(1):145–50.

46. Badaoui B, Rutigliano T, Anselmo A, Vanhee M, Nauwynck H, Giuffra E, Botti S. RNA-sequence analysis of primary alveolar macrophages after in vitro infection with porcine reproductive and respiratory syndrome virus strains of differing virulence. PLoS One. 2014;9(3):e91918.

47. Overend CC, Cui J, Grubman MJ, Garmendia AE. The activation of the IFNbeta induction/signaling pathway in porcine alveolar macrophages by porcine reproductive and respiratory syndrome virus is variable. Vet Res Commun. 2017;41(1):15–22.

48. Huang C, Zhang Q, Feng WH. Regulation and evasion of antiviral immune responses by porcine reproductive and respiratory syndrome virus. Virus Res. 2015;202:101-11.

49. Ait-Ali T, Wilson AD, Westcott DG, Clapperton M, Waterfall M, Mellencamp MA, Drew TW, Bishop SC, Archibald AL. Innate immune responses to replication of porcine reproductive and respiratory syndrome virus in isolated swine alveolar macrophages. Viral Immunol. 2007;20(1):105–18.

50. Choi C, Cho WS, Kim B, Chae C. Expression of interferon-gamma and tumour necrosis factor-alpha in pigs experimentally infected with porcine reproductive and respiratory syndrome virus (PRRSV). J Comp Pathol. 2002; 127(2–3):106–13.

51. Manickam C, Dwivedi V, Patterson R, Papenfuss T, Renukaradhya GJ. Porcine reproductive and respiratory syndrome virus induces pronounced immune modulatory responses at mucosal tissues in the parental vaccine strain VR2332 infected pigs. Vet Microbiol. 2013;162(1):68–77.

52. Truong HM, Lu Z, Kutish GF, Galeota J, Osorio FA, Pattnaik AK. A highly pathogenic porcine reproductive and respiratory syndrome virus generated from an infectious cDNA clone retains the in vivo virulence and transmissibility properties of the parental virus. Virology. 2004;325(2):308–19.

53. Nielsen HS, Liu G, Nielsen J, Oleksiewicz MB, Botner A, Storgaard T, Faaberg KS. Generation of an infectious clone of VR-2332, a highly virulent North American-type isolate of porcine reproductive and respiratory syndrome virus. J Virol. 2003;77(6):3702–11.

54. Berger Rentsch M, Zimmer G. A vesicular stomatitis virus replicon-based bioassay for the rapid and sensitive determination of multi-species type I interferon. PLoS One. 2011;6(10):e25858.

55. Sun D, Khatun A, Kim WI, Cooper V, Cho YI, Wang C, Choi EJ, Yoon KJ. Attempts to enhance cross-protection against porcine reproductive and respiratory syndrome viruses using chimeric viruses containing structural genes from two antigenically distinct strains. Vaccine. 2016;34(36):4335–42.

56. OIE. Porcine reproductive and respiratory syndrome, Terrestrial Manual for Diagnostics and Vaccines, vol. Chapter 2.08.07; 2011.

57. Khatun A, Shabir N, Yoon KJ, Kim WI. Effects of ribavirin on the replication and genetic stability of porcine reproductive and respiratory syndrome virus. BMC Vet Res. 2015;11:21.

58. Halbur PG, Miller LD, Paul PS, Meng XJ, Huffman EL, Andrews JJ. Immunohistochemical identification of porcine reproductive and respiratory syndrome virus (PRRSV) antigen in the heart and lymphoid system of three-week-old colostrum-deprived pigs. Vet Pathol. 1995;32(2):200–4.

59. Winer J, Jung CK, Shackel I, Williams PM. Development and validation of real-time quantitative reverse transcriptase-polymerase chain reaction for monitoring gene expression in cardiac myocytes in vitro. Anal Biochem. 1999;270(1):41–9.

60. Opriessnig T, McKeown NE, Harmon KL, Meng XJ, Halbur PG. Porcine circovirus type 2 infection decreases the efficacy of a modified live porcine reproductive and respiratory syndrome virus vaccine. Clin Vaccine Immunol. 2006;13(8):923–9.

61. Kim WI, Kim JJ, Cha SH, Yoon KJ. Different biological characteristics of wild-type porcine reproductive and respiratory syndrome viruses and vaccine viruses and identification of the corresponding genetic determinants. J Clin Microbiol. 2008;46(5):1758–68.

62. Wu WH, Fang Y, Farwell R, Steffen-Bien M, Rowland RR, Christopher-Hennings J, Nelson EA. A 10-kDa structural protein of porcine reproductive and respiratory syndrome virus encoded by ORF2b. Virology. 2001;287(1):183–91.

63. Niu P, Shabir N, Khatun A, Seo BJ, Gu S, Lee SM, Lim SK, Kim KS, Kim WI. Effect of polymorphisms in the GBP1, Mx1 and CD163 genes on host responses to PRRSV infection in pigs. Vet Microbiol. 2016;182:187–95.

21

Dexamethasone treatment did not exacerbate Seneca Valley virus infection in nursery-age pigs

Alexandra Buckley[1], Nestor Montiel[1,2], Baoqing Guo[3], Vikas Kulshreshtha[1,4], Albert van Geelen[1], Hai Hoang[3], Christopher Rademacher[3], Kyoung-Jin Yoon[3] and Kelly Lager[5]*

Abstract

Background: *Senecavirus A*, commonly known as Seneca Valley virus (SVV), is a picornavirus that has been infrequently associated with porcine idiopathic vesicular disease (PIVD). In late 2014 there were multiple PIVD outbreaks in several states in Brazil and samples from those cases tested positive for SVV. Beginning in July of 2015, multiple cases of PIVD were reported in the United States in which a genetically similar SVV was also detected. These events suggested SVV could induce vesicular disease, which was recently demonstrated with contemporary US isolates that produced mild disease in pigs. It was hypothesized that stressful conditions may exacerbate the expression of clinical disease and the following experiment was performed. Two groups of 9-week-old pigs were given an intranasal SVV challenge with one group receiving an immunosuppressive dose of dexamethasone prior to challenge. After challenge animals were observed for the development of clinical signs and serum and swabs were collected to study viral shedding and antibody production. In addition, pigs were euthanized 2, 4, 6, 8, and 12 days post inoculation (dpi) to demonstrate tissue distribution of virus during acute infection.

Results: Vesicular disease was experimentally induced in both groups with the duration and magnitude of clinical signs similar between groups. During acute infection [0–14 days post infection (dpi)], SVV was detected by PCR in serum, nasal swabs, rectal swabs, various tissues, and in swabs from ruptured vesicles. From 15 to 30 dpi, virus was less consistently detected in nasal and rectal swabs, and absent from most serum samples. Virus neutralizing antibody was detected by 5 dpi and lasted until the end of the study.

Conclusion: Treatment with an immunosuppressive dose of dexamethasone did not drastically alter the clinical disease course of SVV in experimentally infected nursery aged swine. A greater understanding of SVV pathogenesis and factors that could exacerbate disease can help the swine industry with control and prevention strategies directed against this virus.

Keywords: Dexamethasone, Seneca Valley virus (SVV), Vesicular disease, Swine, Nursery-age pigs

Background

Vesicular disease in swine is recognized by the development of vesicles on the feet, snout, and less frequently in the oral cavity [1, 2]. Nonspecific signs include fever, lethargy, anorexia, and lameness. Known viral causes of swine vesicular disease are vesicular stomatitis virus, swine vesicular disease virus, vesicular exanthema of swine virus, and importantly, foot-and-mouth disease virus (FMDV). FMDV causes one of the most highly contagious diseases of livestock that can result in devastating economic losses to the agricultural industry and disruption of the human food supply [3, 4]. Since vesicular diseases of swine are indistinguishable in the field, each case must be treated as if it was an FMDV infection, which triggers a significant response in countries where FMDV is not endemic. Although rare, swine vesicular diseases have occurred in countries without known vesicular viruses or the known causes have been ruled out, resulting in the diagnosis of porcine idiopathic vesicular disease (PIVD).

* Correspondence: kelly.lager@ars.usda.gov
[5]U.S. Department of Agriculture, Virus and Prion Research Unit, National Animal Disease Center, Agricultural Research Service, 1920 Dayton Avenue, PO Box 70, Ames, IA 50010, USA
Full list of author information is available at the end of the article

Senecavirus A, commonly known as Seneca Valley virus (SVV), is a non-enveloped, single-stranded RNA virus in the family *Picornaviridae* first identified as a tissue culture contaminant in 2002 [5]. A retrospective analysis confirmed that since the late 1980s, SVV has been sporadically isolated from swine samples in the United States (US) [6], and detected by PCR in more recent cases of PIVD [7, 8]. Although early attempts to induce vesicular disease with field SVV isolates were unsuccessful [6, 9–11], it was presumed that SVV was the causative agent.

In late 2014, outbreaks of vesicular disease in finishing swine, and in sow farms with concurrent reports of increased neonatal mortality occurred in multiple states in Brazil [12, 13]. As part of the diagnostic investigations ruling out known causes of vesicular disease in swine, SVV was identified by PCR and virus isolation in multiple samples from affected animals. In July 2015, similar outbreaks began in the US and SVV was also detected in those cases [14–17]. Collectively, the association of SVV with vesicular disease in Brazil and the US provided strong support for SVV as the causal agent. This was confirmed with the fulfillment of Koch's postulates in 9-week old pigs using a 2015 SVV isolate from the US [18]. Since that report, vesicular disease was also experimentally reproduced with SVV infection in nursery pigs [19] as well as in finishing-aged swine [20]. Although SVV was rarely detected in North America prior to the 2014/2015 unprecedented emergence of PIVD in Brazil and the United States, it has been detected many times since then in the respective countries as well as recent novel case reports in Canada [21], China [22–24], Thailand [25], and Colombia [26]. Interestingly, viruses from these recent outbreaks are genetically similar sharing > 94% nucleotide identity at the full-length genomic level.

In an early PIVD report there was speculation that stressful events in the field may predispose pigs to SVV clinical disease; e.g., after transportation to slaughter [8]. Similar observations in the 2014/2015 SVV cases supported this assumption which led to the original experiment using an immunosuppressive model to test the hypothesis that administration of a synthetic glucocorticoid would exacerbate the SVV infection in swine. Surprisingly, both non-dexamethasone treated pigs as well as dexamethasone treated pigs developed vesicular disease of comparable severity. The acute phase of the vesicular disease in the non-dexamethasone SVV-challenged pigs was previously reported [18]. This manuscript describes the kinetics of the SVV infection and the comparison between the dexamethasone and non-dexamethasone treated pigs.

Results

Clinical and microscopic observations

All pigs were free from signs of vesicular disease prior to challenge, and all control pigs appeared normal throughout the experiment. One pig in the Dex-SVV group became anorexic at 2 dpi and was removed from the experiment because it was not competitive in a group environment. The pig's health continued to deteriorate and it died 2 days post removal from the group. Although no definitive cause of death was determined, it is believed SVV did not contribute to the illness and death since the only clinical signs recognized in the other pigs was transient lameness.

A mild transient lameness was recognized in 2–3 pigs from both the Dex-SVV and SVV groups on 2 and 3 dpi. No gross abnormalities in behavior or appearance were observed in pigs euthanized on 2, 4, 8, and 12 dpi for necropsy.

The acute lesions for the SVV pigs were previously described [18]. The lesions that developed in the Dex-SVV group were indistinguishable from the SVV group and are briefly described below. At the 4 dpi daily observation, cutaneous lesions were detected in 8/11 Dex-SVV pigs (72.7%) and 7/16 SVA pigs (43.8%). Cutaneous lesions consisted of small vesicles (about 3 mm × 3 mm) and/or erosions first noticed at 4 dpi in the interdigital spaces and coronary bands of one or more feet. At 5 dpi, all Dex-SVV pigs were observed with vesicular lesions and 14/15 SVV pigs had at least one lesion. Lesions were recognized as small, pale or blanched areas of swelling on the coronary band that would grow in size, thicken and become raised (Fig. 1). Usually, the skin would wear away leaving an erosion or ulcer that could coalesce with adjacent lesions. Snout lesions, when present, were mostly recognized as an elliptical erosion (3 mm × 5 mm) that was on the dorsal ridge of the snout which quickly healed. No new coronary band lesions were recognized after 6 dpi at which time the lesions began to heal.

Microscopic examination of coronary band tissue sections from the pigs euthanized revealed lesions in the 4 and 6 dpi pigs. Extensive areas of epidermis, predominantly stratum spinosum, was affected and effaced by multifocal to coalescing vesicles, containing fibrin, edema, necrotic debris and infiltration of lymphocytes and plasma cells (Fig. 2). Occasionally, these vesicles progressed to pustules, which were characterized by degenerate neutrophils admixed with variable amounts of fibrin, cellular and karyorrhectic debris. A few of these inflammatory cells were multifocally observed in the superficial dermis (beneath the affected epidermis).

SVV RNA detection

Selected samples from control animals all tested negative for SVV RNA by PCR. In serum samples of both Dex-SVV ($n = 11$) and SVV ($n = 12$) groups SVV RNA was detected in pigs as early as 1 dpi (Fig. 3). For the Dex-SVV group, peak viremia occurred at 5 dpi with a mean value of 5.7×10^4 genomic copies per microliter (GC/µL) of serum. Peak of viremia was observed at 3 dpi in the SVV group with a mean value of

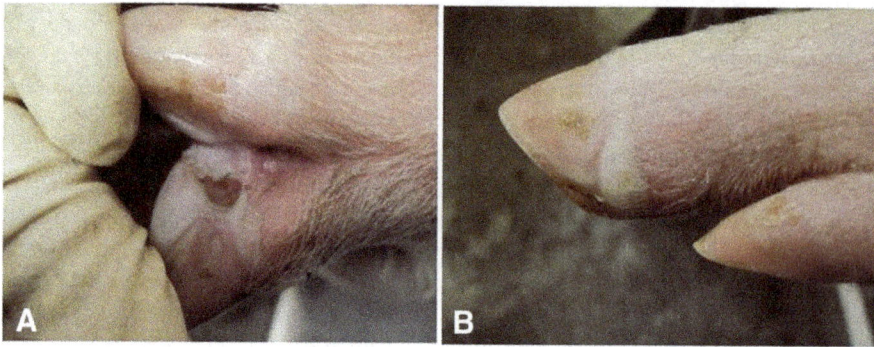

Fig. 1 Vesicular lesions from 9-week-old swine. **a**) Ruptured vesicle in the interdigital space. **b**) Intact vesicle on the lateral coronary band

3.1×10^5 GC/μL of serum. Though different peaks were observed, no statistically significant differences were detected in the magnitude of viremia between Dex-SVV and SVV groups. Mean SVV RNA concentration in serum in both groups decreased over time with minimal amounts detected at 15 (0.6 GC/μL) and 22 dpi (0.3 GC/μL).

In nasal swabs, SVV RNA was detected in one or more pigs at all time-points of the study, with a peak mean value of 2.0×10^5 GC/μL in the Dex-SVV group and 8.8×10^4 GC/μL in the SVV group at 3 dpi (Fig. 3). By 30 dpi, GCs were found at low levels in all animals of the SVV group and only three animals were PCR positive in the Dex-SVV group. There was no statistically significant difference in GC/uL levels between the Dex-SVV and SVV group.

In rectal swabs, SVV RNA levels were approximately 2 to 3 logs lower than those in nasal swabs (Fig. 3). An observable peak occurred at 3 dpi for the Dex-SVV group (9.5×10^2 GC/μL) and at 5 dpi for the SVV group (2.6×10^2 GC/μL) with statistically significant differences between the two groups detected at 3 and 5 dpi ($p < 0.05$). In addition, both experimental groups had a transient increase in SVV RNA levels from 11 dpi to 15 dpi.

SVV RNA was detected in almost every tissue tested in each pig necropsied at 2, 4, 6, 8, and 12 dpi (Fig. 4). In general, the lowest RNA levels were detected in pig 234 (2 dpi) with peak RNA levels detected in pig 235 (4pi) followed by a reduction over time for pigs 236 (6 dpi), 237 (8 dpi), and 238 (12 dpi), respectively. SVV RNA was found at much higher concentrations in the tonsil (7.83×10^5 GC/μL

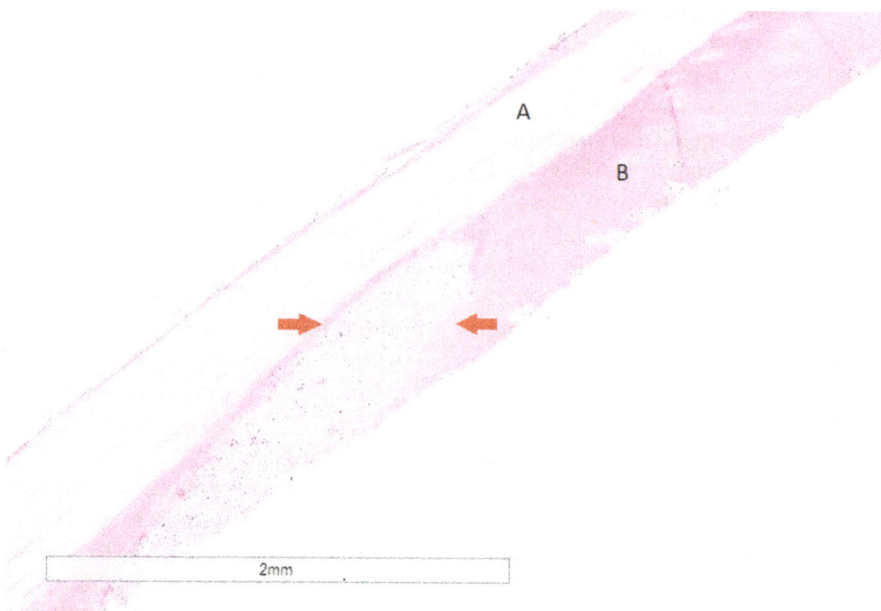

Fig. 2 Microscopic lesions from the coronary band. Cut section of the epidermis with the (**a**) stratum corneum and (**b**) stratum spinosum. Red arrows point to an oval shaped vesicle disrupting the stratum spinosum, which contains small numbers of neutrophils, lymphocytes, varying amounts of fibrin, edema admixed with necrotic and cellular debris

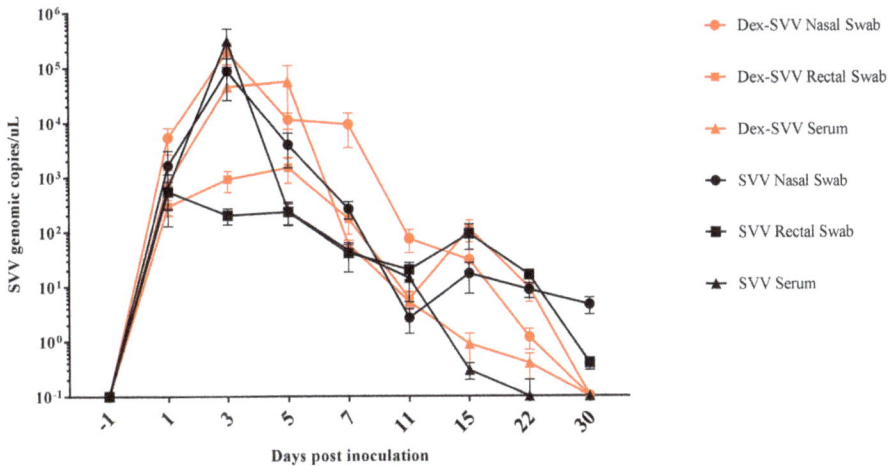

Fig. 3 SVV infection dynamics. Virus shedding in nasal and rectal swabs and viremia levels detected by RT-qPCR. Quantity of viral RNA is expressed as genomic copies per microliter. Error bars represent the standard error of the mean. Red data lines represent the Dex-SVV group and black data lines represent the SVV group. Nasal swabs are designated with a circle, rectal swabs with a square, and serum with triangles. No viral shedding or viremia was detected in control animals

at 4 dpi) and inguinal lymph nodes (6.55×10^5 GC/μL at 6 dpi) compared to the other tissues.

Serological responses

All 0 dpi sera samples were negative for antibodies against SVV by IFA and VN assays. All challenged pigs developed IgG IFA antibodies against SVV by 11 dpi with titers ranging from 1:320 to 1:640. Pigs from both groups had a similar IgG response throughout the study with most having a titer of 1:1280 on 30 dpi (Fig. 5). All but 2 pigs in the Dex-SVV group developed measurable neutralizing antibodies by 5 dpi. VN titers peaked in both groups at 7 dpi. Again, the neutralizing antibody response was similar for both the Dex-SVV and SVV groups over the course of the experiment as shown in Fig. 5. Control pig sera had VN titers ≤1:4.

Discussion

Our hypothesis for this study was that stress induced through an immunosuppressive dose of dexamethasone would exacerbate clinical disease in SVV challenged pigs compared to those not treated with dexamethasone. This experimental question was derived from failed attempts prior to 2016 to reproduce vesicular disease with SVV, and the observations that PIVD cases were associated with times of stress such as transportation of finishing pigs to market [8] and congregations of show pigs. In this study, SVV infection induced prolific vesicular lesions in both the SVV and Dex-SVV treatment groups. The onset, character, and duration of the lesions were similar between groups indicating the dexamethasone treatment did not dramatically alter the clinical disease

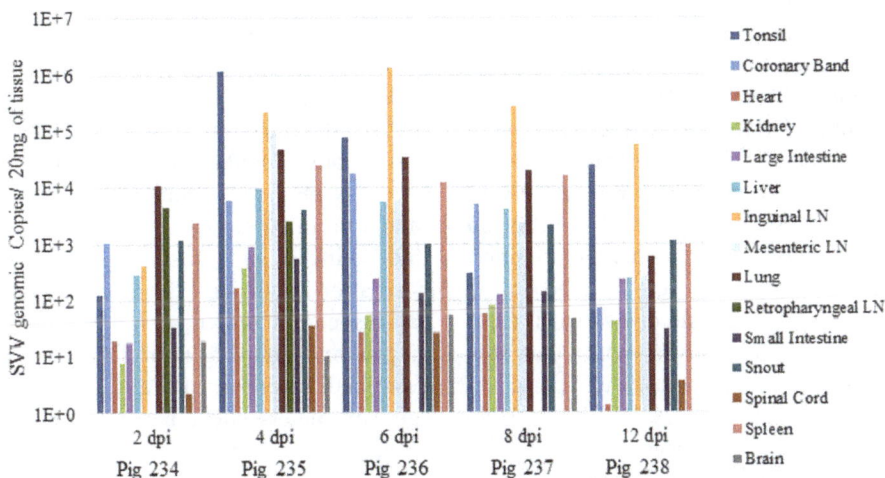

Fig. 4 Tissue distribution of SVV. Viral load present in selected tissues from necropsied pigs on 2, 4, 6, 8, and 12 dpi respectively. Quantity of viral RNA was determined by RT-qPCR and expressed as genomic copies per 20 mg of tissue

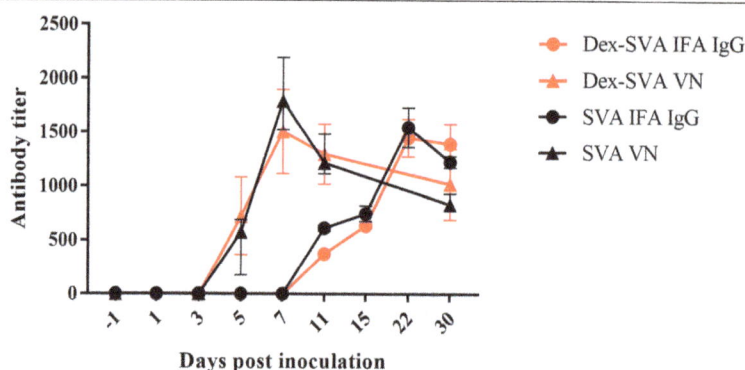

Fig. 5 Antibody response to SVV infection. Indirect immunofluorescence assay measured IgG antibody response to SVV infection. Virus neutralization assay measured neutralizing antibody response to SVV. Red data lines represent the Dex-SVV group and black data lines represent the SVV group. IFA titers are depicted with circles while VN titers are designated with triangles. Error bars represent the standard error of the mean

of pigs after SVV challenge. More pigs developed vesicular lesions earlier in the Dex-SVV group, but by 5 dpi all but one pig in this experiment developed clinical signs. For the purposes of this discussion, both groups will be discussed as one unless noted otherwise.

Development of vesicular disease in both challenge groups was surprising since experimental infections reported before 2016 were unable to induce clinical disease [6, 9–11]. Vesicular disease reproduced in this study was compatible with early field reports of SVV-related PIVD in finishing-aged pigs and sows [12, 13, 17], and similar to recent reports of experimentally induced vesicular lesions with US wild-type virus in 3-week-old pigs [19] and 55 kg pigs [20]. Collectively, this research may provide insight into the kinetics of SVV-associated vesicular disease that can provide evidence for recommendations for diagnostic sampling and possible control strategies.

In this study, lesions were first noted in the interdigital spaces when pigs were observed at 4 dpi. It is likely the onset of foot lesions began at an earlier time-point, given that by 4 dpi some pigs already had both intact and ruptured vesicles on their feet. The incidence of foot lesions peaked about 5–6 dpi and most lesions had resolved in about a week after onset of clinical signs. Our timeline of vesicular lesion development is similar to that reported by Joshi et al. in finishing pigs experimentally inoculated with SVV where lesions were first observed at 4 dpi [20]. In contrast to onset and incidence of coronary band lesions, snout lesions observed in this study were first recognized around 7 dpi in just a few pigs as one or two small elliptical plaques on the dorsal ridge of the snout. In comparison, a higher incidence of snout lesions was reported in older [20] and nursery age pigs [19].

Viremia and viral shedding were measured by RT-PCR to gain a better understanding of SVV infection kinetics. PCR results revealed a rapid onset of viremia with many pigs testing positive for SVV in serum by 1 dpi. Peak RNA

concentration in sera occurred at 3–5 dpi followed by a rapid decrease becoming almost undetectable by 15 dpi. Viral shedding also had a rapid onset with both nasal and fecal swabs testing positive for SVV at 1 dpi. The Dex-SVV group had statistically higher levels of SVV detected by PCR at 3 and 5 dpi in the rectal swabs. Virus was detected longer in nasal swabs than fecal swabs, especially in the SVV group. By 30 dpi most pigs in either group were SVV-negative on rectal swab, but some of the SVV group were still nasal swab positive for viral RNA, though there was no statistical difference between groups. A similar shedding pattern was observed by Joshi et al. who reported that oral, nasal, and fecal swabs became PCR negative by 28 dpi in finishing pigs [20]. Oral swabs were not collected in the present study, but others have reported a greater amount of virus detected and longer shedding compared to fecal or nasal swabs [19, 20].

In the pigs euthanized from 2 to 12 dpi, 4 dpi was the time point with highest detection of SVV nucleic acids in tissues. In general, the tissues with the greatest SVV RNA presence were coronary band epithelia, tonsils, inguinal and mesenteric lymph nodes, lung, liver, and spleen. In addition, most tissues had higher SVV RNA concentrations than serum, which suggests that viral concentrations in the tissues is not solely due to blood contamination. A similar tissue distribution was reported in mid-finishing swine during acute infection in addition to lymph nodes, spleen, intestine, kidneys, and tonsils testing positive by PCR 38 days after challenge [20]. Similarly, diagnostic investigations conducted in sow herds affected by vesicular disease and neonatal mortality in the US and in Brazil described detection of SVV nucleic acids in most tissues of piglets including the small intestine, tonsil, lung, heart, liver, spleen, kidney, myocardium and cerebellum [27, 28]. Previous field case reports have documented detection of SVV in scrapings of ruptured vesicles and ulcerative lesions by PCR in sows and finishing pigs [7, 12, 13, 28]. Based on PCR results in our study, vesicular fluid/swabs had the highest concentration of virus compared to other samples.

The onset of the humoral immune response was similar for both groups with antibody titers first being detected at 5 dpi by VN assay and 11 dpi by IFA assay. The onset and peak of antibody titers was more rapid for neutralizing antibody response compared to IgG measured by IFA assay. Similarly, Joshi et al. detected VN antibody as early as 5 dpi, which was also described by Yang et al., and SVV-specific IgG antibodies by 10 dpi [11, 20]. Chen et al. reported neutralizing antibody response as early as 3 dpi in weaned pigs [19]. Aforementioned studies have credited the early neutralizing antibody response for the quick decline in viremia reported in various aged swine. The duration of a SVV humoral immune response is unknown though unpublished work has demonstrated sterilizing immunity in gilts challenged five months after initial exposure to wild type SVV (Buckley, unpublished observations).

Seneca Valley virus is in the same virus family as FMDV and shares many structural characteristics as well as a similar clinical presentation. Vesicular lesions can develop within 24–48 h of FMDV challenge [2, 3, 29], and are most commonly found on the coronary bands [30]. FMDV can also be detected in serum as soon as 24 h after exposure followed by gradual decline over a couple of weeks [31]. Antibodies against FMDV have been detected by ELISA around 7 dpi (IgM) and 14 dpi (IgG, IgA) in swine either inoculated or in direct contact [32]. Although vesicular lesions contain the highest concentration of FMDV, virus replication does occur in other tissues, e.g., in the tonsil [31]. This study and others have consistently found high concentrations of SVV nucleic acids present in the tonsil not only acutely, but also in convalescent swine [15, 20], which may provide evidence for the tonsil being a primary site of viral replication. The kinetics of the SVV infection are similar to FMDV; however, it is unknown if SVV is as infectious as FMDV. Replication and shedding of SVV from the tonsil into the oropharyngeal cavity probably plays an important role in producing SVV-positive oral secretions, and may contribute to contamination of the nasal passages and the finding of SVV-positive nasal swabs. However, the possible replication of SVV in nasal mucosa has not been ruled out.

Unpublished observations from our laboratory support recent field reports describing environmental contamination with SVV nucleic acid [33], which combined with normal pig behavior such as rooting and inquisitive chewing, could cause false-positives in oral fluid samples. The extended presence of SVV in tonsil tissue may help explain the apparent longer shedding patterns reported for oral fluids and nasal swabs when compared to fecal swabs [19, 20]. Further study will be required under experimental conditions to characterize the duration of shedding of infectious virus from the oral cavity.

Results from Chen et al., Joshi et al., and the present study can certainly be used for diagnostic recommendations for nursery to mid-finishing pigs that will be presented dogmatically for simplicity. Viral RNA can be detected in serum from at least 2–7 days, in feces (rectal swabs) from at least 2–15 days, in nasal swabs from at least 2–15 days, and in oral fluids from at least 2–21 days post infection. During the acute infection viral RNA can be detected in many tissues for at least 2–12 days post infection and can also serve as potential diagnostic samples. Finally, swabs of vesicular lesions may serve as the best sample to collect during clinical disease to reliably detect the presence of SVV due to the large quantities of virus found in that location. Lastly, it is expected that sows infected with SVV will have similar infection kinetics to nursery and mid-finishing swine and sample recommendation listed above would also pertain to mature animals.

Compared to other species, swine are relatively resistant to glucocorticoids [34, 35]; however, their use has induced recrudescence of latent pseudorabies virus in swine indicating glucocorticoids can alter or suppress a pig's immune system [36]. Dexamethasone administration in this study was based on previous studies that successfully induced recrudescence of wild-type and attenuated PRV vaccine [36], and presumably this dosing schedule would be a substitute for stress that would alter the immune system allowing for enhanced vesicular disease. The Dex-SVV group had more pigs observed with vesicular lesions on 4 dpi, but by 5 dpi all except 1 pig was effected in both groups. In addition, there were higher concentrations of SVV detected in rectal swabs on 3 and 5 dpi compared to the SVV group. It is unknown if a higher glucocorticoid dose might have altered the disease course more significantly, but such efforts may not be needed with contemporary SVV isolates based on this experiment and the previous successful reports [19, 20].

The relative ease in which vesicular disease was produced in both groups of pigs in this study, and in the experiments of Chen et al. and Joshi, et al. provides a basis for a variety of questions and assumptions about contemporary SVV isolates, and why attempts to reproduce vesicular disease prior to 2016 were unsuccessful. First, the genetic similarity of current SVV isolates from Brazil, U.S., Canada, China, Thailand, and Colombia suggests a recent common ancestor which might explain the timing of newly recognized outbreaks across separate continents. Second, there may be a mutation in SVV that increased the likelihood that contemporary viruses are more likely to induce vesicular disease than viruses isolated before 2014. If either of these factors played a role in the emergence of the mini SVV epidemics, then there must be some explanation for the relatively recent distribution of the common virus among continents. However, an increased awareness of potential SVV-related vesicular disease may be contributing to the current observations. For example, the increase in reports of clinical disease associated with SVV isolates in China [24] and Thailand [25] that are more distantly related genetically to the contemporary Brazilian and US isolates suggests current clinical cases are not completely dependent on the emergence of a new lineage of virus. Third, there may be a novel cofactor(s) such as another infectious agent allowing disease to develop, which if true, must also have been transmitted to different continents. Fourth, there may be a genetic predisposition to clinical disease that is more prevalent now when compared to pre-2014 swine, and this trait has been slowly building in swine herds in countries reporting recent events. In addition, there may be an age-dependent effect on clinical presentation of disease which makes it more difficult to observe disease in young pigs compared to mature swine. The focal nature of vesicular disease demonstrates a discrete viral tropism for a cell type in a specific anatomical site, i.e., coronary band tissue. Such a condition might involve an increased density of this cell type in older swine, or this tissue may be more prevalent in one genetic line compared to another which would contribute to the presence of gross lesions. Lastly, there is the possibility that two or more of the above factors contributed to the emergence of the mini SVV epidemics.

Conclusions

This study adds to the growing body of knowledge about the pathogenesis of SVV in swine that has focused on understanding the acute infection (clinical disease, viremia, viral shedding, virus distribution in tissues, and immune response). This work has demonstrated the need to investigate the duration of immunity in convalescent swine and the spectrum of immunity against historical and contemporary strains to discern why an apparent change has occurred in the ecology of SVV in swine. The clinical description and kinetics from this study can also help diagnosticians and veterinarians improve strategies to control this disease and differentiate it from other vesicular diseases in swine.

Methods

Cells and virus

A swine testicular (ST) cell line (ATCC® CRL-1746; American Type Culture Collection, Manassas, VA) was cultured at 37 °C and 5% CO_2 in minimum essential medium (MEM, MilliporeSigma, St. Louis, MO) supplemented with 10% fetal bovine serum (FBS, AtlantaBio, Flowery Way, GA), 1x gentamicin, and 1x glutamax (Life Technologies, Carlsbad, CA). NCI-H1299 cells (ATCC® CRL-5803) were cultured under the same conditions in Dulbecco's modified Eagle's medium (DMEM, Millipore-Sigma) supplemented with 10% FBS (AtlantaBio).

SVA15-41901SD was isolated on ST cells from samples collected from a barn of finishing pigs in South Dakota that developed vesicular lesions during the summer of 2015 [15]. The second and third passage of the virus were combined to create a larger volume of working stock virus which was utilized for inoculum and laboratory assays. The stock virus had a titer of 4.75×10^7 plaque-forming units (PFU)/ml and was tested for purity by next-generation-sequencing that only detected SVA.

Animals and experimental design

Forty-nine conventionally-raised weaned pigs were purchased, randomly assigned ear tags and housed at the USDA-ARS-NADC campus in accordance with Institutional Animal Care and Use Committee guidelines (ACUP #2867) until 9 weeks of age at which time 29 pigs received SVV at a dose of 4.75×10^7 PFU/animal via intranasal inoculation. A LMA® MAD Nasal™ Intranasal Mucosal Atomization Device (Teleflex; Morrisville, NC) was used for delivery of about 2.5 mL of the atomized inoculum into each nostril for a total of 5 mL. Twelve pigs (Dex-SVV group) based on sequential ear tag numbers were given an immunosuppressive treatment of dexamethasone (AgriLabs; St. Joseph, MO) intramuscularly in the neck for 5 days prior to challenge as follows: day 1, 2.3 mg/kg of body weight twice a day (BID); days 2–5, 1.1 mg/kg BID. Seventeen pigs (SVV group) did not receive any treatment before challenge. The remaining 20 pigs were unchallenged control pigs (Control group). Each group was housed in a separate animal-biosafety-level 2 room, with at least 17 square feet/pig and constant access to feed and water. One pig each from the SVV group was euthanized on 2, 4, 6, 8, and 12 days post inoculation (dpi). At the time of euthanasia, the animal was physically restrained for the intravenous administration of a barbiturate (Fatal Plus, Vortech Pharmaceuticals, Dearborn, MI) following the manufacturer label dose (1 mL/4.54 kg). At the conclusion of the study the remaining pigs were also euthanized in the same manner.

Sample collection

Progression of clinical disease for each pig was assessed by daily observations for lameness and vesicular lesions. Antemortem sampling consisted of collection of whole blood in serum separation tubes (BD Vacutainer®, Franklin Lakes, NJ), and nasal and rectal swabs. Swabs (Puritan Medical Products, Guilford, ME) were collected and immersed in 3 ml of serum-free MEM (MilliporeSigma). Samples were collected on 0, 1, 3, 5, 7, 11, 15, 23, and 30 dpi from the SVV and Dex-SVV groups. Pigs from the control group were sampled at 0, 7, 15, and 29 dpi. When vesicles were observed, vesicular fluid collection was attempted via aspiration or swabs. Blood tubes were centrifuged to harvest the serum. Sera, vesicular fluids, and swab tubes were stored at − 80 °C until time of testing.

In the pigs euthanized on 2, 4, 6, 8, and 12 dpi, all post-mortem examinations were performed immediately after euthanasia, and tissue specimens collected included snout and coronary band epithelium, tonsil, lymph nodes (retropharyngeal, mesenteric, and/or inguinal), lungs, heart, liver, spleen, kidney, small and large intestine, brain and spinal cord. For each anatomically defined specimen, approximately 10 g of tissue was collected and placed into a self-sealing plastic bag on dry ice for transfer within 2 h to a − 80 °C freezer. Tissues were thawed and 20 mg of each sample resuspended in 5 mL 1x PBS (Thermo Scientific, Waltham, MA) in individual gentleMACS™ M tubes (Miltenyi Biotec, Auburn, CA). Tissue was dissociated using a gentleMACS™ Octo-Dissociator (Miltenyi Biotec, Germany) following the manufacturer's recommendations. After dissociation, tissue suspensions were aliquoted in 2 mL cryogenic vials (Corning, Corning, NY) and frozen at − 80 °C until time of processing. For histology, skin sections from the coronary band region were collected in 10% neutral buffered formalin. Sections were fixed in formalin for 24 h, routinely processed, paraffin-embedded, sectioned at 4-μm thickness, and stained with hematoxylin and eosin.

Seneca Valley virus-specific nucleic acid detection

Samples were extracted using the MagMAX™ Pathogen RNA/DNA kit (Life Technologies, Carlsbad, CA) following manufacturer's recommendations and a MagMAX™ Express instrument 24 (Life Technologies) using program AM1836 (Life Technologies). The viral RNA was eluted in 90 μL of elution buffer. Following extraction, 5 μL of the nucleic acid templates were added to 20 μL of the Path-ID™ Multiplex One-Step RT-PCR reaction master mix (Applied Biosystems, Foster City, CA) for fecal swabs or 20 μL AgPath-ID™ One step RT-PCR kit (Applied Biosystems) for nasal swabs, vesicle fluids, and sera. The primers and probe were designed to target the conserved region between the 5′ untranslated region (5′UTR) and protein L containing nucleotides 602–710 of the SVA genome. The forward primer sequence was 5′-TGCCTTGGATACTGCCTGA TAG-3′, the reverse primer sequence was 5′-GGTG CCAGAGGCTGTATCG-3′ and the probe sequence was 5′-CGACGGCCTAGTCG GTCGGT T-3′. The probe was labeled using 6-FAM™ at the 5′ end, ZEN™ internal quencher, and Iowa Black® quencher at the 3′ end (Integrated DNA Technologies, Coralville, IA). Real-time RT-PCR was performed on an ABI 7500 Fast instrument (Life Technologies) run in standard mode with the following conditions: 1 cycle at 48 °C for 10 min, followed by 1 cycle at 95 °C for 10 min, and 40 cycles of 95 °C for 15 s and 60 °C for 45 s. SVA genome RNA copies were calculated based on a standard RNA transcript overlapping the target region.

Indirect immunofluorescence (IFA) assay

SVV-specific IgG antibody response to SVV challenge was evaluated by an IFA assay. Serum samples were serially diluted 1:2 up to (1:5120). Plates were previously prepared with SVV-infected NCI-H1299 cells overnight and fixed with cold 80% acetone, and stored at − 20 °C. Plates were rehydrated with 200 μL of PBS and 50 μL of diluted serum were added to wells to incubate for 1 h. Plates were washed 3 times and 1:50 diluted anti-swine IgG-FITC antibody (KPL, MD, USA) was added to incubate for 45 min. Again, plates were washed 3 times and wells were observed for fluorescence under a fluorescent microscope. The highest serum dilution with clear and specific staining was considered as the end point (i.e., IFA titer).

Virus neutralization (VN) assay

Serum samples were heat-inactivated at 56 °C for 30 min and serially diluted 1:4 (up to 1:4096) in MEM in a 96-well plate and repeated in quadruplicate. Each diluted serum was mixed with an equal volume of SVA15-41901SD (200 TCID$_{50}$) and incubated for 1 h at 37 °C. One-hundred microliters of each virus-serum mixture was transferred to respective wells of 96-well plates of ST cells grown to a confluent monolayer and replenished with MEM supplemented with 2% FBS. Plates were read for CPE daily for 4 days. VN titers, based on CPE, were recorded as the highest dilution of serum at which the infectivity of SVA-41901SD was completely neutralized in 50% of the inoculated wells. A back titration of the virus was performed during each test.

Statistical analysis

Data analyses and graphic representations were performed by using Microsoft Office Excel 2010 and GraphPad Prism 7.03. Statistical analyses of the data were performed using a mixed linear model (SAS 9.4 for Windows XP, SAS Institute Inc., Cary, NC, USA) for repeated measures and a spatial spherical or autoregressive covariance structure was utilized. Linear combinations of the least squares means estimates for GC/uL were used in a priori contrasts after testing for either a significant ($P < 0.05$) effect of treatment or a significant treatment by time interaction. Comparisons were made between treatment groups at each time-point using 5% level of significance ($P < 0.05$) to assess statistical differences.

Abbreviations
BID: twice a day; CPE: cytopathic effect; DMEM: Dulbecco's modified Eagle's medium; dpi: days post infection; FBS: fetal bovine serum; FMDV: foot-and-mouth disease virus; GC: genomic copies; IFA: indirect immunofluorescence; MEM: minimum essential media; PCR: polymerase chain reaction; PFU: plaque-forming units; PIVD: Porcine idiopathic vesicular disease; ST: swine testicular; SVV: Seneca Valley virus; US: United States; VN: virus neutralization

Acknowledgements
The authors thank Deb Adolphson and Sarah Anderson for technical assistance and Jason Huegel, Justin Miller, and Keiko Sampson for assistance with animal studies.

Funding
This study was supported by USDA-ARS, USDA-APHIS; AB, NM, VK, and AVG were supported in part by an appointment to the ARS-USDA Research Participation Program administered by the Oak Ridge Institute for Science and Education (ORISE) through an interagency agreement between the U.S. Department of Energy (DOE) and USDA. ORISE is managed by ORAU under DOE contract number DE-AC05-06OR23100. Mention of trade names or commercial products in this article is solely for the purpose of providing specific information and does not imply recommendation or endorsement by the USDA, DOE, or ORISE/ORAU. USDA is an equal opportunity provider and employer. Funding agencies did not have a role in the design of the study and collection, analysis, interpretation of data, or writing the manuscript.

Author's contributions
AB, NM, VK, AVG, CR, KJY, and KL contributed to the conception and design of the study. AB, NM, VK, AVG, KL performed the infection trial in pigs. CR, BG, and KJY contributed essential materials. BG, HH, KJY, AB, NM performed laboratory testing of samples. AB, NM, KJY, KL wrote the paper. All authors provided draft edits, read, and approved the final manuscript.

Consent for publication
Not applicable.

Competing interests
The authors declare that they have no competing interests.

Author details
[1]U.S. Department of Agriculture, Oak Ridge Institute for Science and Education and National Animal Disease Center, Ames, IA, USA. [2]Present address: U.S. Department of Agriculture, Avian Viruses Section, Diagnostic Virology Laboratory, National Veterinary Services Laboratories, Animal and Plant Health Inspection Service, Ames, IA, USA. [3]Department of Veterinary Diagnostic and Production Animal Medicine, College of Veterinary Medicine, Iowa State University, Ames, IA, USA. [4]Present address: Toxikon Corporation, Bedford, MA, USA. [5]U.S. Department of Agriculture, Virus and Prion Research Unit, National Animal Disease Center, Agricultural Research Service, 1920 Dayton Avenue, PO Box 70, Ames, IA 50010, USA.

References
1. Stenfeldt C, Diaz-San Segundo F, de Los ST, Rodriguez LL, Arzt J. The pathogenesis of foot-and-mouth disease in pigs. Front Vet Sci. 2016;3:41.
2. Murphy C, Bashiruddin JB, Quan M, Zhang Z, Alexandersen S. Foot-and-mouth disease viral loads in pigs in the early, acute stage of disease. Vet Rec. 2010;166(1):10–4.
3. Grubman MJ, Baxt B. Foot-and-mouth disease. Clin Microbiol Rev. 2004; 17(2):465–93.
4. Leforban Y. Prevention measures against foot-and-mouth disease in Europe in recent years. Vaccine. 1999;17(13–14):1755–9.
5. Hales LM, Knowles NJ, Reddy PS, Xu L, Hay C, Hallenbeck PL. Complete genome sequence analysis of Seneca Valley virus-001, a novel oncolytic picornavirus. The Journal of general virology. 2008;89(Pt 5):1265–75.

6. Knowles NJ, Hales LM, Jones BH, Landgraf JG, House JA, Skele KL, et al. Epidemiology of Seneca Valley virus: identification and characterization of isolates from pigs in the United States. Abstracts of the Northern Lights EUROPIC 2006 – XIV Meeting of the European Study Group on the Molecular Biology of Picornaviruses, Saariselka, Inari, Finland; November 26th – December 1st. 2006;Abstract G2.
7. Singh K, Corner S, Clark SG, Scherba G, Fredrickson R. Seneca Valley virus and vesicular lesions in a pig with idiopathic vesicular disease. J Vet Sci Technol. 2012;3(6):1–3.
8. Pasma T, Davidson S, Shaw SL. Idiopathic vesicular disease in swine in Manitoba. The Canadian veterinary journal La revue veterinaire canadienne. 2008;49(1):84–5.
9. Bracht AJ, O'Hearn ES, Fabian AW, Barrette RW, Sayed A. Real-time reverse transcription PCR assay for detection of Senecavirus a in swine vesicular diagnostic specimens. PLoS One. 2016;11(1):e0146211.
10. Willcocks MM, Locker N, Gomwalk Z, Royall E, Bakhshesh M, Belsham GJ, et al. Structural features of the Seneca Valley virus internal ribosome entry site (IRES) element: a picornavirus with a pestivirus-like IRES. J Virol. 2011;85(9): 4452–61.
11. Yang M, van Bruggen R, Xu W. Generation and diagnostic application of monoclonal antibodies against Seneca Valley virus. Journal of veterinary diagnostic investigation : official publication of the American Association of Veterinary Laboratory Diagnosticians, Inc. 2012;24(1):42–50.
12. Leme RA, Zotti E, Alcantara BK, Oliveira MV, Freitas LA, Alfieri AF, et al. Senecavirus a: an emerging vesicular infection in Brazilian pig herds. Transbound Emerg Dis. 2015;62(6):603–11.
13. Vannucci FA, Linhares DC, Barcellos DE, Lam HC, Collins J, Marthaler D. Identification and complete genome of Seneca Valley virus in vesicular fluid and sera of pigs affected with idiopathic vesicular disease, Brazil. Transbound Emerg Dis. 2015;62(6):589–93.
14. Hause BM, Myers O, Duff J, Hesse RA. Senecavirus a in pigs, United States, 2015. Emerg Infect Dis. 2016;22(7):1323–5.
15. Guo B, Pineyro PE, Rademacher CJ, Zheng Y, Li G, Yuan J, et al. Novel Senecavirus a in swine with vesicular disease, United States, July 2015. Emerg Infect Dis. 2016;22(7):1325–7.
16. Wang L, Prarat M, Hayes J, Zhang Y. Detection and genomic characterization of Senecavirus a, Ohio, USA, 2015. Emerg Infect Dis. 2016;22(7):1321–3.
17. Zhang J, Pineyro P, Chen Q, Zheng Y, Li G, Rademacher C, et al. Full-Length Genome Sequences of Senecavirus A from Recent Idiopathic Vesicular Disease Outbreaks in U.S. Swine. Genome Annouce. 2015;3(6):e01270-15.
18. Montiel N, Buckley A, Guo B, Kulshreshtha V, VanGeelen A, Hoang H, et al. Vesicular disease in 9-week-old pigs experimentally infected with Senecavirus a. Emerg Infect Dis. 2016;22(7):1246–8.
19. Chen Z, Yuan F, Li Y, Shang P, Schroeder R, Lechtenberg K, et al. Construction and characterization of a full-length cDNA infectious clone of emerging porcine Senecavirus a. Virology. 2016;497:111–24.
20. Joshi LR, Fernandes MH, Clement T, Lawson S, Pillatzki A, Resende TP, et al. Pathogenesis of Senecavirus a infection in finishing pigs. J General Virology. 2016;97(12):3267–79.
21. Xu W, Hole K, Goolia M, Pickering B, Salo T, Lung O, et al. Genome wide analysis of the evolution of Senecavirus a from swine clinical material and assembly yard environmental samples. PLoS One. 2017;12(5):e0176964.
22. Qian S, Fan W, Qian P, Chen H, Li X. Isolation and full-genome sequencing of Seneca Valley virus in piglets from China, 2016. Virol J. 2016;13(1):173.
23. Wu Q, Zhao X, Bai Y, Sun B, Xie Q, Ma J. The first identification and complete genome of Senecavirus a affecting pig with idiopathic vesicular disease in China. Transbound Emerg Dis. 2016.
24. Zhao X, Wu Q, Bai Y, Chen G, Zhou L, Wu Z, et al. Phylogenetic and genome analysis of seven senecavirus a isolates in China. Transbound Emerg Dis. 2017.
25. Saeng-Chuto K, Rodtian P, Temeeyasen G, Wegner M, Nilubol D. The first detection of Senecavirus A in pigs in Thailand. Transbound Emerg Dis. 2016:2017.
26. Sun D, Vannucci F, Corzo C, Marthaler DG. Emergence and whole-genome sequence of Senecavirus A in Colombia. Transboundary and emerging diseases: Knutson TP; 2017.
27. Leme RA, Oliveira TE, Alfieri AF, Headley SA, Alfieri AA. Pathological, Immunohistochemical and Molecular Findings Associated with Senecavirus A-Induced Lesions in Neonatal Piglets. Journal of comparative pathology. 2016.
28. Canning P, Canon A, Bates JL, Gerardy K, Linhares DC, Pineyro PE, et al. Neonatal mortality, vesicular lesions and lameness associated with Senecavirus a in a U.S. sow farm. Transbound Emerg Dis. 2016;63(4):373–8.

29. Alexandersen S, Oleksiewicz MB, Donaldson AI. The early pathogenesis of foot-and-mouth disease in pigs infected by contact: a quantitative time-course study using TaqMan RT-PCR. The Journal of general virology. 2001; 82(Pt 4):747–55.

30. Kitching RP, Alexandersen S. Clinical variation in foot and mouth disease: pigs. Rev Sci Tech. 2002;21(3):513–8.

31. Stenfeldt C, Pacheco JM, Rodriguez LL, Arzt J. Early events in the pathogenesis of foot-and-mouth disease in pigs; identification of oropharyngeal tonsils as sites of primary and sustained viral replication. PLoS One. 2014;9(9):e106859.

32. Pacheco JM, Butler JE, Jew J, Ferman GS, Zhu J, Golde WT. IgA antibody response of swine to foot-and-mouth disease virus infection and vaccination. Clin Vaccine Immunol. 2010;17(4):550–8.

33. Joshi LR, Mohr KA, Clement T, Hain KS, Myers B, Yaros J, et al. Detection of the emerging picornavirus Senecavirus a in pigs, mice, and houseflies. J Clin Microbiol. 2016;54(6):1536–45.

34. Roth JA, Flaming KP. Model systems to study immunomodulation in domestic food animals. Adv Vet Sci Comp Med. 1990;35:21–41.

35. Saulnier D, Martinod S, Charley B. Immunomodulatory effects in vivo of recombinant porcine interferon gamma on leukocyte functions of immunosuppressed pigs. Ann Rech Vet. 1991;22(1):1–9.

36. Mengeling WL, Lager KM, Volz DM, Brockmeier SL. Effect of various vaccination procedures on shedding, latency, and reactivation of attenuated and virulent pseudorabies virus in swine. Am J Vet Res. 1992;53(11):2164–73.

22

Identification of novel B-cell epitope in gp85 of subgroup J avian leukosis virus and its application in diagnosis of disease

Kun Qian[1,2,3†], Xue Tian[1†], Hongxia Shao[1,2,3], Jianqiang Ye[1,2,3], Yongxiu Yao[4,5], Venugopal Nair[4,5] and Aijian Qin[1,2,3*] (iD)

Abstract

Background: The gp85 is the main envelope protein of avian leukosis subgroup J (ALV-J) involved in virus neutralization. Here, we mapped the epitope in ALV-J gp85 by ELISA using synthetic peptides and developed epitope based diagnostic methods for ALV-J infection.

Results: The results revealed that monoclonal antibody (mAb) JE9 recognized [83]WDPQEL[88] motif, which was highly conserved in gp85 among different ALV-J strains by homology analysis. Moreover, after evaluation with two hundred and forty sera samples obtained from different chicken farms, the epitope-based peptide ELISA had much higher sensitivity than commercial ELISA kit for antibody detection of ALV-J.

Conclusions: A novel B-cell epitope recognized by the mAb JE9 was identified. The developed peptide-ELISA based on this novel B-cell epitope could be useful in laboratory viral diagnosis, routine surveillance in chicken farms, and also in understanding the pathogenesis of ALV-J.

Keywords: Avian leukosis virus subgroup J, B-cell epitope mapping, Epitope-based peptide ELISA, Antibody detection

Background

Since the first report of avian leukosis virus subgroup J (ALV-J) in the United Kingdom in 1988, the virus has spread rapidly to many countries including China [1–5]. As the most prevalent subgroup currently in China, ALV-J infection causes vascular and visceral neoplasms, decrease egg production, stunted growth, and increased mortality [3, 6]. Compared to the Western countries where ALV-J-associated disease was restricted to the meat-type chickens, the disease in China was also seen in layer flocks and local chicken breeds, possibly due to the extremely high mutation rates [7, 8].

The genome of ALV-J consists of three viral structural genes, *gag*, *pol*, and *env*, which encode the group-specific (gs) antigen, integrase and reverse transcriptase, and

envelop glycoprotein, respectively. The *env* gene encodes two proteins gp85 and gp37, which are synthesized as a single precursor polypeptide. The gp85 protein contains the determinants of ALV subgroup specificity, virus neutralization and receptor binding [9, 10]. Meanwhile, the gp85 is the most variable structural protein which exhibits high diversity in the genome of ALV-J [7, 11, 12]. Therefore, identification of the conserved epitopes in gp85 will facilitate the establishment of serological methods for the detection of ALV-J.

In our previous report, we showed that the gp85-specific mAb JE9 could react with different ALV-J strains but not with other ALV subgroups [13], confirming that the mAb JE9 recognized a conserved antigenic epitope. However, the exact epitope sequence recognised by the mAb JE9 has not been identified. In this study, we identified a conserved linear B-cell epitope recognised by the mAb JE9 using synthetic peptides, and applied it for the diagnosis of ALV-J infection using an epitope-based peptide ELISA. The results in this study will contribute to the understanding of the antigenic structure of gp85 and rational design of vaccines and diagnostic tools.

* Correspondence: aijian@yzu.edu.cn
†Kun Qian and Xue Tian contributed equally to this work.
1Ministry of Education Key Lab for Avian Preventive Medicine, College of Veterinary Medicine, Yangzhou University, No.12 East Wenhui Road, Yangzhou, Jiangsu 225009, People's Republic of China
2The International Joint Laboratory for Cooperation in Agriculture and Agricultural Product Safety, Ministry of Education, Yangzhou University, 225009 Yangzhou, People's Republic of China
Full list of author information is available at the end of the article

Results

Epitope mapping in gp85 protein recognized by mAb JE9

Our preliminary unpublished data using western blot assay showed that the mAb JE9 recognized epitope between the amino acid positions 65 to 155 of gp85 protein. For precise mapping of this epitope, ALV-J-1P (95-125aa), ALV-J-2P (126-155aa) and ALV-J-3P (65-94aa), which covered 65–155 aa of gp85 protein were synthesized. The OD450 values of peptide ELISA revealed that JE9 reacted with ALV-J-3P but not with the other two (Table 1). Subsequently, ALV-J-3P was further truncated into different overlapping peptides described in Table 2. Accordingly, we identified WDPQEL as the target sequence of mAb JE9, which corresponds to 83–88 aa of ALV-gp85 as deletion of 83 W or 88 L disrupted the binding of the peptides with mAb JE9. The results indicated that the peptide ^{83}WDPQEL^{88} was the minimal epitope in the gp85 protein of ALV-J for binding with mAb JE9 (Fig. 1, Table 2).

The epitope is conserved among ALV-J strains

In order to evaluate the conservation of the mAb JE9 defined epitope, alignment analysis was performed with gp85 sequences of 25 ALV-J strains, 6 ALV-A strains, 6 ALV-B strains, 2 ALV-C strains, 1 ALV-D strain, 5 ALV-E strains, and 6 ALV-K strains reported in recent year. As illustrated in Fig. 2, the alignment results showed that the ^{83}WDPQEL^{88} is highly conserved among all ALV-J strains analysed.

Development and optimization of the peptide-ELISA procedure

In order to achieve the best reactivity of the peptide-ELISA, according to the previous experimental results and cost considerations, we choose BSA conjugated polypeptide ALV-J-3P-2 (BSA-3P-2), BSA – C-^{75}QALNTTLPWDPQELDILGSQ^{94}, as ELISA coating antigen. The optimal dilution of reagents was determined at 1 µg/ml for BSA-3P-2 peptide, 8% Normal Rabbit Serum in PBS with 0.05% Tween 20 for blocking reagent, 1:200 for serum sample and 1:10000 for the secondary horseradish peroxidase-conjugated rabbit anti-chicken immunoglobulin. Exposure time was optimally at 37 °C for 25 min. Using these conditions, the best signal with minimum background was obtained in the peptide-ELISA. And the cross reaction of the peptide-ELISA was excluded by sera from other common avian diseases (data not shown).

Table 1 Reactivity of the different synthetic peptides of gp85 with mAb JE9 using ELISA

	ALV-J-1P (95-125aa)	ALV-J-2P (126-155aa)	ALV-J-3P (65-94aa)
PBST	0.055[a]	0.047	0.054
JE9	0.049	0.056	1.236

[a]The mean values of triplicate OD450 detected by Bio-TEK ELISA reader

Determination of the cut-off value

One hundred and seventy SPF serum samples were tested by peptide-ELISA to set cut-off values. The 170 SPF serum samples had mean OD$_{450}$ of 0.087 ± 0.022. Thus, for higher specificity the cut-off value for this peptide-ELISA could be determined at 0.153 by adding three standard deviations to the mean for SPF serum of 0.087. For the test system to be valid, we determined that the positive control OD$_{450}$ should be higher than 0.4 and at least three times higher than the negative control OD$_{450}$. And the negative control OD$_{450}$ should be lower than mean plus 2 standard deviations, 0.131. If the values of positive or negative control were outside these limits, the test was repeated.

Repeatability and reproducibility of the peptide-ELISA

Results from Table 3 determined the repeatability and reproducibility of the diagnostic assay by seven sera samples. The intraplate variation showed CVs from 3.34 to 7.81% (4.89% average), whereas the interplate variation showed CVs from 2.74 to 7.30% (4.14% average). The good reflection of assay precision was proved by the low calculated CV values.

Application of peptide-ELISA to test field samples and comparison with commercial ELISA

To validate the performance of the peptide-ELISA, 240 sera samples derived from different chicken farms were detected by peptide-ELISA and IDEXX ELISA, and IFA served as gold standard method to evaluate the results detected by these two ELISAs. As shown in Table 4, the positivity of peptide-ELISA was 20.42% (49/240), as compared with 4.58% (11/240) for IDEXX ELISA. The 11 positive samples from IDEXX ELISA were all included in the positive results of peptide-ELISA. When the IFA was used as the standard for comparison, we found that the sensitivity of peptide-ELISA and IDEXX ELISA was 85.96% (49/57) and 19.30% (11/57). The specificity of two methods was 95.63% (175/183) and 100% (183/183) respectively (Table 4). The results clearly showed that the peptide-ELISA for ALV-J antibody detection developed in this study had much more sensitivity than commercial ELISA when applied to field sera samples. In order to further verify the specificity of 38 sera samples which were negative in IDEXX ELISA, the IFA was carried out with these sera as primary antibodies. The results in Fig. 3 showed 29 out of 38 sera samples were positive in IFA assay.

Discussion

Invention and development of monoclonal antibody technology provide an easy approach to identify the B-cell epitope which can induce antibody production. The mAb JE9 used in current study is the earliest

Table 2 Epitope mapping of mAb JE9 with synthetic peptides ELISA

Peptide ID	Sequence	Location	Reaction[a]
ALV-J-3p	DLASQTACLIQALNTT LPWDPQELDILGSQ	65–94	+
ALV-J-3p-1	DLASQTACLIQALNTTLPWD	65–84	–
ALV-J-3p-2	QALNTTLPWDPQELDILGSQ	75–94	+
ALV-J-3p-2-1	QALNTTLPWDPQE	75–87	–
ALV-J-3p-2-2	WDPQELDILGSQ	83–94	+
ALV-J-3p-2-3	QALNTTLPWDPQEL	75–88	+
ALV-J-3p-2-4	DPQELDILGSQ	84–94	–

[a]Reaction of peptide coated ELISA with mAb JE9

reported monoclonal antibody specific to the gp85 protein of ALV-J. According to previous reports, the mAb JE9 has good reactivity with many ALV-J isolates all over the world [13]. It suggests that the B-cell epitope recognized by JE9 is a conserved and immunodominant epitope. The results of peptide epitope mapping and homology analysis in Fig. 1 and Fig. 2 proved that the [83]WDPQEL[88] peptide sequence is highly conserved and specific among ALV-J strains.

The peptide-ELISA has been shown to be a sensitive and specific indirect diagnostic tool in virology of a multitude of species, ranging from humans in public health to livestock in the agriculture industry [14–18]. How to choose the peptide sequence is the key point to the peptide-ELISA. The binding ability of the selected polypeptide to the antibody, the length of the polypeptide and the necessity to be conjugated with the carrier are factors that need to be considered. In the results of Fig. 1, when we truncated the 3p-2 peptide to [83]WDPQELDILGSQ[94] and [75]QALNTTLPWDPQEL[88] which contain the minimal conserved epitope, the reactivity of the polypeptide decreased significantly although the OD_{450} value was still positive. The similar

results have been reported in previous study, which described that peptide length alters diagnostic sensitivity and specificity [15]. In the same report, the authors suggested that conjugation of peptides to BSA could improve assay sensitivity. In current study, we got the similar results also. When the peptide was conjugated with BSA, the reactivity of the peptide-ELISA slightly enhanced (data not shown). In another report, B-cell epitope for ALV-J gp85 protein, [134]AEAELRDFI[142], was identified [19]. We combined ALV-J-2P which covered 134-142aa and ALV-J-3P-2 in the same ELISA. Unfortunately, the sensitivity was not improved, but on the contrary the reactivity decreased (data not shown). Thus, we choose BSA conjugated 3p-2 only as the coating antigen in the peptide-ELISA for ALV-J antibody detection.

Full length proteins are normally effective reagents for immunodiagnostics because they contain multiple epitopes. However, in this study the peptide-ELISA for ALV-J antibody detection has much more sensitivity than IDEXX ELISA kit which coated with the whole gp85 of ALV-J expressed by recombinant baculovirus. We speculate that glycosylated gp85 protein expressed by recombinant virus might block the internal antigenic epitope which can be recognized by the positive sera. In addition, there are nine sera samples positive in peptide-ELISA, but negative in IFA assay. This suggests more positive and negative samples need to be tested to revise cut-off value to make the ELISA more precise in the future.

Conclusions

In summary, a novel B-cell epitope, [83]WDPQEL[88], specific for ALV-J gp85 protein recognized by the mAb JE9 was identified in this study. The peptide-ELISA based on this novel B-cell epitope established here was more sensitive for ALV-J antibody than current serological method for ALV-J in IDEXX ELISA. The successful use

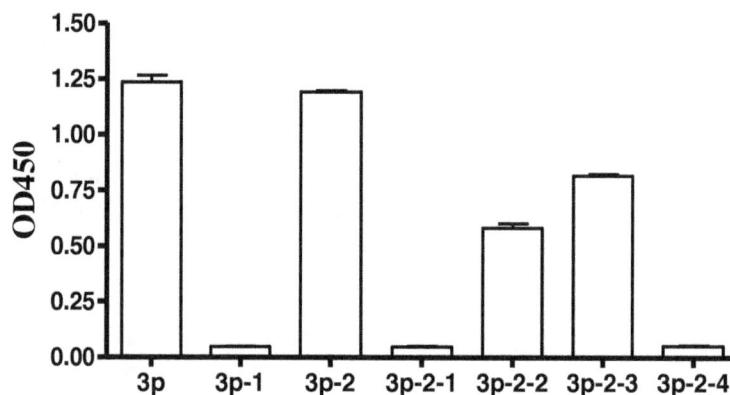

Fig. 1 Reactivity of the different synthetic peptides of gp85 with mAb JE9 using ELISA. The name of each column corresponds to the polypeptide in the Table 2

```
        S - Q T A C L I K A L N T T L P W D P Q E L D I L G S Q M I K  Majority

        150              160              170              180
        . - . . . . . . Q . . . . . . . . . . . . . . . . . . . . . . . .   ALV-J CAUSY01 (HM640945)
        N - . . . . . . Q . . . . . . . . . . . . . . . . . . . . . . . .   ALV-J GX11GL12 (KT598494)
        N - . . . . . . Q T . . . . . . . . . . . . . . . . . . . . . . .   ALV-J JS14XJ02 (KM873186)
        T - . . . . . . Q . . . . . . . . . . . . . . . . V . . . . . V .   ALV-J GX12NN19 (KT598484)
        . - . . . . . . Q T . . . . . . . . . . . . . . . . . . . . . . .   ALV-J GX13YL03 (KT598476)
        . - . . . . . . Q T . . . . . . . . . . . . . . . . . . . . . . .   ALV-J GX13NN13 (KT598480)
        . - . . . . . . Q . . . . . . . . . . . . . . . . . . . . . . E .   ALV-J GX12NN03 (KT598489)
        N - . . . . . . Q T . A . . . . . . . . . . . . . . . . . . . . .   ALV-J JS-J1210 (KC282894)
        . - . . . . . . . . . . . . . . . . . . . . . . . . . . . . . . .   ALV-J SDLWH34J (HQ332146)
        . - . . . . . . . . . . . . . . . . . . . . . . . . . . . . . . .   ALV-J TA081225J (HQ333269)
        . - . . . . . . . . . . . . . . . . . . . . . . . . . . . . . . .   ALV-J SDYC04J (JN388908)
        . - . . . . . . . . . . . . . . . . . . . . . . . . . . . . . . .   ALV-J SQ-J1211 (KC282895)
        . - . . . . . . . . . . . . . . . . . . . . . . . . . . . . . . .   ALV-J WX-J1215 (KC417027)
        . - . . . . . . . . . . . . . . . . . . . . . . . . . . . . . . .   ALV-J SG-1 (KF218958)
        . - . . . . . . Q . . . . . . . . . . . . . . . . . . . . . . . .   ALV-J SCYA759 (KF796640)
        . - . . . . . . . . . . . . . . . . . . . . . . . . . . . . . . .   ALV-J WJ612 (KJ631317)
        . - . . . . . . . . . . . . . . . . . . . . . . . . . . . . . . .   ALV-J WSC912 (KJ631321)
        . - . . . . . . . . . . . . . . . . . . . . . . . . . . . . . . .   ALV-J FJ201306 (KM655820)
        . - . . . . . . . . . . . . . . . . . . . . . . . . . . . . . . .   ALV-J SCYA767 (KF796641)
        . - . . . . . . . . . . . . . . . . . . . . . . . . . . . . . . .   ALV-J WSC512 (KJ631320)
        . - . . . . . . . . . . . . . . . . . . . . . . . . . . . . . . .   ALV-J SD09TA04 (JN378893)
        . - . . . . . . Q T . . . . . . . . . . . . . . . . . . . . . E .   ALV-J PL09DP5-2 (JN378892)
        . - . . . . . . . . . . . . . . . . . . . . . . . . . . . . . . .   ALV-J LYJ195 (KF218957)
        . - . . . . . . . . . . . . . . . . . . . . . . . . . . . . . . .   ALV-J HuB09XT02 (JN378889)
        . - . . . . . . Q T . . . . . . . . . . . . . . . . . . . . . E .   ALV-J GL09DP02 (JN378887)
        . - . . . . . . . . . . . . . . . . . . . . . . . . . . . . . . .   ALV-J GD1401J (KP317564)
        . - . . . . . . . . . . . . . . . . . . . . . . . . . . . . . . .   ALV-J DJ146 (KF218959)
        . - . . . . . . Q T . . . . . . . . . . . . . . . . . . . . . . .   ALV-J CAUXT01 (HM640947)
        . - . . . . . . . . . . . . . . . . . . . . . . . . . . . . . . .   ALV-J HPRS103 (Z46390)
        T - . . . . . . Q S . . . . . . . . . . . . . . . R . . . . . . .   ALV-J ADOL-7501 (AY027920)
        Y R K V S . . L L K . . V S M W N E . P . . Q L . . . . S L P   ALV-A SDAU09E2 (HM452342)
        Y R K V S . . L L K . . V S M W D E . P . . Q L . . . . S L P   ALV-A MQNCSU (DQ365814)
        Y R K V S . . L L K . . V S M W D E . P . . Q L . . . . S L P   ALV-A MAV-1 (L10922)
        Y R K V S . . L Q K . . V S M W D E . P . . Q L . . . . S L P   ALV-A DPRE32 (KM434201)
        Y R K V S . . L L K . . V S . L D E . S . . Q L . . . . S L P   ALV-B SDAU09C2 (HM446005)
        Y Q K V S . . L L K . . V S . L D E . S . . Q L . . . . S L P   ALV-B Schmidt-Ruppin B (AF052428)
        Y R K V S . . L L K . . V S . L D E . P . . Q L . . . . S L P   ALV-B MAV-2 (L10924)
        Y R K V S . . L L K . . V S M W D E . P . . Q L . . . . S L P   ALV-C RSV-PragueC (J02342)
        Y R K V S . . L L K . . V S M W D E . P . . Q L . . . . S L P   ALV-C (AF033808)
        Y R K V S . . L L K . . V S M W D E . P . . Q L . . . . S L P   ALV-D RSV Schmidt-RuppinD (D10652)
        Y R K V S . . L L K . . V S M W D E . P . . Q L . . . . S L P   ALV-E ev-1 (AY013303)
        Y R K V S . . L L K . . V S M W D E . P . . Q L . . . . S L P   ALV-E-B11 (KC610517)
        Y R K V S . . L S K . . V P M W D E . P . . Q L . . . . S L P   ALV-K tw-3593 (HM582658)
        Y R K V S . . L L K . . V S M L D E . P . . Q L . . . . S L P   ALV-K SD110503R (KF738251)
        Y R K V S . . L L K . . V S M W D E . P . . Q L . . . . S L P   ALV-K PDRC-3249 (EU070902)
        Y R K V S . . L L K . . V S . L D E . P . . Q L . . . . S L P   ALV-K km_5844 (AB670312)
        Y R K V S . . L L K . . V P M L D E . P . . Q L . . . . S L P   ALV-K JS11C1 (KF746200)
        Y R K V S . . L L K . . V S . L D E . P . . Q L . . . . S L P   ALV-K GDFX0602 (KP686143)
```

Fig. 2 Alignment of the epitopes motif with 51 ALV strains. The GenBank accession numbers of the ALV strains used are indicated in parentheses. The homologous sequences of different ALV strains corresponding to the identified epitope are boxed. Identical residues are indicated by ".". "-" indicates that there was no corresponding amino acid at this position

of this peptide in detecting ALV-J in the clinical serum samples suggests that the epitope-based peptide ELISA could possibly be used as a serologic reagent in the diagnosis of ALV-J infection and it will contribute to the rational design of vaccines by further understanding of the antigenic structure of gp85.

Methods

Cells, antibody and ELISA kit

The pcDNA-env-DF1 cell line [10] were maintained in Dulbecco's modified Eagle medium (DMEM; GIBCO, Shanghai China) supplemented with 5% fatal bovine serum (FBS), 100 U/mL of penicillin, 100 g/mL of streptomycin and 100 µg/mL of Zeocin at 37 °C in a 5% CO_2 atmosphere. ALV-J gp85 specific monoclonal antibody, JE9, was kept in our lab [13]. The antibodies of chicken sera samples were determined by ALV-J antibody ELISA (IDEXX, Beijing, China) according to the manufacturer's protocol.

Table 3 Intraplate and interplate variation of the peptide-ELISA

Sample No.	1	2	3	4	5	6	7
Intraplate CV (%)	3.58	5.81	3.34	3.67	4.20	7.81	5.81
Interplate CV (%)	2.74	4.06	2.79	2.88	4.13	7.30	5.07

Table 4 Evaluation of the developed peptide-ELISA for detection ALV-J antibodies with clinical serum samples

		Peptide-ELISA		IDEXX ELISA[a]	
		Positive	Negative	Positive	Negative
IFA	Positive	49	8	11	46
	Negative	8	175	0	183

[a]ALV-J antibody ELISA kit obtained from IDEXX Inc., Beijing China

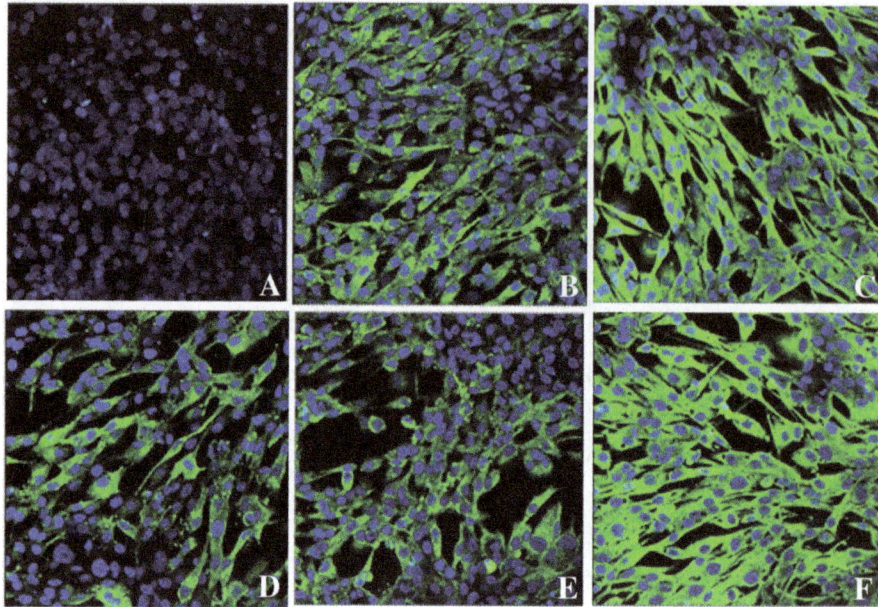

Fig. 3 The result of partial serum samples evaluated with IFA. Panel **a** and **b** were negative and positive control respectively; Panel **c** to **f** were partial positive results by serum samples which were negative in IDEXX ELISA, but positive in peptide-ELISA

Sera samples collection

The 170 SPF chicken sera samples were kindly provided by Spirax Ferrer Poultry Science and Technology Co.Ltd., Jinan, China. The other 240 chicken sera samples were collected from chicken farms with the eradication programme of ALV being conducted in Jiangsu and Shandong Province in China. All experiments complied with institutional animal care guidelines and were approved by the Animal Care Committee of Yangzhou University.

Peptide design and synthesis

The 65–155 amino acids of gp85 protein were equally divided into three segments, ALV-J-1P (95-125aa), ALV-J-2P (126-155aa), and ALV-J-3P (65-94aa), respectively. A series of peptides of ALV-J-3P for mAb epitope mapping was designed in Table 2. All of the peptides were synthesized by Shanghai Jietai Biotech Company. The crude peptides were purified by semi-preparative HPLC on a Beckman System Gold with a reverse-phase C18 column, resulting in purity greater than 95%, checked by analytical HPLC on a Shimadzu system. The BSA conjugated polypeptide (BSA-3P-2) was also synthesized and purified by Shanghai Jietai Biotech Company.

Homology analysis

The conservation of the sequence that contained the B-cell epitope of gp85 protein was analysed with sequences of different ALV strains using MegAlign software version 7.1.0 (DNAstar, Madison, USA). The

sequences of the ALV strains that were used as reference were downloaded from the GenBank database. Analysis of the antigenic index and the surface probability of the ALV gp85 protein was also performed using Protean software version 7.1.0 (DNAstar, Madison, USA).

Peptide-ELISA procedure

The peptide-ELISA procedure was performed as previously described with minor modifications [17]. Maxisorp ELISA plates (NUNC, Thermo, Shanghai) were coated overnight at 4 °C with 100 µl per well of a 1 µg/ml peptide solution in 0.1 M carbonate/bicarbonate buffer, pH 9.6. Next morning, 400 µl per well of 8% Normal Rabbit Serum (Biyuntian, Nantong, China) in PBS was added for blocking, and the plate was incubated for 2 h at 37 °C followed by three washes with PBS supplemented with 0.05% Tween 20. Reagent layers were removed by striking plates repeatedly, bottom up, on a stack of absorbent tissue.

After blocking and washing, the plate was incubated with 100 µl per well of chicken sera diluted in 8% Normal Rabbit Serum in PBS with 0.05% Tween 20 at 37 °C for 30 min. Unless stated otherwise, sera were assayed at a 1:200 dilution. After being washed, plates were incubated with 100 µl per well of horseradish peroxidase-conjugated rabbit anti-chicken immunoglobulin (SIGMA, Shanghai, China) at 37 °C for 45 min, diluted 1:10000 in 8% Normal Rabbit Serum in PBS with 0.05% Tween 20. Following a final wash, plates were developed for approximately 25 min at 37 °C with 100 µl per well of TMB Turbo substrate (Pierce, Thermo, USA), and were stopped with 100 μl per well of 2 M sulphuric acid. Absorbance at OD450 was determined using a standard ELISA plate reader (Bio-TEK, Vermont, USA).

For storage, ELISA plates were coated with peptide and blocked as described above. Excess blocking reagent was removed, and the plates were dried with bottom up and the wells exposed to circulating air, for 4 h at room temperature. Plates were then stored in Vacuum packaging at 4 °C. The ELISA protocol for pre-coated plates was the same as described above, starting with the addition of the primary antibody.

Preliminary precision assessment

Preliminary precision assessment was performed as previously described [20]. Coefficients of variation (CVs) for 3 replicates of a total of 7 specimens were run on the same plate and calculated for repeatability (intraplate variation). For reproducibility, CVs were obtained by using the same specimens run in triplicate in two different plates (interplate variation).

Indirect immunofluorescence assay (IFA)

The protocol was the same as that of our previous report [21]. Briefly, the pcDNA-env-DF1 cells were fixed with 4% paraformaldehyde in PBS for 20 min at room temperature, permeabilized with 0.25% Triton X-100 for 5 min, washed with PBS, blocked with 2% BSA for 30 min, and incubated with chicken serum (1:20 dilution) in PBS for 45 min at room temperature. The cells were then washed in PBS and incubated with rabbit anti-chicken IgG conjugated with FITC (Sigma, Shanghai, China) and stained with 10 μg/mL of Hoechst 33342 dye (SIGMA, Shanghai, China) at room temperature for an additional 30 min. The pictures were captured with a Leica SP2 confocal microscope.

Statistical analysis

Statistical analysis including CV calculation was performed by using GraphPad Prism version 5.0 software (GraphPad Software, SanDiego, CA).

Abbreviations

ALV-J: Avian leukosis virus subgroup J; BSA: Bovine serum albumin; DMEM: Dulbecco's modified Eagle's medium; ELISA: Enzyme-linked immunosorbent assay; env: envelop;; FBS: Foetal bovine serum; FITC: Fluorescein isothiocyanate; mAb: Monoclonal antibody

Funding

The research was supported by the National key research and development program (Grant No. 2016YFD0500803), the National Natural Science Foundation of China (Grant No. 31772734), NCFC-RCUK-BBSRC (Grant No. 31761133002), Special Foundation for State Basic Research Program of China (2013FY113300-4), the Priority Academic Program Development of Jiangsu Higher Education Institutions and the Jiangsu Co-innovation Centre for the Prevention and Control of Important Animal Infectious Diseases and Zoonoses.

Authors' contributions

KQ and AQ designed the project. KQ and XT carried out the experiments, analysed the data, and drafted the manuscript. AQ supervised all the experiments and participated in the data analysis. HS, JY, YY and VN discussed and prepared the final report. All of the authors have read and approved the final manuscript.

Consent for publication

Not applicable.

Competing interests

The authors declare that they have no competing interests.

Author details

[1]Ministry of Education Key Lab for Avian Preventive Medicine, College of Veterinary Medicine, Yangzhou University, No.12 East Wenhui Road, Yangzhou, Jiangsu 225009, People's Republic of China. [2]The International Joint Laboratory for Cooperation in Agriculture and Agricultural Product Safety, Ministry of Education, Yangzhou University, 225009 Yangzhou, People's Republic of China. [3]Jiangsu Key Lab of Zoonosis, No.12 East Wenhui Road, Yangzhou, Jiangsu 225009, People's Republic of China. [4]Avian Oncogenic Virus Group, The Pirbright Institute, Ash Road, Pirbright, Surrey GU24 0NF, UK. [5]The UK-China Centre of Excellence for Research on Avian Diseases, 169 Huanghe 2nd Road, Binzhou, Shandong, People's Republic of China.

References

1. Payne LN, Nair V. The long view: 40 years of avian leukosis research. Avian pathology: journal of the WVPA. 2012;41(1):11–9.
2. Wu X, Qian K, Qin A, Shen H, Wang P, Jin W, Eltahir YM. Recombinant avian leukosis viruses of subgroup J isolated from field infected commercial layer chickens with hemangioma and myeloid leukosis possess an insertion in the E element. Vet Res Commun. 2010;34(7):619–32.
3. Sun S, Cui Z. Epidemiological and pathological studies of subgroup J avian leukosis virus infections in Chinese local "yellow" chickens. Avian pathology: journal of the WVPA. 2007;36(3):221–6.
4. Wang CH, Juan YW. Occurrence of subgroup J avian leukosis virus in Taiwan. Avian pathology: journal of the WVPA. 2002;31(5):435–9.
5. Sung HW, Kim JH, Reddy S, Fadly A. Isolation of subgroup J avian leukosis virus in Korea. Journal of veterinary science. 2002;3(2):71–4.
6. Wang G, Jiang Y, Yu L, Wang Y, Zhao X, Cheng Z. Avian leukosis virus subgroup J associated with the outbreak of erythroblastosis in chickens in China. Virol J. 2013;10:92.
7. Wang P, Lin L, Li H, Yang Y, Huang T, Wei P. Diversity and evolution analysis of glycoprotein GP85 from avian leukosis virus subgroup J isolates from chickens of different genetic backgrounds during 1989–2016: Coexistence of five extremely different clusters. Arch Virol. 2018;163(2):377–89.
8. Shen Y, Cai L, Wang Y, Wei R, He M, Wang S, Wang G, Cheng Z. Genetic mutations of avian leukosis virus subgroup J strains extended their host range. The Journal of general virology. 2014;95(Pt 3):691–9.
9. Venugopal K. Avian leukosis virus subgroup J: a rapidly evolving group of oncogenic retroviruses. Res Vet Sci. 1999;67(2):113–9.
10. Mei M, Ye J, Qin A, Wang L, Hu X, Qian K, Shao H. Identification of novel viral receptors with cell line expressing viral receptor-binding protein. Sci Reports. 2015;5:7935.
11. Dong X, Meng F, Hu T, Ju S, Li Y, Sun P, Wang Y, Chen W, Zhang F, Su H, et al. Dynamic Co-evolution and Interaction of Avian Leukosis Virus Genetic Variants and Host Immune Responses. Front Microbiol. 2017;8:1168.
12. Silva RF, Fadly AM, Hunt HD. Hypervariability in the envelope genes of subgroup J avian leukosis viruses obtained from different farms in the United States. Virology. 2000;272(1):106–11.

13. Qin A, Lee LF, Fadly A, Hunt H, Cui Z. Development and characterization of monoclonal antibodies to subgroup J avian leukosis virus. Avian Dis. 2001; 45(4):938–45.

14. Gao M, Zhang R, Li M, Li S, Cao Y, Ma B, Wang J. An ELISA based on the repeated foot-and-mouth disease virus 3B epitope peptide can distinguish infected and vaccinated cattle. Appl Microbiol Biotechnol. 2012;93(3):1271–9.

15. Dubois ME, Hammarlund E, Slifka MK. Optimization of peptide-based ELISA for serological diagnostics: a retrospective study of human monkeypox infection. Vector Borne Zoonotic Dis. 2012;12(5):400–9.

16. Tian H, Chen Y, Wu J, Shang Y, Liu X. Serodiagnosis of sheeppox and goatpox using an indirect ELISA based on synthetic peptide targeting for the major antigen P32. Virol J. 2010;7:245.

17. Oleksiewicz MB, Stadejek T, Mackiewicz Z, Porowski M, Pejsak Z. Discriminating between serological responses to European-genotype live vaccine and European-genotype field strains of porcine reproductive and respiratory syndrome virus (PRRSV) by peptide ELISA. J Virol Methods. 2005; 129(2):134–44.

18. Wang CH, Hong CC, Seak JC. An ELISA for antibodies against infectious bronchitis virus using an S1 spike polypeptide. Vet Microbiol. 2002;85(4): 333–42.

19. Li X, Zhu H, Wang Q, Sun J, Gao Y, Qi X, Wang Y, Gao H, Gao Y, Wang X. Identification of a novel B-cell epitope specific for avian leukosis virus subgroup J gp85 protein. Arch Virol. 2015;160(4):995–1004.

20. Lang A, de Vries M, Feineis S, Muller E, Osterrieder N, Damiani AM: Development of a peptide ELISA for discrimination between serological responses to equine herpesvirus type 1 and 4. J Virol Methods 2013, 193(2): 667–673.

21. Qian K, Gao AJ, Zhu MY, Shao HX, Jin WJ, Ye JQ, Qin AJ. Genistein inhibits the replication of avian leucosis virus subgroup J in DF-1 cells. Virus Res. 2014;192:114–20.

Permissions

The contributors of this book come from diverse backgrounds, making this book a truly international effort. This book will bring forth new frontiers with its revolutionizing research information and detailed analysis of the nascent developments around the world.

We would like to thank all the contributing authors for lending their expertise to make the book truly unique. They have played a crucial role in the development of this book. Without their invaluable contributions this book wouldn't have been possible. They have made vital efforts to compile up to date information on the varied aspects of this subject to make this book a valuable addition to the collection of many professionals and students.

This book was conceptualized with the vision of imparting up-to-date information and advanced data in this field. To ensure the same, a matchless editorial board was set up. Every individual on the board went through rigorous rounds of assessment to prove their worth. After which they invested a large part of their time researching and compiling the most relevant data for our readers.

The editorial board has been involved in producing this book since its inception. They have spent rigorous hours researching and exploring the diverse topics which have resulted in the successful publishing of this book. They have passed on their knowledge of decades through this book. To expedite this challenging task, the publisher supported the team at every step. A small team of assistant editors was also appointed to further simplify the editing procedure and attain best results for the readers.

Apart from the editorial board, the designing team has also invested a significant amount of their time in understanding the subject and creating the most relevant covers. They scrutinized every image to scout for the most suitable representation of the subject and create an appropriate cover for the book.

The publishing team has been an ardent support to the editorial, designing and production team. Their endless efforts to recruit the best for this project, has resulted in the accomplishment of this book. They are a veteran in the field of academics and their pool of knowledge is as vast as their experience in printing. Their expertise and guidance has proved useful at every step. Their uncompromising quality standards have made this book an exceptional effort. Their encouragement from time to time has been an inspiration for everyone.

The publisher and the editorial board hope that this book will prove to be a valuable piece of knowledge for researchers, students, practitioners and scholars across the globe.

List of Contributors

Mutien Garigliany, Gautier Gilliaux, Sandra Jolly, Tomas Casanova, Calixte Bayrou, Etienne Lévy, Dominique Cassart and Daniel Desmecht
Department of Morphology and Pathology, University of Liège, Liège, Belgium

Dominique Peeters and Kris Gommeren
Department of Clinical Sciences, University of Liège, Liège, Belgium

Thomas Fett and Axel Mauroy
Department of Infectious and Parasitic Diseases, Centre for Fundamental and Applied Research for Animals & Health, Faculty of Veterinary Medicine, University of Liège, Liège, Belgium

Luc Poncelet
Laboratory of Anatomy, Biomechanics and Organogenesis, Faculty of Medicine, Free University of Brussels, Brussels, Belgium

Daria Dziewulska, Tomasz Stenzel, Marcin Śmiałek, Bartłomiej Tykałowski and Andrzej Koncki
Department of Poultry Diseases, Faculty of Veterinary Medicine, University of Warmia and Mazury in Olsztyn, ul. Oczapowskiego 13/14, 10-719 Olsztyn, Poland

Waqas Ashraf, Sunaina Haris, Ameena Mobeen, Muhammad Farooq, Muhammad Asif and Qaiser Mahmood Khan
National Institute for Biotechnology and Genetic Engineering (NIBGE), Faisalabad, Pakistan
Pakistan Institute of Engineering and Applied Sciences (PIEAS), Islamabad, Pakistan

Hermann Unger
Animal Production and Health Section, Joint FAO/IAEA Division of Nuclear Techniques in Food and Agriculture, Vienna, Austria

Maarten Haspeslagh, Mireia Jordana Garcia, Lieven E. M. Vlaminck and Ann M. Martens
Department of Surgery and Anaesthesiology of Domestic Animals, Faculty of Veterinary Medicine, Ghent University, Salisburylaan 133, 9820 Merelbeke, Belgium

M. Perez Andrés, Ponder Julia, Wünschmann Arno, Vander Waal Kimberly, Alvarez Julio and Willette Michelle
University of Minnesota, St. Paul, MN, USA

Alba Ana
University of Minnesota, St. Paul, MN, USA
Univ of Minnesota College of Veterinary Medicine, 1920 Fitch Avenue, St. Paul, MN 55108, USA

Puig Pedro
Universitat Autònoma de Barcelona, Cerdanyola del Vallès, Barcelona, Spain

Hiroshi Bannai, Akihiro Ochi, Manabu Nemoto, Koji Tsujimura, Takashi Yamanaka and Takashi Kondo
Equine Research Institute, Japan Racing Association, 1400-4 Shiba, Shimotsuke, Tochigi 329-0412, Japan

Juliano Ribeiro
Graduate Program in Cellular and Molecular Biology, Federal University of Parana, Curitiba, Paraná 81531-990, Brazil

Claudia Staudacher
Zoonoses Control Center, City Secretary of Health, Curitiba, Paraná 80060-130, Brazil

Camila Marinelli Martins and Fernando Ferreira
Department of Preventive Veterinary Medicine and Animal Health, University of São Paulo, São Paulo 05508-270, Brazil

Leila Sabrina Ullmann and João Pessoa Araujo Jr
UNESP – Univ. Estadual Paulista, Campus de Botucatu, Institute of Biotechnology, Botucatu, São Paulo, Botucatu, São Paulo 18607-440, Brazil

Alexander Welker Biondo
Department of Veterinary Medicine, Federal University of Paraná, Rua dos Funcionários, 1540, Curitiba, Paraná 80035-050, Brazil

M. C. M. de Jong and K. Frankena
Quantitative Veterinary Epidemiology, Wageningen University & Research, Droevendaalsesteeg 1, 6708 PB Wageningen, The Netherlands

W. Molla
Quantitative Veterinary Epidemiology, Wageningen University & Research, Droevendaalsesteeg 1, 6708 PB Wageningen, The Netherlands
Faculty of Veterinary Medicine, University of Gondar, Gondar, Ethiopia

Mar Melero and Jose M. Sánchez-Vizcaíno
VISAVET Center and Animal Health Department, Veterinary School, Complutense University of Madrid, Avda. Puerta del Hierro s/n, 28040 Madrid, Spain

Consuelo Rubio-Guerri
VISAVET Center and Animal Health Department, Veterinary School, Complutense University of Madrid, Avda. Puerta del Hierro s/n, 28040 Madrid, Spain
Fundación Oceanografic de la Comunitat Valenciana, C/. Eduardo Primo Yúfera (Científic) 1B, 46013 Valencia, Spain

Jose L. Crespo-Picazo
Fundación Oceanografic de la Comunitat Valenciana, C/. Eduardo Primo Yúfera (Científic) 1B, 46013 Valencia, Spain

Daniel García-Párraga
Fundación Oceanografic de la Comunitat Valenciana, C/. Eduardo Primo Yúfera (Científic) 1B, 46013 Valencia, Spain
Veterinary Services, Avanqua Oceanogràfic S.L., C/ Eduardo Primo Yúfera (Científic) 1B, 46013 Valencia, Spain

M. Ángeles Jiménez
Medicine and Surgery Department (Anatomic Pathology), Veterinary School, Complutense University of Madrid, 28040 Madrid, Spain

Josué Díaz-Delgado, Eva Sierra and Manuel Arbelo
Unit of Histology and Veterinary Pathology, Institute for Animal Health, Veterinary School, University of Las Palmas de Gran Canaria, Trasmontaña, s, /n 35416 Arucas (Las Palmas), Canary Islands, Spain

Edwige N. Bellière and Fernando Esperón
National Institute for Agricultural and Food Research and Technology, Ctra. de Algete a El Casar s/n, 28130 Madrid, Spain

S. Beck, D. J. Hicks and A. Núñez
Pathology Department, Animal and Plant Health Agency, Weybridge, UK

P. Gunawardena
Department of Veterinary Pathobiology, University of Peradeniya, Peradeniya, Sri Lanka

A. Ortiz-Pelaez
Animal and Plant Health Agency, Weybridge, UK

D. A. Marston, A. R. Fooks and D. L. Horton
Wildlife Zoonoses and Vector Borne Diseases Research Group, Animal and Plant Health Agency, Weybridge, UK

Steven Lawson, Xiaodong Liu, Aaron Singrey, Travis Clement, Kyle Hain, Julie Nelson, Jane Christopher-Hennings and Eric A. Nelson
Veterinary & Biomedical Sciences Department, South Dakota State University, Brookings, SD, USA

Faten Okda
Veterinary & Biomedical Sciences Department, South Dakota State University, Brookings, SD, USA
National Research Center, Giza, Egypt

Chen Yuan, En Zhang, Lulu Huang, Jialu Wang and Qian Yang
MOE Joint International Research Laboratory of Animal Health and Food Safety, College of veterinary medicine, Nanjing Agricultural University, Weigang 1, Nanjing, Jiangsu 210095, People's Republic of China

Jianchang Wang, Jinfeng Wang and Libing Liu
Center of Inspection and Quarantine, Hebei Entry-Exit Inspection and Quarantine Bureau, No.318 Hepingxilu Road, Shijiazhuang, Hebei Province 050051, People's Republic of China

Ruiwen Li and Wanzhe Yuan
College of Veterinary Medicine, Agricultural University of Hebei, No.38 Lingyusi Street, Baoding, Hebei 071001, People's Republic of China

María Elena Trujillo-Ortega, Rolando Beltrán-Figueroa and Alicia Sotomayor-González
Departamento de Medicina y Zootecnia de Cerdos, Facultad de MedicinaVeterinaria y Zootecnia, Universidad Nacional Autónoma de México, Mexico City 04510, Mexico

José F. Becerra-Hernández, Rosa Elena Sarmiento-Silva, Erika N. Hernández-Villegas and Montserrat Elemi García-Hernández
Departamento de Microbiología e Inmunología, Facultad de Medicina Veterinaria y Zootecnia, Universidad Nacional Autónoma de México, Mexico City 04510, Mexico

Mireya Juárez-Ramírez
Departamento de Patología, Facultad de Medicina Veterinaria y Zootecnia, Universidad Nacional Autónoma de México, Mexico City 04510, Mexico

Francesca Rizzo, Simona Zoppi, Alessandro Dondo, Serena Robetto, Riccardo Orusa and Maria Lucia Mandola
Istituto zooprofilattico sperimentale del Piemonte, Liguria e Valle d'Aosta, Via Bologna 148, 10148 Torino, Italy

Angelika Lander, Kathryn M. Edenborough and Andreas Kurth
Robert Koch Institute, Seestraße 10, 13353 Berlin, Germany

Roberto Toffoli and Paola Culasso
Chirosphera, via Tetti Barbiere 11, 10026 Santena, TO, Italy

Sergio Rosati and Luigi Bertolotti
Department of Veterinary Science, Largo Paolo Braccini 2, 10095 Grugliasco, TO, Italy

Van Phan Le
Department of Microbiology and Infectious Disease, Faculty of Veterinary Medicine, Vietnam National University of Agriculture, Hanoi, Vietnam

Thi Thu Hang Vu
Research and Development Laboratory, Rural Technology Development JSC, Hung Yen, Vietnam

Hong-Quan Duong
Institute of Research and Development, Duy Tan University, Danang, Vietnam

Van Thai Than
Department of Microbiology, Chung-Ang University College of Medicine, Seoul, South Korea

Daesub Song
College of Pharmacy, Korea University, Sejong, South Korea

Libing Liu and Jinfeng Wang
Center of Inspection and Quarantine, Hebei Entry-Exit Inspection and Quarantine Bureau, Shijiazhuang 050051, People's Republic of China

Ruihan Shi and Jianchang Wang
Center of Inspection and Quarantine, Hebei Entry-Exit Inspection and Quarantine Bureau, Shijiazhuang 050051, People's Republic of China
Hebei Academy of Science and Technology for Inspection and Quarantine, Shijiazhuang 050051, People's Republic of China

Wanzhe Yuan
College of Veterinary Medicine, Agricultural University of Hebei, No.38 Lingyusi Street, Baoding, Hebei 071001, People's Republic of China

Qingan Han and Ruoxi Zhang
Hebei Animal Disease Control Center, Shijiazhuang 050050, People's Republic of China

Mi Lin
State Key Laboratory of Veterinary Etiological Biology, Lanzhou Veterinary Research Institute, Chinese Academy of Agricultural Sciences, Lanzhou 730046, People's Republic of China

Kwang-Soo Lyoo
Korea Zoonosis Research Institute, Chonbuk National University, Iksan, South Korea

Sun-Woo Yoon and Hye Kwon Kim
Infectious Disease Research Center, Korea Research Institute of Bioscience and Biotechnology, Daejeon 305-806, South Korea

Min-Chul Jung and Dae Gwin Jeong
Infectious Disease Research Center, Korea Research Institute of Bioscience and Biotechnology, Daejeon 305-806, South Korea
University of Science and Technology (UST), Daejeon, South Korea

Naouel Feknous and Hamza Khaled
LBRA, Institute of Veterinary Sciences, Saad Dahlab University, Soumaa Road, BP 270, 09000 Blida, Algeria

Brigitte Cay, Jean-Baptiste Hanon and Marylène Tignon
Sciensano, Infectious animal diseases directorate, Service of enzootic, vector-borne and bee diseases, Groeselenberg 99, 1180 Brussels, Belgium

Abdallah Bouyoucef
ENSV, National superior veterinary school, Bab ezzouar, El allia, Algeria

Amina Khatun, Salik Nazki, Myoun-Sik Yang, Bumseok Kim and Won-Il Kim
College of Veterinary Medicine, Chonbuk National University, 79 Gobong-ro, Iksan, Jeonbuk, Korea

Nadeem Shabir
College of Veterinary Medicine, Chonbuk National University, 79 Gobong-ro, Iksan, Jeonbuk, Korea
Division of Animal Biotechnology, Faculty of Veterinary Sciences and Animal Husbandry, Sher-e-Kashmir University of Agricultural Sciences and Technology of Kashmir, Srinagar, India

Suna Gu and Sang-Myoung Lee
College of Environmental & Biosource Science, Division of Biotechnology, Chonbuk National University, Iksan, South Korea

Tai-Young Hur
Dairy Science Division, National Institute of Animal Science, Rural Development Administration, Cheonan 31000, South Korea

Alexandra Buckley and Albert van Geelen
U.S. Department of Agriculture, Oak Ridge Institute for Science and Education and National Animal Disease Center, Ames, IA, USA

Vikas Kulshreshtha
U.S. Department of Agriculture, Oak Ridge Institute for Science and Education and National Animal Disease Center, Ames, IA, USA
Toxikon Corporation, Bedford, MA, USA

Nestor Montiel
U.S. Department of Agriculture, Oak Ridge Institute for Science and Education and National Animal Disease Center, Ames, IA, USA

U.S. Department of Agriculture, Avian Viruses Section, Diagnostic Virology Laboratory, National Veterinary Services Laboratories, Animal and Plant Health Inspection Service, Ames, IA, USA

Baoqing Guo, Hai Hoang, Kyoung-Jin Yoon and Christopher Rademacher
Department of Veterinary Diagnostic and Production Animal Medicine, College of Veterinary Medicine, Iowa State University, Ames, IA, USA

Kelly Lager
U.S. Department of Agriculture, Virus and Prion Research Unit, National Animal Disease Center, Agricultural Research Service, 1920 Dayton Avenue, Ames, IA 50010, USA

Xue Tian
Ministry of Education Key Lab for Avian Preventive Medicine, College of Veterinary Medicine, Yangzhou University, No.12 East Wenhui Road, Yangzhou, Jiangsu 225009, People's Republic of China

Kun Qian, Hongxia Shao, Jianqiang Ye and Aijian Qin
Ministry of Education Key Lab for Avian Preventive Medicine, College of Veterinary Medicine, Yangzhou University, No.12 East Wenhui Road, Yangzhou, Jiangsu 225009, People's Republic of China
The International Joint Laboratory for Cooperation in Agriculture and Agricultural Product Safety, Ministry of Education, Yangzhou University, 225009 Yangzhou, People's Republic of China
Jiangsu Key Lab of Zoonosis, No.12 East Wenhui Road, Yangzhou, Jiangsu 225009, People's Republic of China

Yongxiu Yao and Venugopal Nair
Avian Oncogenic Virus Group, The Pirbright Institute, Ash Road, Pirbright, Surrey GU24 0NF, UK
The UK-China Centre of Excellence for Research on Avian Diseases, 169 Huanghe 2nd Road, Binzhou, Shandong, People's Republic of China

Index

www.ingramcontent.com/pod-product-compliance
Lightning Source LLC
Chambersburg PA
CBHW082035190326
41458CB00010B/3377